Wolfram Koch, M

A Chemist's Gu
Density Functional Theory

Second Edition

Further Reading from Wiley-VCH and John Wiley & Sons

P. Comba/T. W. Hambley
Molecular Modeling of Inorganic Compounds, Second Edition
2000, approx. 250 pages with approx. 200 figures and a CD-ROM with an interactive tutorial. Wiley-VCH.
ISBN 3-527-29915-7

H.-D. Höltje/G. Folkers
Molcular Modeling. Basic Priniciples and Applications
1997, 206 pages. Wiley-VCH.
ISBN 3-527-29384-1

F. Jensen
Introduction to Computational Chemistry
1998, 454 pages. Wiley.
ISBN 0-471-98425-6

K. B. Lipkowitz/D. B. Boyd (Eds.)
Reviews in Computational Chemistry, Vol. 13
1999, 384 pages. Wiley.
ISBN 0-471-33135-X

M. F. Schlecht
Molecular Modeling on the PC
1998, 763 pages. Wiley-VCH.
ISBN 0-471-18467-1

P. von Schleyer (Ed.)
Encyclopedia of Computational Chemistry
1998, 3580 pages. Wiley.
ISBN 0-471-96588-X

J. Zupan/J. Gasteiger
Neural Networks in Chemistry and Drug Design
1999, 400 pages. Wiley-VCH.
ISBNs 3-527-29779-0 (Softcover), 3-527-29778-2 (Hardcover)

Wolfram Koch, Max C. Holthausen

A Chemist's Guide to Density Functional Theory

Second Edition

Weinheim · New York · Chichester · Brisbane · Singapore · Toronto

Prof. Dr. Wolfram Koch
Gesellschaft Deutscher Chemiker
(German Chemical Society)
Varrentrappstraße 40–42
D-60486 Frankfurt
Germany

Dr. Max C. Holthausen
Fachbereich Chemie
Philipps-Universität Marburg
Hans-Meerwein-Straße
D-35032 Marburg
Germany

1st Edition 2000
2nd Edition 2001
1st Reprint 2002
2nd Reprint 2003
3rd Reprint 2004
4th Reprint 2007
5th Reprint 2008

Library of Congress Card No.: applied for

British Library Cataloguing-in-Publication Data:
A catalogue record for this book
is available from the British Library

Die Deutsche Bibliothek – CIP-Cataloguing-in-Publication Data:
A catalogue record for this book is available from the Deutsche Bibliothek
ISBN 978-3-527-30422-6 (Hardcover)
978-3-527-30372-4 (Softcover)

Composition: Text- und Software-Service Manuela Treindl, D-93059 Regensburg

Foreword

It is a truism that in the past decade density functional theory has made its way from a peripheral position in quantum chemistry to center stage. Of course the often excellent accuracy of the DFT based methods has provided the primary driving force of this development. When one adds to this the computational economy of the calculations, the choice for DFT appears natural and practical. So DFT has conquered the rational minds of the quantum chemists and computational chemists, but has it also won their hearts? To many, the success of DFT appeared somewhat miraculous, and maybe even unjust and unjustified. Unjust in view of the easy achievement of accuracy that was so hard to come by in the wave function based methods. And unjustified it appeared to those who doubted the soundness of the theoretical foundations. There has been misunderstanding concerning the status of the one-determinantal approach of Kohn and Sham, which superficially appeared to preclude the incorporation of correlation effects. There has been uneasiness about the molecular orbitals of the Kohn-Sham model, which chemists used qualitatively as they always have used orbitals but which in the physics literature were sometimes denoted as mathematical constructs devoid of physical (let alone chemical) meaning.

Against this background the Chemist's Guide to DFT is very timely. It brings in the second part of the book the reader up to date with the most recent successes and failures of the density functionals currently in use. The literature in this field is exploding in such a manner that it is extremely useful to have a comprehensive overview available. In particular the extensive coverage of property evaluation, which has very recently been enormously stimulated by the time-dependent DFT methods, will be of great benefit to many (computational) chemists. But I wish to emphasize in particular the good service the authors have done to the chemistry community by elaborating in the first part of the book on the approach that DFT takes to the physics of electron correlation. A full appreciation of DFT is only gained through an understanding of how the theory, in spite of working with an orbital model and a single determinantal wave function for a model system of noninteracting electrons, still achieves to incorporate electron correlation. The authors justly put emphasis on the pictorial approach, by way of Fermi and Coulomb correlation holes, to understanding exchange and correlation. The present success of DFT proves that modelling of these holes, even if rather crudely, can provide very good energetics. It is also in the simple physical language of shape and extent (localized or delocalized) of these holes that we can understand where the problems of that modelling with only local input (local density, gradient, Laplacian, etc.) arise. It is because of the well equilibrated treatment of physical principles and chemical applications that this book does a good and very timely service to the computational and quantum chemists as well as to the chemistry community at large. I am happy to recommend it to this audience.

EVERT JAN BAERENDS, Amsterdam
October 1999

Preface

This book has been written by chemists for chemists. In particular, it has not been written by genuine theoretical chemists but by chemists who are primarily interested in solving chemical problems and in using computational methods for addressing the many exciting questions that arise in modern chemistry. This is important to realize right from the start because our background of course determined how we approached this project. Density functional theory is a fairly recent player in the computational chemistry arena. WK, the senior author of this book remembers very well his first encounter with this new approach to tackle electronic structure problems. It was only some ten years back, when he got a paper to review for the *Journal of Chemical Physics* where the authors employed this method for solving some chemical problems. He had a pretty hard time to understand what the authors really did and how much the results were worth, because the paper used a language so different from conventional wave function based ab initio theory that he was used to. A few years later we became interested in transition-metal chemistry, the reactivity of coordinatively unsaturated open-shell species in mind. During a stay with Margareta Blomberg and Per Siegbahn at the University of Stockholm, leading researchers in this field then already for a decade, MCH was supposed to learn the tricks essential to cope with the application of highly correlated multireference wave function based methods to tackle such systems. So he did – yet, what he took home was the feeling that our problems could not be solved for the next decade with this methodology, but that there might be something to learn about density functional theory (DFT) instead. It did not take long and DFT became the major computational workhorse in our group. We share this kind of experience with many fellow computational chemists around the globe. Starting from the late eighties and early nineties approximate density functional theory enjoyed a meteoric rise in computational chemistry, a success story without precedent in this area. In the Figure below we show the number of publications where the phrases 'DFT' or 'density functional theory' appear in the title or abstract from a Chemical Abstracts search covering the years from 1990 to 1999. The graph speaks for itself.

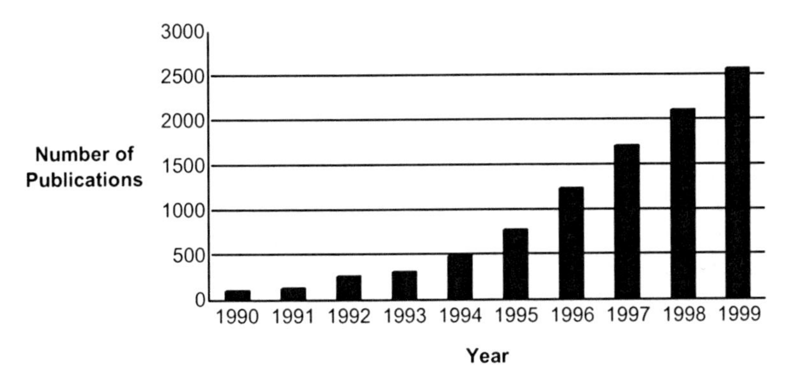

This stunning progress was mainly fueled by the development of new functionals – gradient-corrected functionals and most notably hybrid functionals such as B3LYP – which cured many of the deficiencies that had plagued the major model functional used back then, i. e., the local density approximation. Their subsequent implementation in the popular quantum chemistry codes additionally catalyzed this process, which is steadily gaining momentum. The most visible documentation that computational methods in general and density functional theory in particular finally lost their 'new kid on the block' image is the award of the 1998 Noble Prize in chemistry to two exceptional protagonists of this genre, John Pople and Walter Kohn.

Many experimental chemists use sophisticated spectroscopic techniques on a regular basis, even though they are not experts in the field, and probably never need to be. In a similar manner, more and more chemists start to use approximate density functional theory and take advantage of black box implementations in modern programs without caring too much about the theoretical foundations and – more critically – limitations of the method. In the case of spectroscopy, this partial unawareness is probably just due to a lack of time or motivation since almost any level of education required seems to be well covered by textbooks. In computational chemistry, however, the lack of digestible sources tailored for the needs of chemists is serious. Everyone trying to supplement a course in computational chemistry with pointers to the literature well suited for amateurs in density functional theory has probably had this experience. Certainly, there is a vast and fast growing literature on density functional theory including many review articles, monographs, books containing collections of high-level contributions and also text books. Indeed, some of these were very influential in advancing density functional theory in chemistry and we just mention what is probably the most prominent example, namely Parr's and Yang's '*Density-Functional Theory of Atoms and Molecules*' which appeared in 1989, just when density functional theory started to lift off. Still, many of these are either addressing primarily the physics community or present only specific aspects of the theory. What is not available is a text book, something like Tim Clark's '*A Handbook of Computational Chemistry*', which takes a chemist, who is interested but new to the field, by the hand and guides him or her through basic theoretical and related technical aspects at an easy to understand level. This is precisely the gap we are attempting to fill with the present book. Our main motivation to embark on the endeavor of this project was to provide the many users of standard codes with the kind of background knowledge necessary to master the many possibilities and to critically assess the quality obtained from such applications. Consequently, we are neither concentrating on all the important theoretical difficulties still related to density functional theory nor do we attempt to exhaustively review all the literature of important applications. Intentionally we sacrifice the purists' theoretical standpoint and a broad coverage of fields of applications in favor of a pragmatic point of view. However, we did our best to include as many theoretical aspects and relevant examples from the literature as possible to encourage the interested readers to catch up with the progress in this rapidly developing field. In collecting the references we tried to be as up-to-date as possible, with the consequence that older studies are not always cited but can be traced back through the more recent investigations included in the bibliography. The literature was covered through the fall of 1999.

However, due to the huge amount of relevant papers appearing in a large variety of journals, certainly not all papers that should have come to our attention actually did and we apologize at this point to anyone whose contribution we might have missed. One more point: we have written this book dwelling from our own background. Hence, the subjects covered in this book, particularly in the second part, mirror to some extent the areas of interest of the authors. As a consequence, some chemically relevant domains of density functional theory are not mentioned in the following chapters. We want to make clear that this does not imply that we assign a reduced importance to these fields, rather it reflects our own lack of experience in these areas. The reader will, for example, search in vain for an exposition of density functional based ab initio molecular dynamics (Car-Parrinello) methods, for an assessment of the use of DFT as a basis for qualitative models such as soft- and hardness or Fukui functions, an introduction into the treatment of solvent effects or the rapidly growing field of combining density functional methods with empirical force fields, i. e., QM/MM hybrid techniques and probably many more areas.

The book is organized as follows. In the first part, consisting of Chapters 1 through 7, we give a systematic introduction to the theoretical background and the technical aspects of density functional theory. Even though we have attempted to give a mostly self-contained exposition, we assume the reader has at least some basic knowledge of molecular quantum mechanics and the related mathematical concepts. The second part, Chapters 8 to 13 presents a careful evaluation of the predictive power that can be expected from today's density functional techniques for important atomic and molecular properties as well as examples of some selected areas of application. Of course, also the selection of these examples was governed by our own preferences and cannot cover all important areas where density functional methods are being successfully applied. The main thrust here is to convey a general feeling about the versatility but also the limitations of current density functional theory.

For any comments, hints, corrections, or questions, or to receive a list of misprints and corrections please drop a message at **DFT-Guide@chemie.uni-marburg.de.**

Many colleagues and friends contributed important input at various stages of the preparation of this book, by making available preprints prior to publication, by discussions about several subjects over the internet, or by critically reading parts of the manuscript. In particular we express our thanks to V. Barone, M. Bühl, C. J. Cramer, A. Fiedler, M. Filatov, F. Haase, J. N. Harvey, V. G. Malkin, P. Nachtigall, G. Schreckenbach, D. Schröder, G. E. Scuseria, Philipp Spuhler, M. Vener, and R. Windiks. Further, we would like to thank Margareta Blomberg and Per Siegbahn for their warm hospitality and patience as open minded experts and their early inspiring encouragement to explore the pragmatic alternatives to rigorous conventional ab initio theory. WK also wants to thank his former and present diploma and doctoral students who helped to clarify many of the concepts by asking challenging questions and always created a stimulating atmosphere. In particular we are grateful to A. Pfletschinger and N. Sändig for performing some of the calculations used in this book. Brian Yates went through the exercise of reading the whole manuscript and helped to clarify the discussion and to correct some of our 'Germish'. He did a great job – thanks a lot, Brian – of course any remaining errors are our sole responsibility. Last but certainly not least we are greatly indebted to Evert Jan Baerends who not only contributed many enlightening discussions on the theoretical aspects and provided preprints, but who also volunteered to write the Foreword for this book and to Paul von Ragué Schleyer for providing thoughtful comments. MCH is grateful to Joachim Sauer and Walter Thiel for support, and to the Fonds der Chemischen Industrie for a Liebig fellowship, which allowed him to concentrate on this enterprise free of financial concerns. At Wiley-VCH we thank R. Wengenmayr for his competent assistance in all technical questions and his patience. The victims that suffered most from sacrificing our weekends and spare time to the progress of this book were certainly our families and we owe our wives Christina and Sophia, and WK's daughters Juliana and Leora a deep thank you for their endurance and understanding.

WOLFRAM KOCH, Frankfurt am Main
MAX C. HOLTHAUSEN, Berlin
November 1999

Preface to the second edition

Due to the large demand, a second edition of this book had to be prepared only about one year after the original text appeared. In the present edition we have corrected all errors that came to our attention and we have included new references where appropriate. The discussion has been brought up-to-date at various places in order to document significant recent developments.

WOLFRAM KOCH, Frankfurt am Main
MAX C. HOLTHAUSEN, Marburg
April 2001

Contents

PART A

The Definition of the Model

What is density functional theory? The first part of this book is devoted to this question and we will try in the following seven chapters to give the reader a guided tour through the current state of the art of approximate density functional theory. We will try to lift some of the secrets veiling that magic black box, which, after being fed with only the charge density of a system somewhat miraculously cranks out its energy and other ground state properties. Density functional theory is rooted in quantum mechanics and we will therefore start by introducing or better refreshing some elementary concepts from basic molecular quantum mechanics, centered around the classical Hartree-Fock approximation. Since modern density functional theory is often discussed in relation to the Hartree-Fock model and the corresponding extensions to it, a solid appreciation of the related physics is a crucial ingredient for a deeper understanding of the things to come. We then comment on the very early contributions of Thomas and Fermi as well as Slater, who used the electron density as a basic variable more out of intuition than out of solid physical arguments. We go on and develop the red line that connects the seminal theorems of Hohenberg and Kohn through the realization of this concept by Kohn and Sham to the currently popular approximate exchange-correlation functionals. The concept of the exchange-correlation hole, which is rarely discussed in detail in standard quantum chemical textbooks holds a prominent place in our exposition. We believe that grasping its characteristics helps a lot in order to acquire a more pictorial and less abstract comprehension of the theory. This intellectual exercise is therefore well worth the effort. Next to the theory, which – according to our credo – we present in a down-to-earth like fashion without going into all the many intricacies which theoretical physicists make a living of, we devote a large fraction of this part to very practical aspects of density functional theory, such as basis sets, numerical integration techniques, etc. While it is neither possible nor desirable for the average user of density functional methods to apprehend all the technicalities inherent to the implementation of the theory, the reader should nevertheless become aware of some of the problems and develop a feeling of how a solution can be realized.

1 Elementary Quantum Chemistry

In this introductory chapter we will review some of the fundamental aspects of electronic structure theory in order to lay the foundations for the theoretical discussion on density functional theory (DFT) presented in later parts of this book. Our exposition of the material will be kept as brief as possible and for a deeper understanding the reader is encouraged to consult any modern textbook on molecular quantum chemistry, such as Szabo and Ostlund, 1982, McWeeny, 1992, Atkins and Friedman, 1997, or Jensen, 1999. After introducing the Schrödinger equation with the molecular Hamilton operator, important concepts such as the antisymmetry of the electronic wave function and the resulting Fermi correlation, the Slater determinant as a wave function for non-interacting fermions and the Hartree-Fock approximation are presented. The exchange and correlation energies as emerging from the Hartree-Fock picture are defined, the concepts of dynamical and nondynamical electron correlation are discussed and the dissociating hydrogen molecule is introduced as a prototype example.

1.1 The Schrödinger Equation

The ultimate goal of most quantum chemical approaches is the – approximate – solution of the time-independent, non-relativistic Schrödinger equation

$$\hat{H}\Psi_i(\vec{x}_1,\vec{x}_2,\ldots,\vec{x}_N,\vec{R}_1,\vec{R}_2,\ldots,\vec{R}_M) = E_i\Psi_i(\vec{x}_1,\vec{x}_2,\ldots,\vec{x}_N,\vec{R}_1,\vec{R}_2,\ldots,\vec{R}_M) \qquad (1\text{-}1)$$

where $\hat{H}$ is the Hamilton operator for a molecular system consisting of M nuclei and N electrons in the absence of magnetic or electric fields. $\hat{H}$ is a differential operator representing the total energy:

$$\hat{H} = -\frac{1}{2}\sum_{i=1}^{N}\nabla_i^2 - \frac{1}{2}\sum_{A=1}^{M}\frac{1}{M_A}\nabla_A^2 - \sum_{i=1}^{N}\sum_{A=1}^{M}\frac{Z_A}{r_{iA}} + \sum_{i=1}^{N}\sum_{j>i}^{N}\frac{1}{r_{ij}} + \sum_{A=1}^{M}\sum_{B>A}^{M}\frac{Z_A Z_B}{R_{AB}} \qquad (1\text{-}2)$$

Here, A and B run over the M nuclei while i and j denote the N electrons in the system. The first two terms describe the kinetic energy of the electrons and nuclei respectively, where the Laplacian operator ∇_q^2 is defined as a sum of differential operators (in cartesian coordinates)

$$\nabla_q^2 = \frac{\partial^2}{\partial x_q^2} + \frac{\partial^2}{\partial y_q^2} + \frac{\partial^2}{\partial z_q^2} \qquad (1\text{-}3)$$

and M_A is the mass of nucleus A in multiples of the mass of an electron (atomic units, see below). The remaining three terms define the potential part of the Hamiltonian and represent the attractive electrostatic interaction between the nuclei and the electrons and the repulsive potential due to the electron-electron and nucleus-nucleus interactions, respectively. r_{pq} (and similarly R_{pq}) is the distance between the particles p and q, i. e., $r_{pq} = |\vec{r}_p - \vec{r}_q|$. $\Psi_i(\vec{x}_1, \vec{x}_2, \ldots, \vec{x}_N, \vec{R}_1, \vec{R}_2, \ldots, \vec{R}_M)$ stands for the wave function of the i'th state of the system, which depends on the 3N spatial coordinates $\{\vec{r}_i\}$, and the N spin coordinates[1] $\{s_i\}$ of the electrons, which are collectively termed $\{\vec{x}_i\}$ and the 3M spatial coordinates of the nuclei, $\{\vec{R}_I\}$. The wave function Ψ_i contains all information that can possibly be known about the quantum system at hand. Finally, E_i is the numerical value of the energy of the state described by Ψ_i.

All equations given in this text appear in a very compact form, without any fundamental physical constants. We achieve this by employing the so-called system of atomic units, which is particularly adapted for working with atoms and molecules. In this system, physical quantities are expressed as multiples of fundamental constants and, if necessary, as combinations of such constants. The mass of an electron, m_e, the modulus of its charge, $|e|$, Planck's constant h divided by 2π, $\hbar$, and $4\pi\varepsilon_0$, the permittivity of the vacuum, are all set to unity. Mass, charge, action etc. are then expressed as multiples of these constants, which can therefore be dropped from all equations. The definitions of atomic units used in this book and their relations to the corresponding SI units are summarized in Table 1-1.

Table 1-1. Atomic units.

Quantity	Atomic unit	Value in SI units	Symbol (name)
mass	rest mass of electron	9.1094×10^{-31} kg	m_e
charge	elementary charge	1.6022×10^{-19} C	e
action	Planck's constant/2π	1.0546×10^{-34} J s	$\hbar$
length	$4\pi\varepsilon_0 \hbar / m_e e^2$	5.2918×10^{-11} m	a_0 (bohr)
energy	$\hbar^2 / m_e a_0^2$	4.3597×10^{-18} J	E_h (hartree)

Note that the unit of energy, 1 hartree, corresponds to twice the ionization energy of a hydrogen atom, or, equivalently, that the exact total energy of an H atom equals –0.5 E_h. Thus, 1 hartree corresponds to 27.211 eV or 627.51 kcal/mol.[2]

The Schrödinger equation can be further simplified if we take advantage of the significant differences between the masses of nuclei and electrons. Even the lightest of all nuclei, the proton (^{1}H), weighs roughly 1800 times more than an electron, and for a typical nucleus such as carbon the mass ratio well exceeds 20,000. Thus, the nuclei move much slower than the electrons. The practical consequence is that we can – at least to a good approximation – take the extreme point of view and consider the electrons as moving in the field of fixed

---

[1] Remember from basic quantum mechanics that to completely describe an electron its spin needs to be specified in addition to the spatial coordinates. The spin coordinates can only assume the values ±½; the possible values of the spin functions $\alpha(s)$ and $\beta(s)$ are: $\alpha(½) = \beta(-½) = 1$ and $\alpha(-½) = \beta(½) = 0$.

[2] We use kcal/mol rather than kJ/mol throughout the book. 1 kcal/mol = 4.184 kJ/mol.

nuclei. This is the famous *Born-Oppenheimer* or clamped-nuclei approximation. Of course, if the nuclei are fixed in space and do not move, their kinetic energy is zero and the potential energy due to nucleus-nucleus repulsion is merely a constant. Thus, the complete Hamiltonian given in equation (1-2) reduces to the so-called electronic Hamiltonian

$$\hat{H}_{elec} = -\frac{1}{2}\sum_{i=1}^{N}\nabla_i^2 - \sum_{i=1}^{N}\sum_{A=1}^{M}\frac{Z_A}{r_{iA}} + \sum_{i=1}^{N}\sum_{j>i}^{N}\frac{1}{r_{ij}} = \hat{T} + \hat{V}_{Ne} + \hat{V}_{ee}. \quad (1\text{-}4)$$

The solution of the Schrödinger equation with $\hat{H}_{elec}$ is the electronic wave function Ψ_{elec} and the electronic energy E_{elec}. Ψ_{elec} depends on the electron coordinates, while the nuclear coordinates enter only parametrically and do not explicitly appear in Ψ_{elec}. The total energy E_{tot} is then the sum of E_{elec} and the constant nuclear repulsion term, $E_{nuc} = \sum_{A=1}^{M}\sum_{B>A}^{M}\frac{Z_A Z_B}{r_{AB}}$, i. e.,

$$\hat{H}_{elec}\Psi_{elec} = E_{elec}\Psi_{elec} \quad (1\text{-}5)$$

and

$$E_{tot} = E_{elec} + E_{nuc}. \quad (1\text{-}6)$$

The attractive potential exerted on the electrons due to the nuclei – the expectation value of the second operator $\hat{V}_{Ne}$ in equation (1-4) – is also often termed the external potential, V_{ext}, in density functional theory, even though the external potential is not necessarily limited to the nuclear field but may include external magnetic or electric fields etc. From now on we will only consider the electronic problem of equations (1-4) – (1-6) and the subscript 'elec' will be dropped.

The wave function Ψ itself is not observable. A physical interpretation can only be associated with the square of the wave function in that

$$\left|\Psi(\vec{x}_1, \vec{x}_2, \ldots, \vec{x}_N)\right|^2 d\vec{x}_1 d\vec{x}_2 \ldots d\vec{x}_N \quad (1\text{-}7)$$

represents the probability that electrons 1, 2, ..., N are found simultaneously in volume elements $d\vec{x}_1 d\vec{x}_2 \ldots d\vec{x}_N$. Since electrons are indistinguishable, this probability must not change if the coordinates of any two electrons (here i and j) are switched, viz.,

$$\left|\Psi(\vec{x}_1, \vec{x}_2, \ldots, \vec{x}_i, \vec{x}_j, \ldots, \vec{x}_N)\right|^2 = \left|\Psi(\vec{x}_1, \vec{x}_2, \ldots, \vec{x}_j, \vec{x}_i, \ldots, \vec{x}_N)\right|^2. \quad (1\text{-}8)$$

Thus, the two wave functions can at most differ by a unimodular complex number $e^{i\phi}$. It can be shown that the only possibilities occurring in nature are that either the two functions are identical (symmetric wave function, applies to particles called *bosons* which have integer spin, including zero) or that the interchange leads to a sign change (antisymmetric wave function, applies to *fermions*, whose spin is half-integral). Electrons are fermions with spin = ½ and Ψ must therefore be antisymmetric with respect to interchange of the spatial and spin coordinates of any two electrons:

$$\Psi(\vec{x}_1, \vec{x}_2, \ldots, \vec{x}_i, \vec{x}_j, \ldots, \vec{x}_N) = -\Psi(\vec{x}_1, \vec{x}_2, \ldots, \vec{x}_j, \vec{x}_i, \ldots, \vec{x}_N). \tag{1-9}$$

We will soon encounter the enormous consequences of this antisymmetry principle, which represents the quantum-mechanical generalization of *Pauli's exclusion principle* ('no two electrons can occupy the same state'). A logical consequence of the probability interpretation of the wave function is that the integral of equation (1-7) over the full range of all variables equals one. In other words, the probability of finding the N electrons anywhere in space must be exactly unity,

$$\int \cdots \int \left|\Psi(\vec{x}_1, \vec{x}_2, \ldots, \vec{x}_N)\right|^2 d\vec{x}_1 d\vec{x}_2 \ldots d\vec{x}_N = 1. \tag{1-10}$$

A wave function which satisfies equation (1-10) is said to be *normalized*. In the following we will deal exclusively with normalized wave functions.

1.2 The Variational Principle

What we need to do in order to solve the Schrödinger equation (1-5) for an arbitrary molecule is first to set up the specific Hamilton operator of the target system. To this end we need to know those parts of the Hamiltonian $\hat{H}$ that are specific for the system at hand. Inspection of equation (1-4) reveals that the only information that depends on the actual molecule is the number of electrons in the system, N, and the external potential V_{ext}. The latter is in our cases completely determined through the positions and charges of all nuclei in the molecule. All the remaining parts, such as the operators representing the kinetic energy or the electron-electron repulsion, are independent of the particular molecule we are looking at. In the second step we have to find the eigenfunctions Ψ_i and corresponding eigenvalues E_i of $\hat{H}$. Once the Ψ_i are determined, all properties of interest can be obtained by applying the appropriate operators to the wave function. Unfortunately, this simple and innocuous-looking program is of hardly any practical relevance, since apart from a few, trivial exceptions, no strategy to solve the Schrödinger equation exactly for atomic and molecular systems is known.

Nevertheless, the situation is not completely hopeless. There is a recipe for systematically approaching the wave function of the ground state Ψ_0, i. e., the state which delivers the lowest energy E_0. This is the *variational principle*, which holds a very prominent place in all quantum-chemical applications. We recall from standard quantum mechanics that the expectation value of a particular observable represented by the appropriate operator $\hat{O}$ using any, possibly complex, wave function Ψ_{trial} that is normalized according to equation (1-10) is given by

$$\langle \hat{O} \rangle = \int \cdots \int \Psi^*_{trial} \, \hat{O} \, \Psi_{trial} d\vec{x}_1 d\vec{x}_2 \ldots d\vec{x}_N \equiv \left\langle \Psi_{trial} \middle| \hat{O} \middle| \Psi_{trial} \right\rangle \quad (1\text{-}11)$$

where we introduce the very convenient *bracket notation* for integrals first used by Dirac, 1958, and often used in quantum chemistry. The star in Ψ^*_{trial} indicates the complex-conjugate of Ψ_{trial}.

The variational principle now states that the energy computed via equation (1-11) as the expectation value of the Hamilton operator $\hat{H}$ from any guessed Ψ_{trial} will be an *upper bound* to the true energy of the ground state, i. e.,

$$\left\langle \Psi_{trial} \middle| \hat{H} \middle| \Psi_{trial} \right\rangle = E_{trial} \geq E_0 = \left\langle \Psi_0 \middle| \hat{H} \middle| \Psi_0 \right\rangle \quad (1\text{-}12)$$

where the equality holds if and only if Ψ_{trial} is identical to Ψ_0. The proof of equation (1-12) is straightforward and can be found in almost any quantum chemistry textbook.

Before we continue let us briefly pause, because in equations (1-11) and (1-12) we encounter for the first time the main mathematical concept of density functional theory. A rule such as that given through (1-11) or (1-12), which assigns a number, e. g., E_{trial}, to a function, e. g., Ψ_{trial}, is called a *functional*. This is to be contrasted with the much more familiar concept of a function, which is the mapping of one number onto another number. Phrased differently, we can say that a functional is a function whose argument is itself a function. To distinguish a functional from a function in writing, one usually employs square brackets for the argument. Hence, f(x) is a function of the variable x while F[f] is a functional of the function f. Recall that a function needs a number as input and also delivers a number:

$$x \xrightarrow{f(x)} y \, .$$

For example, $f(x) = x^2 + 1$. Then, for $x = 2$, the function delivers $y = 5$. On the other hand, a functional needs a function as input, but again delivers a number:

$$f(x) \xrightarrow{F[f(x)]} y \, .$$

For example, if we define $F[f] = \int_0^1 [f(x)]^2 \, dx$ and use f(x) as defined above as input, this functional delivers $F[f(x) = x^2 + 1] = 28/15$. If, instead we choose $f(x) = 2x^2+1$, the result is $F[f(x) = 2x^2 + 1] = 47/15$.

Expectation values such as $\langle \hat{O} \rangle$ in equation (1-11) are obviously functionals, since the value of $\langle \hat{O} \rangle$ depends on the function Ψ_{trial} inserted.

Coming back to the variational principle, the strategy for finding the ground state energy and wave function should be clear by now: we need to minimize the functional $E[\Psi]$ by searching through *all acceptable N-electron wave functions*. Acceptable means in this context that the trial functions must fulfill certain requirements which ensure that these functions make physical sense. For example, to be eligible as a wave function, Ψ must be continuous everywhere and be quadratic integrable. If these conditions are not fulfilled the normalization of equation (1-10) would be impossible. The function[3] which gives the lowest energy will be Ψ_0 and the energy will be the true ground state energy E_0. This recipe can be compactly expressed as

$$E_0 = \min_{\Psi \to N} E[\Psi] = \min_{\Psi \to N} \left\langle \Psi \middle| \hat{T} + \hat{V}_{Ne} + \hat{V}_{ee} \middle| \Psi \right\rangle \tag{1-13}$$

where $\Psi \to N$ indicates that Ψ is an allowed N-electron wave function. While such a search over *all eligible* functions is obviously not possible, we can apply the variational principle as well to subsets of all possible functions. One usually chooses these subsets such that the minimization in equation (1-13) can be done in some algebraic scheme. The result will be the best approximation to the exact wave function that can be obtained from this particular subset. It is important to realize that by restricting the search to a subset the exact wave function itself cannot be identified (unless the exact wave function is included in the subset, which is rather improbable). A typical example is the Hartree-Fock approximation discussed below, where the subset consists of all antisymmetric products (Slater determinants) composed of N spin orbitals.

Let us summarize what we have shown so far: once N and V_{ext} (uniquely determined by Z_A and R_A) are known, we can construct $\hat{H}$. Through the prescription given in equation (1-13) we can then – at least in principle – obtain the ground state wave function, which in turn enables the determination of the ground state energy and of all other properties of the system. Pictorially, this can be expressed as

$$\{N, Z_A, R_A\} \Rightarrow \hat{H} \Rightarrow \Psi_0 \Rightarrow E_0 \text{ (and all other properties).}$$

Thus, N and V_{ext} completely and uniquely determine Ψ_0 and E_0. We say that the *ground state energy is a functional of the number of electrons* N *and the nuclear potential* V_{ext},

$$E_0 = E\left[N, V_{ext}\right]. \tag{1-14}$$

1.3 The Hartree-Fock Approximation

In this and the following sections we will introduce the *Hartree-Fock (HF) approximation* and some of the fundamental concepts intimately connected with it, such as exchange, self-interaction, dynamical and non-dynamical electron correlation. We will meet many of these terms again in our later discussions on related topics in the framework of DFT. The HF

[3] In general there can be more than one function associated with the same energy. If the lowest energy results from n functions, this energy is said to be n-fold degenerate.

approximation is not only the corner stone of almost all conventional, i. e., wave function based quantum chemical methods, it is also of great conceptual importance. An understanding of the physics behind this approximation will thus be of great help in our later analysis of various aspects of density functional theory. In what follows we will concentrate on the interpretation of the HF scheme rather than on a detailed outline how the relevant expressions are being derived. An excellent source for an in-depth discussion of many aspects of the HF approximation and more sophisticated techniques related to it is the book by Szabo and Ostlund, 1982.

As discussed above, it is impossible to solve equation (1-13) by searching through all acceptable N-electron wave functions. We need to define a suitable subset, which offers a physically reasonable approximation to the exact wave function without being unmanageable in practice. In the Hartree-Fock scheme the simplest, yet physically sound approximation to the complicated many-electron wave function is utilized. It consists of approximating the N-electron wave function by an *antisymmetrized* product[4] of N one-electron wave functions $\chi_i(\vec{x}_i)$. This product is usually referred to as a *Slater determinant*, Φ_{SD}:

$$\Psi_0 \approx \Phi_{SD} = \frac{1}{\sqrt{N!}} \begin{vmatrix} \chi_1(\vec{x}_1) & \chi_2(\vec{x}_1) & \cdots & \chi_N(\vec{x}_1) \\ \chi_1(\vec{x}_2) & \chi_2(\vec{x}_2) & & \chi_N(\vec{x}_2) \\ \vdots & \vdots & & \vdots \\ \chi_1(\vec{x}_N) & \chi_2(\vec{x}_N) & \cdots & \chi_N(\vec{x}_N) \end{vmatrix} \tag{1-15}$$

or using a convenient short-hand notation, where only the diagonal elements are given:

$$\Phi_{SD} = \frac{1}{\sqrt{N!}} \det\{\chi_1(\vec{x}_1)\ \ \chi_2(\vec{x}_2) \ldots \chi_N(\vec{x}_N)\}. \tag{1-16}$$

The one-electron functions $\chi_i(\vec{x}_i)$ are called *spin orbitals*, and are composed of a spatial orbital $\phi_i(\vec{r})$ and one of the two spin functions, $\alpha(s)$ or $\beta(s)$.

$$\chi(\vec{x}) = \phi(\vec{r})\ \sigma(s), \quad \sigma = \alpha, \beta\,. \tag{1-17}$$

The spin functions have the important property that they are orthonormal, i. e., $\langle\alpha|\alpha\rangle = \langle\beta|\beta\rangle = 1$ and $\langle\alpha|\beta\rangle = \langle\beta|\alpha\rangle = 0$. For computational convenience, the spin orbitals themselves are usually chosen to be orthonormal also:

[4] A simple product $\Xi = \chi_1(\vec{x}_1)\,\chi_2(\vec{x}_2)\ldots\chi_i(\vec{x}_i)\,\chi_j(\vec{x}_j)\cdots\chi_N(\vec{x}_N)$ is not acceptable as a model wave function for fermions because it assigns a particular one-electron function to a particular electron (for example χ_1 to x_1) and hence violates the fact that electrons are indistinguishable. In addition, $\chi_1(\vec{x}_1)\,\chi_2(\vec{x}_2)\ldots\chi_i(\vec{x}_i)\,\chi_j(\vec{x}_j)\ldots\chi_N(\vec{x}_N) \neq -\chi_1(\vec{x}_1)\,\chi_2(\vec{x}_2)\ldots\chi_i(\vec{x}_j)\,\chi_j(\vec{x}_i)\ldots\chi_N(\vec{x}_N)$, i. e. such a product is not antisymmetric with respect to particle interchange.

$$\int \chi_i^*(\vec{x}) \; \chi_i(\vec{x}) \; d\vec{x} = \langle \chi_i \mid \chi_j \rangle = \delta_{ij} \qquad (1\text{-}18)$$

where we have used the Kronecker delta symbol δ_{ij} which equals 1 for i = j and 0 otherwise. Spin orbitals carry the usual physical interpretation that $|\chi(\vec{x})|^2 d\vec{x}$ represents the probability of finding the electron with spin given by σ within the volume element $d\vec{r}$. The $(N!)^{-1/2}$ prefactor ensures that Φ_{SD} fulfills the normalization condition, equation (1-10). The Slater determinant of equation (1-15) is indeed antisymmetric, since a determinant changes sign upon exchange of two rows or two columns. However, we want to reiterate at this point that replacing the true N-electron wave function Ψ_{exact} by a single Slater determinant Φ_{SD} represents a fairly drastic approximation.

Now that we have decided on the form of the wave function the next step is to use the variational principle in order to find the *best* Slater determinant, i. e., that one particular Φ_{SD} which yields the lowest energy. The only flexibility in a Slater determinant is provided by the spin orbitals. In the Hartree-Fock approach the spin orbitals $\{\chi_i\}$ are now varied under the constraint that they remain orthonormal such that the energy obtained from the corresponding Slater determinant is minimal

$$E_{HF} = \min_{\Phi_{SD} \to N} E[\Phi_{SD}]. \qquad (1\text{-}19)$$

The expectation value of the Hamilton operator with a Slater determinant can be derived by expanding the determinant and constructing the individual terms with respect to the various parts in the Hamiltonian. The derivation is not very complicated and can again be found in all relevant textbooks. We just give here the final result; the HF energy is given by

$$E_{HF} = \langle \Phi_{SD} | \hat{H} | \Phi_{SD} \rangle = \sum_i^N (i \mid \hat{h} \mid i) + \frac{1}{2} \sum_i^N \sum_j^N (ii \mid jj) - (ij \mid ji) \qquad (1\text{-}20)$$

where

$$(i \mid \hat{h} \mid i) = \int \chi_i^*(\vec{x}_1) \left\{ -\frac{1}{2} \nabla^2 - \sum_A^M \frac{Z_A}{r_{1A}} \right\} \chi_i(\vec{x}_1) \, d\vec{x}_1 \qquad (1\text{-}21)$$

defines the contribution due to the kinetic energy and the electron-nucleus attraction and

$$(ii \mid jj) = \iint |\chi_i(\vec{x}_1)|^2 \frac{1}{r_{12}} |\chi_j(\vec{x}_2)|^2 \, d\vec{x}_1 d\vec{x}_2 \qquad (1\text{-}22)$$

$$(ij \mid ji) = \iint \chi_i(\vec{x}_1) \, \chi_j^*(\vec{x}_1) \frac{1}{r_{12}} \chi_j(\vec{x}_2) \, \chi_i^*(\vec{x}_2) \, d\vec{x}_1 d\vec{x}_2 \qquad (1\text{-}23)$$

are the so-called *Coulomb* and *exchange* integrals, respectively, which represent the interaction between two electrons as discussed in more detail below.

E_{HF} from equation (1-20) is obviously a functional of the spin orbitals, $E_{HF} = E[\{\chi_i\}]$. Thus, the variational freedom in this expression is in the choice of the orbitals. In addition, the constraint that the $\{\chi_i\}$ remain orthonormal must be satisfied throughout the minimization, which introduces the *Lagrangian multipliers* ε_i in the resulting equations. These equations (1-24) represent the *Hartree-Fock equations*, which determine the 'best' spin orbitals, i. e., those $\{\chi_i\}$ for which E_{HF} attains its lowest value (for a detailed derivation see Szabo and Ostlund, 1982)

$$\hat{f}\,\chi_i = \varepsilon_i\,\chi_i\,,\ i = 1, 2, \ldots, N. \tag{1-24}$$

These N equations have the appearance of eigenvalue equations, where the Lagrangian multipliers ε_i are the eigenvalues of the operator $\hat{f}$. The ε_i have the physical interpretation of *orbital energies*. The Fock operator $\hat{f}$ is an effective one-electron operator defined as

$$\hat{f}_i = -\frac{1}{2}\nabla_i^2 - \sum_A^M \frac{Z_A}{r_{iA}} + V_{HF}(i). \tag{1-25}$$

The first two terms are the kinetic energy and the potential energy due to the electron-nucleus attraction. $V_{HF}(i)$ is the *Hartree-Fock potential*. It is the average repulsive potential experienced by the i'th electron due to the remaining N-1 electrons. Thus, the complicated two-electron repulsion operator $1/r_{ij}$ in the Hamiltonian is replaced by the simple one-electron operator $V_{HF}(i)$ where the electron-electron repulsion is taken into account only in an average way. Explicitly, V_{HF} has the following two components:

$$V_{HF}(\vec{x}_1) = \sum_j^N \left(\hat{J}_j(\vec{x}_1) - \hat{K}_j(\vec{x}_1)\right). \tag{1-26}$$

The Coulomb operator $\hat{J}$ is defined as

$$\hat{J}_j(\vec{x}_1) = \int \left|\chi_j(\vec{x}_2)\right|^2 \frac{1}{r_{12}}\, d\vec{x}_2 \tag{1-27}$$

and represents the potential that an electron at position $\vec{x}_1$ experiences due to the average charge distribution of another electron in spin orbital χ_j. Remember that $|\chi_j(\vec{x}_2)|^2\, d\vec{x}_2$ represents the probability that the electron is within the volume element $d\vec{x}_2$. Thus the Coulomb repulsion corresponding to a particular distance between the reference electron at $\vec{x}_1$ and another one at position $\vec{x}_2$ is weighted by the probability that the other electron is at this point in space. Finally, this interaction is integrated over all space and spin coordinates. Since the result of application of $\hat{J}_j(\vec{x}_1)$ on a spin orbital $\chi_i(\vec{x}_1)$ depends solely on the value of χ_i at position $\vec{x}_1$, this operator and the corresponding potential are called *local.*

The second term in equation (1-26) is the exchange contribution to the HF potential. The exchange operator $\hat{K}$ has no classical interpretation and can only be defined through its effect when operating on a spin orbital:

$$\hat{K}_j(\vec{x}_1)\,\chi_i(\vec{x}_1) = \int \chi_j^*(\vec{x}_2)\frac{1}{r_{12}}\,\chi_i(\vec{x}_2)\,d\vec{x}_2\,\chi_j(\vec{x}_1)\,. \qquad (1\text{-}28)$$

As evident from the above definition, $\hat{K}_j(\vec{x}_1)$ leads to an exchange of the variables in the two spin orbitals. Furthermore, the result of operating with $\hat{K}_j(\vec{x}_1)$ on $\chi_i(\vec{x}_1)$ depends on the value of χ_i on all points in space, since χ_i is now related to $\vec{x}_2$, the variable over which we integrate. Consequently, this operator and the corresponding exchange potential are called *non-local.* It is important to realize that the occurrence of the exchange term is entirely due to the antisymmetry of the Slater determinant and applies to all fermions, be they charged or neutral. The $1/r_{12}$ operator is spin independent. Thus the integration over the spin coordinate in equation (1-28) can be separated and we have the integral over the product of two different spin orbitals χ_i and χ_j which both depend on the same coordinate $\vec{x}_2$. Because spin functions are orthonormal, it follows that exchange contributions exist *only for electrons of like spin*, because in the case of antiparallel spins, the integrand would contain a factor $\langle\alpha(s_2)|\beta(s_2)\rangle$ (or $\langle\beta(s_2)|\alpha(s_2)\rangle$) which is zero and thus makes the whole integral vanish.

It can easily be shown from their definitions that the expectation values of $\hat{J}_j(\vec{x}_1)$ and $\hat{K}_j(\vec{x}_1)$ are the Coulomb and exchange integrals given in equations (1-22) and (1-23), respectively, presented above. There is one more thing that we need to emphasize: in the double summation in equation (1-20) the term i = j is allowed. This means that if i = j, the integral (1-22) describes the Coulomb interaction of the charge distribution of one electron with itself. As a consequence, even if we compute the energy of a one-electron system, such as the hydrogen atom, where there is definitely no electron-electron repulsion, equation (1-22) would nevertheless give a non-zero result. This *self-interaction* is obviously physical nonsense. However, the exchange term takes perfect care of this: for i = j, the Coulomb and exchange integrals are identical and both reduce to $\iint |\chi_i(\vec{x}_1)|^2 \frac{1}{r_{12}} |\chi_i(\vec{x}_2)|^2\, d\vec{x}_1 d\vec{x}_2$. Since they enter equation (1-20) with opposite signs the self-interaction is exactly cancelled. As we will soon see, the self-interaction problem, so elegantly solved in the HF scheme, and the representation of the exchange energy, constitute major obstacles in density functional approaches. Finally, we should note that because the Fock operator depends through the HF potential on the spin orbitals, i. e., on the very solutions of the eigenvalue problem that needs to be solved, equation (1-24) is not a regular eigenvalue problem that can be solved in a closed form. Rather, we have here a *pseudo-eigenvalue problem* that has to be worked out iteratively. The technique used is called the *self-consistent field* (SCF) procedure since the orbitals are derived from their own effective potential. Very briefly, this technique starts with a 'guessed' set of orbitals, with which the HF equations are solved. The resulting new set of orbitals is then used in the next iteration and so on until the input and output orbitals differ by less than a predetermined threshold. For the sake of completeness we also point out that the Hartree-Fock SCF problem is usually solved through the introduction of a finite basis set to expand the molecular orbitals. We will have to discuss all these aspects in much more detail in the context of the Kohn-Sham equations in later chapters.

Finally, we should note Koopmans' theorem (Koopmans, 1934) which provides a physical interpretation of the orbital energies ε from equation (1-24): it states that the orbital energy ε_i obtained from Hartree-Fock theory is an approximation of minus the ionization energy associated with the removal of an electron from that particular orbital χ_i, i. e., $\varepsilon_i \approx E_N - E^i_{N-1} = -IE(i)$. The simple proof of this theorem can be found in any quantum chemistry textbook.

An important consequence of the only approximate treatment of the electron-electron repulsion is that the true wave function of a many electron system is never a single Slater determinant! We may ask now: if Φ_{SD} is not the exact wave function of N interacting electrons, is there any other (necessarily artificial model) system of which it is the correct wave function? The answer is 'Yes'; it can easily be shown that a Slater determinant is indeed an eigenfunction of a Hamilton operator defined as the sum of the Fock operators of equation (1-25)

$$\hat{H}_{HF}\Phi_{SD} = E^0_{HF}\Phi_{SD} = \sum_i^N \hat{f}_i\Phi_{SD} = \sum_i^N \varepsilon_i\Phi_{SD}. \tag{1-29}$$

Since the Fock operator is a effective one-electron operator, equation (1-29) describes a system of N electrons which do not interact among themselves but experience an effective potential V_{HF}. In other words, the Slater determinant is the exact wave function of N *non-interacting* particles moving in the field of the effective potential V_{HF}.[5] It will not take long before we will meet again the idea of non-interacting systems in the discussion of the Kohn-Sham approach to density functional theory.

1.4 The Restricted and Unrestricted Hartree-Fock Models

Frequently we are dealing with the special but common situation that the system has an even number of electrons which are all paired to give an overall singlet, so-called *closed-shell* systems. The vast majority of all 'normal' compounds, such as water, methane or most other ground state species in organic or inorganic chemistry, belongs to this class. In these

[5] Strictly speaking, this statement applies only to closed-shell systems of non-degenerate point group symmetry, otherwise the wave function consists of a linear combination of a few Slater determinants.

instances the Hartree-Fock solution is usually characterized by having doubly occupied spatial orbitals, i. e., two spin orbitals χ_p and χ_q share the *same spatial orbital* ϕ_p connected with an α and a β spin function, respectively and have the same orbital energy. If we impose this double occupancy right from the start, we arrive at the *restricted* Hartree-Fock approximation, RHF for short. Situations where the RHF picture is inadequate are provided by any system containing an odd number of electrons (the methyl radical or even the hydrogen atom with its single electron fall into this category) or by systems with an even number of electrons, but where not all of these electrons occupy pair-wise one spatial orbital – i. e., *open-shell* situations, such as the triplet ground states of methylene, CH_2 ($\tilde{X}^3B_1$) or the oxygen molecule ($X^3\Sigma_g^-$). There are two possibilities for how one can treat such species within the Hartree-Fock approximation. Either we stay as closely as possible to the RHF picture and doubly occupy all spatial orbitals with the only exception being the explicitly singly occupied ones, or we completely abandon the notion of doubly occupied spatial orbitals and allow each spin orbital to have its own spatial part. The former is the *restricted open-shell* HF scheme (ROHF) while the latter is the much more popular *unrestricted* Hartree-Fock variant (UHF). In UHF the α and β orbitals do not share the same effective potential but experience different potentials, V_{HF}^{α} and V_{HF}^{β}. As a consequence, the α- and β-orbitals differ in their spatial characteristics and have different orbital energies. The UHF scheme affords equations that are much simpler than their ROHF counterparts. Particularly, the ROHF wave function is usually composed not of a single Slater determinant, but corresponds to a limited linear combination of a few determinants where the expansion coefficients are determined by the symmetry of the state. On the other hand, in the UHF scheme we are always dealing with single-determinantal wave functions. However, the major disadvantage of the UHF technique is that unlike the true and also the ROHF wave function, a UHF Slater determinant is no longer an eigenfunction of the total spin operator, $\hat{S}^2$. The more the $\langle \hat{S}^2 \rangle$ expectation value of a Slater determinant deviates from the correct value – i. e., S(S+1) where S is the spin quantum number representing the total spin of the system – the more this unrestricted determinant is contaminated by functions corresponding to states of higher spin multiplicity and the less physically meaningful it obviously gets.

1.5 Electron Correlation

As we have seen in the preceding section a single Slater determinant Φ_{SD} as an approximate wave function captures a significant portion of the physics of a many electron system. However, it never corresponds to the exact wave function. Thus, owing to the variational principle, E_{HF} is necessarily always larger (i. e., less negative) than the exact (within the Born-Oppenheimer approximation and neglecting relativistic effects) ground state energy E_0. The difference between these two energies is, following Löwdin, 1959, called the *correlation energy*

$$E_C^{HF} = E_0 - E_{HF}. \qquad (1\text{-}30)$$

E_C^{HF} is a negative quantity because E_0 and $E_{HF} < 0$ and $|E_0| > |E_{HF}|$. It is a measure for the error introduced through the HF scheme. The development of methods to determine the correlation contributions accurately and efficiently is still a highly active research area in conventional quantum chemistry. Electron correlation is mainly caused by the instantaneous repulsion of the electrons, which is not covered by the effective HF potential. Pictorially speaking, the electrons get often too close to each other in the Hartree-Fock scheme, because the electrostatic interaction is treated in only an average manner. As a consequence, the electron-electron repulsion term is too large resulting in E_{HF} being above E_0. This part of the correlation energy is directly connected to the $1/r_{12}$ term controlling the electron-electron repulsion in the Hamiltonian and is obviously the larger the smaller the distance r_{12} between electrons 1 and 2 is. It is usually called *dynamical* electron correlation because it is related to the actual movements of the individual electrons and is known to be a short range effect. The second main contribution to E_C^{HF} is the *non-dynamical* or static correlation. It is related to the fact that in certain circumstances the ground state Slater determinant is not a good approximation to the true ground state, because there are other Slater determinants with comparable energies. A typical example is provided by one of the famous laboratories of quantum chemistry, the H_2 molecule. At the equilibrium distance the RHF scheme provides a good approximation to the H_2 molecule. The correlation error, which is almost exclusively due to dynamical correlation is small and amounts to only 0.04 E_h. However, as we stretch the bond the correlation gets larger and in the limit of very large distances converges to some 0.25 E_h as evident from Figure 1-1, which displays the computed (RHF and UHF) as well as the exact potential curves for the ground state of the hydrogen molecule.

Obviously, this cannot be dynamical correlation because at $r_{HH} \rightarrow \infty$ we have two independent hydrogen atoms with only one electron at each center and no electron-electron interaction whatsoever (because $1/r_{HH} \rightarrow 0$). To understand this wrong dissociation behavior

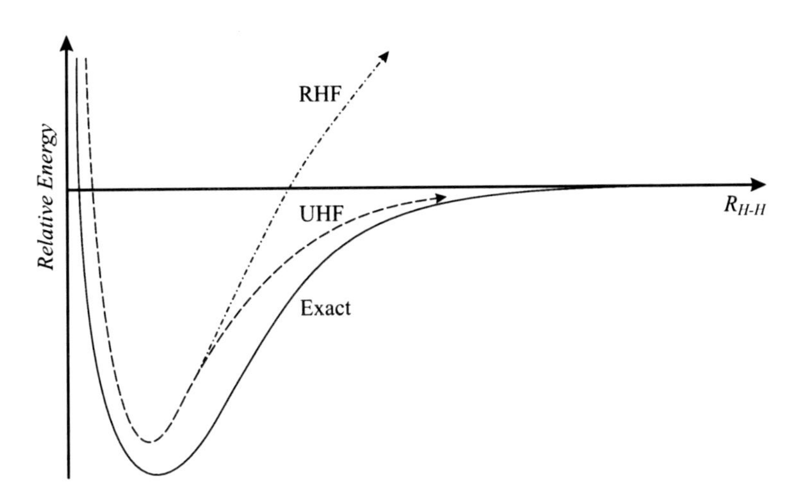

Figure 1-1. Potential curves for H_2.

in the HF picture let us recall from basic quantum mechanics that the HF ground state wave function of the H_2 molecule is the Slater determinant where the bonding σ orbital is doubly occupied

$$\Phi_{GS} = \frac{1}{\sqrt{2}} \det \{\sigma_g(\vec{r}_1)\, \alpha(s_1)\, \sigma_g(\vec{r}_2)\, \beta(s_2)\}. \tag{1-31}$$

Using the simplest picture (and neglecting the effect of overlap on the normalization), this doubly occupied σ_g spatial molecular orbital can be thought of as being the symmetric linear combination of the two 1s atomic orbitals on the 'left' and 'right' hydrogens, H_L and H_R

$$\sigma_g = \frac{1}{\sqrt{2}} \{1s_L + 1s_R\}. \tag{1-32}$$

If we expand the determinant (1-31) in terms of the atomic orbitals (1-32) we get (implicitly assuming that in the determinants the first term is always associated with the coordinates $\vec{r}_1$ and s_1 and the second with $\vec{r}_2$ and s_2)

$$\Phi_{GS} = \frac{1}{2}[\det\{1s_L\alpha\ 1s_R\beta\} + \det\{1s_L\beta\ 1s_R\alpha\} + \det\{1s_L\alpha\ 1s_L\beta\} + \det\{1s_R\alpha\ 1s_R\beta\}]. \tag{1-33}$$

or, pictorially $(H^{\uparrow} \cdots H^{\downarrow}) + (H^{\downarrow} \cdots H^{\uparrow}) + (H^{-\uparrow\downarrow} \cdots H^{+}) + (H^{+} \cdots H^{-\uparrow\downarrow})$

We see that there is an equal probability that the two spin paired electrons in that orbital are shared between the two protons $(H^{\uparrow} \cdots H^{\downarrow}) + (H^{\downarrow} \cdots H^{\uparrow})$, as indicated by the first two terms in equation (1-33) or that both electrons are on one nucleus, giving rise to a hydrogen anion while the other is a mere proton $(H^{-\uparrow\downarrow} \cdots H^{+}) + (H^{+} \cdots H^{-\uparrow\downarrow})$, given by the third and fourth term. While the inclusion of these latter, ionic terms is perfectly adequate for a description at the equilibrium distance, it is not suited at all for the dissociation limit, where the weight of the ionic contributions must of course be zero in order to give the correct asymptotic wave function consisting of two isolated hydrogen atoms

$$\Phi_{DISS} = \frac{1}{\sqrt{2}} \left[\det\{1s_L\alpha\ 1s_R\beta\} + \det\{1s_L\beta\ 1s_R\alpha\}\right]. \tag{1-34}$$

The fact that the HF wave function even at large internuclear distances consists of 50 % of ionic terms, even though H_2 dissociates into two neutral hydrogen atoms, leads to an overestimation of the interaction energy and finally to the large error in the dissociation energy. Another way of looking at this phenomenon is to recognize that to construct the correct expression (1-34) from the molecular orbitals, we have to include the determinant (1-36) composed of the orbital resulting from the antisymmetric linear combination of the 1s atomic orbitals, i. e., the σ_u antibonding orbital,

$$\sigma_u = \frac{1}{\sqrt{2}}\left\{ 1s_L - 1s_R \right\}, \qquad (1\text{-}35)$$

$$\Phi_{AS} = \frac{1}{\sqrt{2}} \det\left\{ \sigma_u(\vec{r}_1)\,\alpha(s_1)\; \sigma_u(\vec{r}_2)\,\beta(s_2) \right\}. \qquad (1\text{-}36)$$

If, as in the RHF scheme, only one of the two determinants is used and the other is completely neglected the picture cannot be complete. Indeed, in terms of determinants constructed from molecular orbitals the qualitatively correct wave function for $r_{HH} \rightarrow \infty$ is

$$\Phi_{DISS} = \frac{1}{\sqrt{2}} \left\{ \Phi_{GS} + \Phi_{AS} \right\} \qquad (1\text{-}37)$$

where both determinants enter with equal weight. This kind of non-dynamical correlation is often also referred to as *left-right* correlation, because it describes the effect that if one electron is at the left nucleus, the other will most likely be at the right one. Obviously, unlike the dynamical correlation discussed before, these non-dynamical contributions are a long range effect and, as in the H_2 case discussed above, become the more important the more the bond is stretched (Cook and Karplus, 1987). However, we also see from Figure 1-1 that using the unrestricted (UHF) scheme rather than RHF cures the problem of the wrong dissociation energy. At an H-H distance of some 1.24 Å an unrestricted solution lower than the RHF one appears and develops into a reasonable potential curve. However, there is no such thing as a free lunch and the price to be paid here is that the resulting UHF wave function no longer resembles the H_2 singlet ground state. At large internuclear distances it actually converges to a physically unreasonable 1:1 mixture between a singlet (S = 0, hence S (S + 1) = 0) and a triplet (S = 1, hence S (S + 1) = 2) as indicated by the expectation value of the $\hat{S}^2$ operator, $\langle \hat{S}^2 \rangle = 1$. The correct energy emerges because the UHF wave function breaks the inversion symmetry inherent to a homonuclear diatomic such as H_2 and localizes one electron with spin down at one nucleus and the second one with opposite spin at the other nucleus. For details, see Szabo and Ostlund, 1982.

Finally, we want to point out that E_C^{HF} is not restricted to the direct contributions connected to the electron-electron interaction. As this quantity measures the difference between the expectation value of $\hat{H}$ with a Slater determinant $\langle \Phi_{SD} | \hat{T} + \hat{V}_{Ne} + \hat{V}_{ee} | \Phi_{SD} \rangle$ and the correct energy obtained from the true wave function Ψ_0, it should come as no surprise that there are also correlation contributions due to the kinetic energy or even the nuclear-electron term. If, for example, the average distance between the electrons is too small at the Hartree-Fock level, this automatically will lead to a kinetic energy that is too large and a nuclear-electron attraction which is too small (i. e., too strong). These indirect contributions can get quite significant and in some cases even constitute the decisive part of E_C^{HF} (Baerends and Gritsenko, 1997). We will see in Chapter 5 that the definition of electron correlation that emerges from the Kohn-Sham formalism of density functional theory is in many aspects similar to the classical one based on the HF scheme discussed at this point, but that there are also some significant differences. Some quantitative data to corroborate the statements of this section can be found in Table 5-1.

In the context of traditional wave function based ab initio quantum chemistry a large variety of computational schemes to deal with the electron correlation problem has been devised during the years. Since we will meet some of these techniques in our forthcoming discussion on the applicability of density functional theory as compared to these conventional techniques, we now briefly mention (but do not explain) the most popular ones.[6] Electron correlation can most economically be accounted for through second order perturbation theory due to Møller and Plesset. This frequently used level is abbreviated MP2. MP4, i. e., Møller-Plesset perturbation theory to fourth order is also often used. This technique is more accurate but also significantly more costly than MP2: while MP2 formally scales with the fifth power of the system size, MP4 scales as $O(m^7)$; m being a measure of the molecular size. For comparison, the formal scaling of Hartree-Fock calculations is $O(m^4)$.[7] Other popular methods are based on configuration interaction (CI), quadratic CI (QCI) and coupled cluster approaches (CC). In principle the exact wave functions and energies of all states of the system could be obtained by these techniques. Of course, in real applications some kind of approximation has to be used. The most common among these are methods known as CISD, QCISD and CCSD, where 'SD' stands for single and double excitations. Even more sophisticated are extensions to QCISD and CCSD where triple excitations are also accounted for through a perturbative treatment, leading to methods called QCISD(T) and CCSD(T), respectively. These last two methods are among the most accurate, but also most expensive (formal scaling is also m^7) computational wave function based techniques generally available.

[6] There is a vast literature on these methods. For a concise but very instructive overview we recommend Bartlett and Stanton, 1995.

[7] The real scaling is significantly smaller, usually between $O(m^2)$ and $O(m^3)$, depending on the system size.

2 Electron Density and Hole Functions

In this chapter we make first contact with the electron density. We will discuss some of its properties and then extend our discussion to the closely related concept of the pair density. We will recognize that the latter contains all information needed to describe the exchange and correlation effects in atoms and molecules. An appealing avenue to visualize and understand these effects is provided by the concept of the exchange-correlation hole which emerges naturally from the pair density. This important concept, which will be of great use in later parts of this book, will finally be used to discuss from a different point of view why the restricted Hartree-Fock approach so badly fails to correctly describe the dissociation of the hydrogen molecule.

2.1 The Electron Density

The probability interpretation from equation (1-7) of the wave function leads directly to the central quantity of this book, the electron density $\rho(\vec{r})$. It is defined as the following multiple integral over the spin coordinates of all electrons and over all but one of the spatial variables

$$\rho(\vec{r}) = N\int\cdots\int\left|\Psi(\vec{x}_1,\vec{x}_2,\ldots,\vec{x}_N)\right|^2 ds_1 d\vec{x}_2\ldots d\vec{x}_N . \tag{2-1}$$

$\rho(\vec{r})$ determines the probability of finding any of the N electrons within the volume element $d\vec{r}_1$ but with arbitrary spin while the other N-1 electrons have arbitrary positions and spin in the state represented by Ψ. Strictly speaking $\rho(\vec{r})$ is a probability density, but calling it the electron density is common practice. It should be noted that the multiple integral as such represents the probability that one particular electron is within the volume element $d\vec{r}_1$. However, since electrons are indistinguishable the probability of finding any electron at this position is just N times the probability for one particular electron. Clearly, $\rho(\vec{r})$ is a non-negative function of only the three spatial variables which vanishes at infinity and integrates to the total number of electrons:

$$\rho(\vec{r}\rightarrow\infty) = 0 , \tag{2-2}$$

$$\int\rho(\vec{r})\, d\vec{r}_1 = N . \tag{2-3}$$

Unlike the wave function, the electron density is an observable and can be measured experimentally, e. g. by X-ray diffraction. One of its important features is that at any position of an atom, $\rho(\vec{r})$ exhibits a maximum with a finite value, due to the attractive force exerted by the positive charge of the nuclei. However, at these positions the gradient of the density has a discontinuity and a *cusp* results. This cusp is a consequence of the singularity in the $-\frac{Z_A}{r_{iA}}$ part in the Hamiltonian as $r_{iA}\rightarrow 0$. Actually, it has long been recognized that

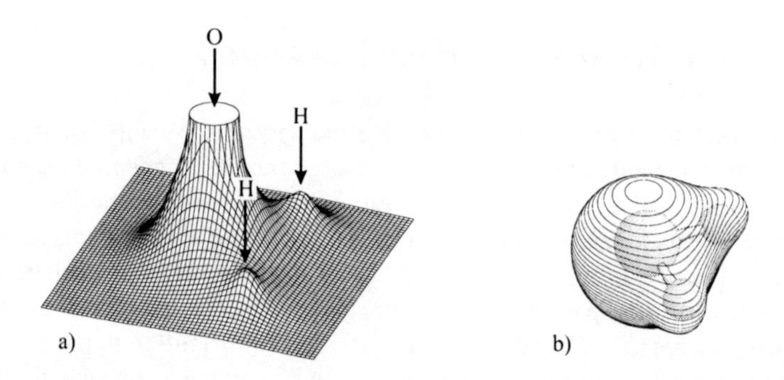

Figure 2-1. Representations of the electron density of the water molecule: (a) relief map showing values of $\rho(r)$ projected onto the plane, which contains the nuclei (large values near the oxygen atom are cut out); (b) three dimensional molecular shape represented by an envelope of constant electron density (0.001 a.u.).

the properties of the cusp are intimately related to the nuclear charge Z of the nucleus according to

$$\lim_{r_{iA}\to 0}\left[\frac{\partial}{\partial r}+2Z_A\right]\bar{\rho}(\vec{r})=0 \tag{2-4}$$

where $\bar{\rho}(\vec{r})$ is the spherical average of $\rho(\vec{r})$. Among the other properties of the density, we mention its asymptotic exponential decay for large distances from all nuclei

$$\rho(\vec{r}) \propto \exp[-2\sqrt{2I}\,|\vec{r}|] \tag{2-5}$$

where I is the exact first ionization energy of the system.

As a typical example we illustrate in Figure 2-1 the electron density of the water molecule in two different representations. In complete analogy, $\rho(\vec{x})$ extends the electron density to the spin-dependent probability of finding any of the N electrons within the volume element $d\vec{r}_1$ and having a spin defined by the spin coordinate s.

2.2 The Pair Density

The concept of electron density, which provides an answer to the question 'how likely is it to find one electron of arbitrary spin within a particular volume element while all other electrons may be anywhere' can now be extended to the probability of finding not one but a pair of two electrons with spins σ_1 and σ_2 simultaneously within two volume elements $d\vec{r}_1$ and $d\vec{r}_2$, while the remaining N-2 electrons have arbitrary positions and spins. The quantity which contains this information is the *pair density* $\rho_2(\vec{x}_1,\vec{x}_2)$, which is defined as

$$\rho_2(\vec{x}_1,\vec{x}_2)=N(N-1)\int\cdots\int\left|\Psi(\vec{x}_1,\vec{x}_2,\ldots,\vec{x}_N)\right|^2 d\vec{x}_3\ldots d\vec{x}_N\,. \tag{2-6}$$

This quantity is of great importance, since it actually contains all information about electron correlation, as we will see presently. Like the density, the pair density is also a non-negative quantity. It is symmetric in the coordinates and normalized to the total number of non-distinct pairs, i. e., N(N-1).[8] Obviously, if electrons were identical, classical particles that do not interact at all, such as for example billiard balls of one color, the probability of finding one electron at a particular point of coordinate-spin space would be completely independent of the position and spin of the second electron. Since in our model we view electrons as idealized mass points with no volume, this would even include the possibility that both electrons are simultaneously found in the same volume element. In this case the pair density would reduce to a simple product of the individual probabilities, i.e.,

$$\rho_2(\vec{x}_1,\vec{x}_2) = \frac{N-1}{N}\rho(\vec{x}_1)\,\rho(\vec{x}_2)\,. \tag{2-7}$$

The (N−1)/N factor enters because the particles are identical and not distinguishable. Pictorially speaking, the probability that any of the N electrons is at $\vec{x}_1$ is given by $\rho(\vec{x}_1)$. The probability that another electron is simultaneously at $\vec{x}_2$ is only (N−1)/N $\rho(\vec{x}_2)$ because the electron at $\vec{x}_1$ cannot at the same time be at $\vec{x}_2$ and the probability must be reduced accordingly.

However, billiard balls are a pretty bad model for electrons. First of all, as discussed above, electrons are fermions and therefore have an antisymmetric wave function. Second, they are charged particles and interact through the Coulomb repulsion; they try to stay away from each other as much as possible. Both of these properties heavily influence the pair density and we will now enter an in-depth discussion of these effects. Let us begin with an exposition of the consequences of the antisymmetry of the wave function. This is most easily done if we introduce the concept of the *reduced density matrix* for two electrons, which we call γ_2. This is a simple generalization of $\rho_2(\vec{x}_1,\vec{x}_2)$ given above according to

$$\gamma_2(\vec{x}_1,\vec{x}_2;\vec{x}'_1,\vec{x}'_2) = \\ N(N-1)\int\cdots\int \Psi(\vec{x}_1,\vec{x}_2,\vec{x}_3,\ldots,\vec{x}_N)\,\Psi^*(\vec{x}'_1,\vec{x}'_2,\vec{x}_3,\ldots,\vec{x}_N)\,d\vec{x}_3\ldots d\vec{x}_N\,. \tag{2-8}$$

When going from ρ_2 to γ_2 we prime those variables in the second factor which are not included in the integration. The two sets of independent and continuous variables, i. e., $\vec{x}_1,\vec{x}_2$ and $\vec{x}'_1,\vec{x}'_2$, define the value of $\gamma_2(\vec{x}_1,\vec{x}_2;\vec{x}'_1,\vec{x}'_2)$ which is the motivation for calling this quantity a matrix (for more information on reduced density matrices see in particular Davidson, 1976, or McWeeny, 1992). If we now interchange the variables $\vec{x}_1$ and $\vec{x}_2$ (or $\vec{x}'_1$ and $\vec{x}'_2$), γ_2 will change sign because of the antisymmetry of Ψ:

[8] This is the normalization adopted for example by McWeeny, 1967, 1992. One also finds 1/2 N(N-1) as prefactor, which corresponds to a normalization to the distinct number of pairs, e. g. Löwdin, 1959 or Parr and Yang, 1989.

$$\gamma_2(\vec{x}_1,\vec{x}_2;\vec{x}'_1,\vec{x}'_2) = -\gamma_2(\vec{x}_2,\vec{x}_1;\vec{x}'_1,\vec{x}'_2). \quad (2\text{-}9)$$

It should be obvious that the diagonal elements of this 'matrix' (i. e., for $\vec{x}_1 = \vec{x}'_1$ and $\vec{x}_2 = \vec{x}'_2$) bring us back to our pair density $\rho_2(\vec{x}_1,\vec{x}_2)$ defined above. If we now look at the special situation that $\vec{x}_1 = \vec{x}_2$, that is the probability that two electrons with the *same spin* are found within the same volume element, we find that

$$\rho_2(\vec{x}_1,\vec{x}_1) = -\rho_2(\vec{x}_1,\vec{x}_1). \quad (2\text{-}10)$$

This can only be true if $\rho_2(\vec{x}_1,\vec{x}_1) = 0$. In other words, this result tells us that *the probability of finding two electrons with the same spin at the same point in space is exactly zero.* Hence, electrons of like spin do not move independently from each other. It is important to realize that this kind of correlation is in no way connected to the charge of the electrons but is a direct consequence of the Pauli principle. It applies equally well to neutral fermions and – also this is very important to keep in mind – does not hold if the two electrons have different spin. This effect is known as *exchange* or *Fermi* correlation. As we will show below, this kind of correlation is included in the Hartree-Fock approach due to the antisymmetry of a Slater determinant and therefore has nothing to do with the correlation energy E_C^{HF} discussed in the previous chapter.

Next, let us explore the consequences of the charge of the electrons on the pair density. Here it is the electrostatic repulsion, which manifests itself through the $1/r_{12}$ term in the Hamiltonian, which prevents the electrons from coming too close to each other. This effect is of course independent of the spin. Usually it is this effect which is called simply electron correlation and in Section 1.4 we have made use of this convention. If we want to make the distinction from the Fermi correlation, the electrostatic effects are known under the label *Coulomb correlation.*

It can easily be shown that the HF approximation discussed in Chapter 1 does include the Fermi-correlation, but completely neglects the Coulomb part. To demonstrate this, we analyze the Hartree-Fock pair density for a two-electron system with the two spatial orbitals ϕ_1 and ϕ_2 and spin functions σ_1 and σ_2

$$\rho_2^{HF}(\vec{x}_1,\vec{x}_2) = \left[\det\{\phi_1(\vec{r}_1)\sigma_1(s_1)\ \phi_2(\vec{r}_2)\sigma_2(s_2)\}\right]^2 \quad (2\text{-}11)$$

which after squaring the expanded determinants becomes

$$\begin{aligned}\rho_2^{HF}(\vec{x}_1,\vec{x}_2) = {} & \phi_1(\vec{r}_1)^2\,\phi_2(\vec{r}_2)^2\,\sigma_1(s_1)^2\,\sigma_2(s_2)^2 \\ & +\phi_1(\vec{r}_2)^2\,\phi_2(\vec{r}_1)^2\,\sigma_1(s_2)^2\,\sigma_2(s_1)^2 \\ & -2\,\phi_1(\vec{r}_1)\,\phi_2(\vec{r}_1)\,\phi_1(\vec{r}_2)\,\phi_2(\vec{r}_2)\,\sigma_1(s_1)\,\sigma_2(s_1)\,\sigma_1(s_2)\,\sigma_2(s_2)\end{aligned} \quad (2\text{-}12)$$

The spin-independent probability of finding one electron at $\vec{r}_1$ and the other simultaneously at $\vec{r}_2$ is obtained by integrating over the spins. Since the spin functions are orthonormal (recall Section 1.3) this integration simply yields 1 for the first two terms. Furthermore, the first and second term in equation (2-12) are identical because electrons are indistinguishable and therefore it does not matter which of the electrons – 'number 1' or 'number 2' – is associated with the first or the second orbital. If, however, $\sigma_1 \neq \sigma_2$, i. e., the electrons' spins are *antiparallel*, the last of the three terms in equation (2-12) will vanish due to the orthonormality of the spin functions, $<\alpha(s_1) \mid \beta(s_1)> = 0$. This finally leads to $\rho_2^{HF,\sigma_1 \neq \sigma_2}(\vec{r}_1, \vec{r}_2) = \rho(\vec{r}_1)\,\rho(\vec{r}_2)$ which corresponds to the completely uncorrelated situation.[9] Note that $\rho_2^{HF,\sigma_1 \neq \sigma_2}(\vec{r}_1, \vec{r}_2)$ does not necessarily vanish even for $\vec{r}_1 = \vec{r}_2$. On the other hand, if $\sigma_1 = \sigma_2$, i. e., the electrons' spins are *parallel*, the last term in equation (2-12) will not vanish but yields $<\sigma(s_i) \mid \sigma(s_i)> = 1$ ($\sigma = \alpha, \beta$). Hence, $\rho_2^{HF,\sigma_1 = \sigma_2}(\vec{r}_1, \vec{r}_2)$ does not reduce to the simple, uncorrelated product of individual probabilities. Rather, for $\vec{r}_1 = \vec{r}_2$, the third term exactly cancels the first two and we indeed arrive at $\rho_2^{HF}(\vec{x}_1, \vec{x}_1) = 0$. Thus, we re-derived the conclusions from the end of the preceding chapter that the correlation due to the antisymmetry of the wave function is covered by the HF scheme – after all no surprise since a Slater determinant is antisymmetric in the coordinates of any two electrons. Electrons of antiparallel spins though move in a completely uncorrelated fashion and Coulomb correlation is not present at the Hartree-Fock level, as discussed in the previous chapter.

It is now convenient to express the influence of the Fermi and Coulomb correlation on the pair density by separating the pair density into two parts, i. e. the simple product of independent densities and the remainder, brought about by Fermi and Coulomb effects and accounting for the (N-1)/N normalization

$$\rho_2(\vec{x}_1, \vec{x}_2) = \rho(\vec{x}_1)\,\rho(\vec{x}_2)\left[1 + f(\vec{x}_1; \vec{x}_2)\right]. \tag{2-13}$$

Here, $f(\vec{x}_1; \vec{x}_2)$ is sometimes called the *correlation factor*. Consequently, $f(\vec{x}_1; \vec{x}_2) = 0$ defines the completely uncorrelated case. However, note that in this case, i. e., for $f(\vec{x}_1; \vec{x}_2) = 0$, $\rho_2(\vec{x}_1, \vec{x}_2)$ is normalized to the wrong number of pairs, since $\iint \rho_2^{HF,\sigma_1 \neq \sigma_2}(\vec{r}_1, \vec{r}_2)\, d\vec{r}_1\, d\vec{r}_2 = \iint \rho(\vec{r}_1)\rho(\vec{r}_2)\, d\vec{r}_1\, d\vec{r}_2 = N^2$ rather than N(N-1) and therefore contains the unphysical self-interaction. We now go one step further and define the *conditional probability* $\Omega(\vec{x}_2; \vec{x}_1)$. This is the probability of finding any electron at position 2 in coordinate-spin space if there is one already known to be at position 1

$$\Omega(\vec{x}_2; \vec{x}_1) = \frac{\rho_2(\vec{x}_1, \vec{x}_2)}{\rho(\vec{x}_1)}. \tag{2-14}$$

The conditional density obviously integrates to N-1 electrons, containing all electrons but the reference electron at $\vec{x}_1$

$$\int \Omega(\vec{x}_2; \vec{x}_1)\, d\vec{x}_2 = N - 1. \tag{2-15}$$

[9] The (N-1)/N factor of equation (2-7) disappears because the two electrons in question have different spin.

The difference between $\Omega(\vec{x}_2;\vec{x}_1)$ and the uncorrelated probability of finding an electron at $\vec{x}_2$ describes the change in conditional probability caused by the correction for self-interaction, exchange and Coulomb correlation, compared to the completely uncorrelated situation:

$$h_{XC}(\vec{x}_1;\vec{x}_2) = \frac{\rho_2(\vec{x}_1,\vec{x}_2)}{\rho(\vec{x}_1)} - \rho(\vec{x}_2) = \rho(\vec{x}_2)f(\vec{x}_1;\vec{x}_2). \tag{2-16}$$

Since correlation typically leads to a depletion of the electron density at $\vec{x}_2$ as compared to the independent particle situation, the quantity $h_{XC}(\vec{x}_1;\vec{x}_2)$ is called the *exchange-correlation hole* which usually has a negative sign, in particular in the vicinity of the reference electron. In addition, if we integrate equation (2-16), recalling from equation (2-15) that $\Omega(\vec{x}_2;\vec{x}_1)$ integrates to N-1, while $\int \rho(\vec{x}_2)d\vec{x}_2 = N$ we immediately see the important result that the *exchange-correlation hole contains exactly the charge of one electron*

$$\int h_{XC}(\vec{x}_1;\vec{x}_2)d\vec{x}_2 = -1. \tag{2-17}$$

The concept of the exchange-correlation hole is widely used in density functional theory and its most relevant properties are the subject of the following section.

2.3 Fermi and Coulomb Holes

The idea of the exchange-correlation hole function allows a very pictorial and intuitively appealing access to an understanding of how exchange and Coulomb correlation affects the electron distribution in an atom or molecule. In this context we can imagine the electron digging a hole around itself such that the probability of finding another electron nearby is diminished. As the hole density usually has a negative sign, the electrostatic interaction of the necessarily positive electron density at a certain position with the exchange-correlation hole surrounding it is attractive. Using the new concepts introduced so far, it is worthwhile to take a fresh look at the expectation value of $\hat{V}_{ee}$, the electron-electron repulsion term in the Hamiltonian, equation (1-4), which corresponds to the potential energy due to the electrostatic repulsion of the electrons, E_{ee}. This interaction depends on the distance between two electrons weighted by the probability that this distance will occur. Thus, we can express E_{ee} in terms of the spin-independent equivalent of the pair density (i.e., where we have integrated over the spin coordinates) which contains just this information

$$E_{ee} = \left\langle \Psi \left| \sum_{i}^{N}\sum_{j>i}^{N}\frac{1}{r_{ij}} \right| \Psi \right\rangle = \frac{1}{2}\iint \frac{\rho_2(\vec{r}_1,\vec{r}_2)}{r_{12}} d\vec{r}_1 d\vec{r}_2. \tag{2-18}$$

Using $\rho_2(\vec{r}_1,\vec{r}_2) = \rho(\vec{r}_1)\rho(\vec{r}_2) + \rho(\vec{r}_1)h_{XC}(\vec{r}_1;\vec{r}_2)$ (cf. the spin-integrated analog of equation 2-16) we can split E_{ee} in two contributions which can be easily interpreted,

$$E_{ee} = \frac{1}{2}\iint \frac{\rho(\vec{r}_1)\rho(\vec{r}_2)}{r_{12}} d\vec{r}_1 d\vec{r}_2 + \frac{1}{2}\iint \frac{\rho(\vec{r}_1) h_{XC}(\vec{r}_1;\vec{r}_2)}{r_{12}} d\vec{r}_1 d\vec{r}_2 . \tag{2-19}$$

The first term is $J[\rho]$, the classical electrostatic energy of a charge distribution with itself. Again, it is important to realize that $J[\rho]$ contains also the unphysical self-interaction as already alluded to in Chapter 1. This can most easily be illustrated by considering a one-electron system: with only one electron there obviously cannot be any electron-electron Coulomb interaction. Nevertheless, even in these cases $J[\rho] \neq 0$. The second term is the energy of interaction between the charge density and the charge distribution of the exchange-correlation hole. It includes the correction for the self-interaction as well as all contributions of quantum-mechanical correlation effects. It should be obvious by now why the hole functions are so useful for discussing exchange and correlation effects. The more we know about the characteristics of h_{XC} and the better the approximate hole functions we use in our calculations resemble the true ones, the more accurate results we can expect.

The exchange-correlation hole can formally be split into the *Fermi hole*, $h_X^{\sigma_1=\sigma_2}(\vec{r}_1,\vec{r}_2)$ and the *Coulomb hole* $h_C^{\sigma_1,\sigma_2}(\vec{r}_1,\vec{r}_2)$,

$$h_{XC}(\vec{r}_1;\vec{r}_2) = h_X^{\sigma_1=\sigma_2}(\vec{r}_1;\vec{r}_2) + h_C^{\sigma_1,\sigma_2}(\vec{r}_1,\vec{r}_2) \tag{2-20}$$

where the former is the hole in the probability density of electrons due to the Pauli principle, i. e., the antisymmetry of the wave function and applies only to electrons with the *same spin*. The latter has contributions for electrons of either spin and is the hole resulting from the $1/r_{12}$ electrostatic interaction. These definitions are motivated by the HF picture where the Fermi hole is accounted for through the use of a single Slater determinant whereas the Coulomb hole is neglected. Even though the separation of h_{XC} into an exchange and a correlation contribution is convenient, we must keep in mind that only the *total hole* has a real physical meaning. In the following we will discuss some of the properties of these individual holes.

2.3.1 The Fermi Hole

First of all we note that the Fermi hole – which is due to the antisymmetry of the wave function – dominates by far the Coulomb hole. Second, another, very important property of the Fermi hole is that it, just like the total hole, integrates to –1

$$\int h_X(\vec{r}_1;\vec{r}_2) d\vec{r}_2 = -1 . \tag{2-21}$$

This is easy to understand because it means that the conditional probability for electrons of spin σ integrates to N_σ - 1 instead of N_σ because there is one electron of the same spin σ

already known to be at $\vec{r}_1$. Hence, this electron is removed from the distribution. By this removal of one charge, the Fermi hole also takes care of the self-interaction problem. Further, due to the Pauli principle which ensures that two electrons of the same spin cannot be at the same position in space, the Fermi hole has to become equal to minus the density of electrons with this spin at the position of the reference electron for $\vec{r}_2 \rightarrow \vec{r}_1$,

$$h_X(\vec{r}_2 \rightarrow \vec{r}_1; \vec{r}_1) = -\rho(\vec{r}_1). \tag{2-22}$$

What can we say about the shape of the Fermi hole? First, it can be shown that h_X is negative everywhere,

$$h_X(\vec{r}_1; \vec{r}_2) < 0. \tag{2-23}$$

Second, if we recall the definition, equation (2-16), and modify it for the exchange-only case

$$h_X(\vec{r}_1; \vec{r}_2) = \rho(\vec{r}_2) f_X(\vec{r}_1; \vec{r}_2) \tag{2-24}$$

we see that the actual shape depends not only on the Fermi correlation factor but also on the density at $\vec{r}_2$. As a consequence, it will certainly not be spherically symmetric. Usually, the exchange hole is largest around the probe electron. However, if, for example, the reference electron is at a position $\vec{r}_1$ located far away from the atom or molecule, the Fermi hole will be only slowly varying for $\vec{r}_2$ being within regions of appreciable electron density. Around the reference electron it will be almost negligible because $\rho(\vec{r}_2 \rightarrow \vec{r}_1)$ will be small. In a way, the Fermi hole 'stays behind' when the reference electron goes outside the regions of normal electron density. There are also situations where the Fermi hole actually tends to be delocalized. Let us again use the ground state of the H_2 molecule as a simple but very instructive example. Here, there are only two electrons with spin paired and the only duty of the Fermi hole in this case is to cancel the self-interaction. Thus, for the α (or equivalently the β) electron, $h_X^{\alpha}(\vec{r}_1; \vec{r}_2)$ is equal to minus the α- (or β-) density which equals half the total density. Thus, this hole is half the squared σ_g molecular orbital in H_2. It is therefore delocalized over the whole molecule, representing a charge depletion of half an electron from the vicinity of each nucleus. Note that the Fermi hole is in this case completely independent of the location of the reference electron. Even for $r_{HH} \rightarrow \infty$, where there should be one electron at each center, the Fermi hole will still remove only *half* of the density from the location of the reference electron. As a consequence, the attraction of the reference electron to the nucleus will be partially screened and its density will therefore be too diffuse. This is exactly what happens in the HF scheme, where the total hole is approximated solely by the Fermi hole. The too diffuse density causes a severe underestimation of the electron-nuclear attraction accompanied by a too low kinetic energy and an electron-electron repulsion which is also too low. All these effects contribute to the large error in the Hartree-Fock dissociation curve, as we will show in more detail in Section 5.3.2 (see also Baerends and Gritsenko, 1997).

2.3.2 The Coulomb Hole

From equations (2-17) and (2-21) it is obvious that the Coulomb hole must be normalized to zero, i. e. the integral over all space contains no charge:

$$\int h_C(\vec{r}_1;\vec{r}_2)\,d\vec{r}_2 = 0\,. \tag{2-25}$$

This makes good physical sense since for electrons of unlike spin the probability of finding an electron of spin σ *anywhere in space* is of course the total number of electrons of this spin, i. e., N_σ. This result is independent of the positions of electrons with spin $\sigma' \neq \sigma$. Also, there is no need for a self-interaction correction. The Coulomb hole will be negative and largest at the position of the reference electron since it originates from the $1/r_{12}$ electrostatic interaction which keeps electrons apart. Since there is a finite probability that two electrons of different spin (and approximated as volume-less point charges) can be found within the same volume element $h_C(\vec{r}_1;\vec{r}_1)$ has no predetermined value at vanishing interelectronic distance, unlike the Fermi hole which for $\vec{r}_2 \rightarrow \vec{r}_1$ approaches $-\rho(\vec{r}_1)$. However, what $h_C(\vec{r}_1;\vec{r}_1)$ mirrors for $\vec{r}_2 \rightarrow \vec{r}_1$ is the cusp condition that we already met in the discussion of the density. Because no two electrons of parallel spin can occupy the same point in space, the cusp condition occurs only for electrons of antiparallel spin. Since the Coulomb hole integrates to zero it must also be positive in some regions. In other words, density is taken away from areas close to the reference electron and is piled up in regions farther away from it. How does the Coulomb hole look like in our H_2 laboratory molecule? Of course, this is a particularly simple case, because there are no parallel spin contributions and the Coulomb hole refers only to the interaction of electrons of antiparallel spins. If one of the two electrons is found at, say, the left proton, the probability to find the other one at the right nucleus will be higher and vice versa. The larger the distance between the two H atoms the more pronounced this effect will be. Thus, also the Coulomb hole will be delocalized with a negative part at the nucleus where the reference electron sits and a positive part, i. e., a build-up of charge at the other nucleus. At the extreme, when $r_{HH} \rightarrow \infty$, the Coulomb hole will remove half an electron from the nucleus where the reference electron is positioned and build a charge of half an electron at the other nucleus. Unlike the Fermi hole, which for H_2 was found to be completely independent of where the reference electron is located, it should be clear from the foregoing discussion that the Coulomb hole has to switch abruptly if the reference electron moves from the left to the right nucleus.

To visualize the above discussion we show in Figure 2-2 the Fermi, Coulomb and total exchange-correlation holes for H_2 at various distances. The probe electron is placed at 0.3 bohr to the left of the right proton. We see immediately – in particular for large r – that while both components of the hole are *delocalized*, the sum of the two, i. e., the total hole, is *localized* at the proton of the reference electron. At large distances the Coulomb hole is negative at the right proton and adds to the Fermi hole, while at the left proton the Coulomb hole is positive and exactly cancels the Fermi hole. The result is a total hole that removes exactly one electron from the right proton as it should in order to yield two undisturbed hydrogen atoms in the dissociation limit. An important observation here is that neither of the two individual components of the total hole is anywhere close to a reasonable representation of the E_{XC} total hole. *Only if the Fermi and Coulomb holes are combined, the correct shape emerges.*

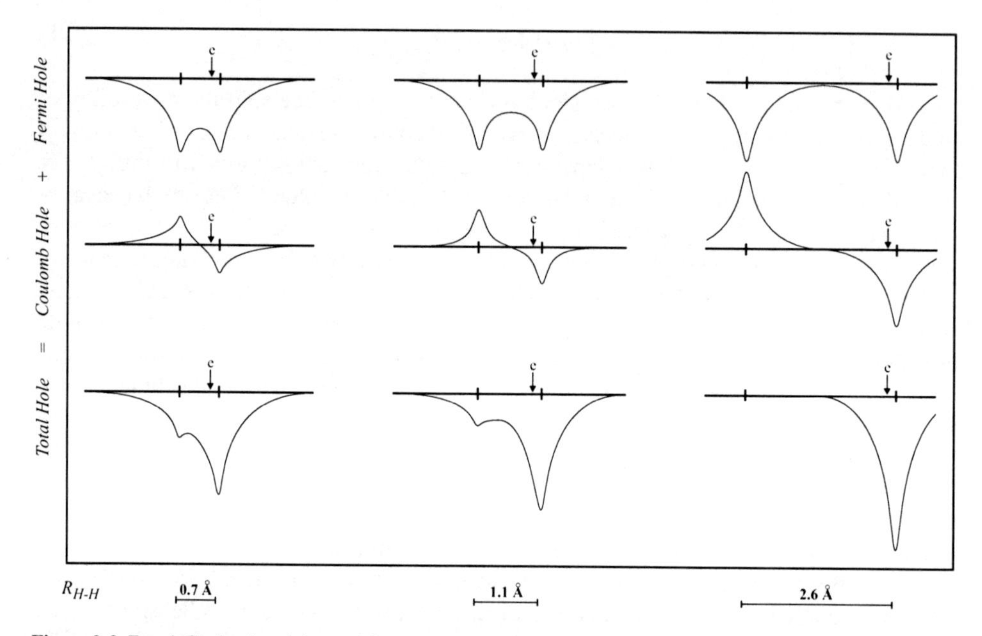

Figure 2-2. Fermi, Coulomb and the resulting total exchange-correlation holes for H_2 at three different internuclear distances; the position of the probe electron is marked with an arrow (adapted from Baerends and Gritsenko, *J. Phys. Chem. A*, **101**, 5390 (1997), with permission by the American Chemical Society).

3 The Electron Density as the Basic Variable: Early Attempts

In this section we will approach the question which is at the very heart of density functional theory: can we possibly replace the complicated N-electron wave function with its dependence on 3N spatial plus N spin variables by a simpler quantity, such as the electron density? After using plausibility arguments to demonstrate that this seems to be a sensible thing to do, we introduce two early realizations of this idea, the Thomas-Fermi model and Slater's approximation of Hartree-Fock exchange defining the X_α method. The discussion in this chapter will prepare us for the next steps, where we will encounter physically sound reasons why the density is really all we need.

3.1 Does it Make Sense?

The conventional approach to quantum chemistry uses the wave function Ψ as the central quantity. The reason is that once we know Ψ (or a good approximation to it) we have access to all information that can be known about this particular state of our target system. A typical example of this approach is the Hartree-Fock approximation that we expounded in Chapter 1. There is, however, a severe problem. The wave function is a very complicated quantity that cannot be probed experimentally and that depends on 4N variables, three spatial and one spin variable for each of the N electrons. The systems we are interested in in chemistry, biology and material science contain many atoms and many more electrons. Thus, any wave function based treatment will soon reach an unmanageable size. This not only makes a computational treatment very difficult if not impossible but also reduces the possibility of a descriptive understanding and renders this approach inaccessible to intuition. On the other hand, the Hamilton operator $\hat{H}$ contains only operators that act on one ($\hat{T}$ and $\hat{V}_{Ne}$) or at most two ($\hat{V}_{ee}$) particles at a time, independent of the size of the system. Hence, one may wonder whether the complicated wave function is really needed for obtaining the energy and other properties of interest or whether it contains redundant and irrelevant information with regard to this purpose and we can get away with a less complicated quantity as the central variable. This is indeed the case! First, it is fairly straightforward to show that the Schrödinger equation can be rewritten in terms of the reduced one- and two-particle density matrices[10] and we end up with an equation that depends on 8, rather than 4N variables, independent of the system size (McWeeny, 1992, Kryachko and Ludeña, 1990). However, in the present context we are not going to follow that avenue any further, since it actually represents a detour from our real goal, i. e., density functional theory. Rather, we want to use the electron density $\rho(\vec{r})$ as discussed in the previous chapter – a quantity that depends only on the three spatial variables and is therefore an object in

[10] The former is a generalization of the electron density in the same spirit as $\gamma_2(\vec{x}_1, \vec{x}_2; \vec{x}'_1, \vec{x}'_2)$ is a generalization of $\rho_2(\vec{x}_1, \vec{x}_2)$, see Section 2.2.

the three-dimensional physical space – as a means to reach a solution to the Schrödinger equation. That this endeavor has some chances of success can be deduced from the following sequence of observations: recall that in Section 1.2 we arrived at the conclusion that the Hamilton operator of any atomic or molecular system is uniquely defined by N, the number of electrons, R_A, the position of the nuclei in space, and Z_A the charges of the nuclei. We went on by saying that once $\hat{H}$ is known, we can – of course only in principle – solve the Schrödinger equation. We also showed in the discussion of the properties of the electron density $\rho(\vec{r})$ in Section 2.1 that

(i) $\int \rho(\vec{r}_1)\, d\vec{r}_1 = N$, i. e., the density integrates to the number of electrons,

(ii) $\rho(\vec{r})$ has maxima, that are actually even cusps, only at the positions R_A of the nuclei, and that

(iii) $\lim\limits_{r_{iA}\to 0}\left[\frac{\partial}{\partial r} + 2Z_A\right]\bar{\rho}(\vec{r}) = 0$, i. e.,
the density at the position of the nucleus contains information about the nuclear charge Z.

Thus, the electron density already provides all the ingredients that we identified as being necessary for setting up the system specific Hamiltonian and it seems at least very plausible that in fact $\rho(\vec{r})$ suffices for a complete determination of all molecular properties (of course, this does not relieve us from the task of actually solving the corresponding Schrödinger equation and all the difficulties related to this). As noted by Handy, 1994, these very simple and beautifully intuitive arguments in favor of density functional theory are attributed to E. B. Wilson. So the answer to the question posed in the caption to this section is certainly a loud and clear 'Yes'.

3.2 The Thomas-Fermi Model

Actually, the first attempts to use the electron density rather than the wave function for obtaining information about atomic and molecular systems are almost as old as is quantum mechanics itself and date back to the early work of Thomas, 1927 and Fermi, 1927. In the present context, their approach is of only historical interest. We therefore refrain from an in-depth discussion of the Thomas-Fermi model and restrict ourselves to a brief summary of the conclusions important to the general discussion of DFT. The reader interested in learning more about this approach is encouraged to consult the rich review literature on this subject, for example by March, 1975, 1992 or by Parr and Yang, 1989.

At the center of the approach taken by Thomas and Fermi is a quantum statistical model of electrons which, in its original formulation, takes into account only the kinetic energy while treating the nuclear-electron and electron-electron contributions in a completely classical way. In their model Thomas and Fermi arrive at the following, very simple expression for the kinetic energy based on the uniform electron gas, a fictitious model system of constant electron density (more information on the uniform electron gas will be given in Section 6.4):

$$T_{TF}[\rho(\vec{r})] = \frac{3}{10}(3\pi^2)^{2/3}\int \rho^{5/3}(\vec{r})d\vec{r}\,. \tag{3-1}$$

If this is combined with the classical expression for the nuclear-electron attractive potential and the electron-electron repulsive potential we have the famous Thomas-Fermi expression for the energy of an atom,

$$E_{TF}[\rho(\vec{r})] = \frac{3}{10}(3\pi^2)^{2/3}\int \rho^{5/3}(\vec{r})d\vec{r} - Z\int\frac{\rho(\vec{r})}{r}\,d\vec{r} + \frac{1}{2}\int\int\frac{\rho(\vec{r}_1)\rho(\vec{r}_2)}{r_{12}}\,d\vec{r}_1 d\vec{r}_2\,. \tag{3-2}$$

The importance of this equation is not so much how well it is able to really describe the energy of an atom (actually it is only of limited use in that respect because T_{TF} is only a very coarse approximation to the true kinetic energy and exchange and correlation effects are completely neglected), but that the energy is given completely in terms of the electron density $\rho(\vec{r})$. Thus, we have the first example of a genuine density functional for the energy! In other words, equation (3-2) is a prescription for how to map a density $\rho(\vec{r})$ onto an energy E without any additional information required. In particular no recourse to the wave function is taken. Now that we have a functional expressing the energy in terms of the density, the next important step is to find a strategy for how the correct density that we need to insert into (3-2) can be identified. To this end, the Thomas-Fermi model employs the variational principle. It is *assumed* that the ground state of the system is connected to the electron density for which the energy according to equation (3-2) is minimized under the constraint of $\int \rho(\vec{r}_1)\,d\vec{r}_1 = N$. Note, at this point we do not know either whether expressing the energy as a functional of $\rho(\vec{r})$ is physically justified or whether a procedure employing the variational principle on the density is really allowed in this context. Thus, for the time being the only right of existence of the Thomas-Fermi model is that it seems reasonable!

3.3 Slater's Approximation of Hartree-Fock Exchange

Let us introduce another early example by Slater, 1951, where the electron density is exploited as the central quantity. This approach was originally constructed not with density functional theory in mind, but as an approximation to the non-local and complicated exchange contribution of the Hartree-Fock scheme. We have seen in the previous chapter that the exchange contribution stemming from the antisymmetry of the wave function can be expressed as the interaction between the charge density of spin σ and the Fermi hole of the same spin

$$E_X = \frac{1}{2}\int\int\frac{\rho(\vec{r}_1)h_X(\vec{r}_1;\vec{r}_2)}{r_{12}}\,d\vec{r}_1 d\vec{r}_2\,. \tag{3-3}$$

Hence, if we can construct a simple but reasonable approximation to the Fermi hole, the solution of equation (3-3) can be made considerably easier. Slater's idea was to assume that the exchange hole is spherically symmetric and centered around the reference electron at $\vec{r}_1$. We further assume that within the sphere the exchange hole density is constant, having minus the value of $\rho(\vec{r}_1)$, while outside it is zero. Since the Fermi hole is known to contain exactly one elementary charge (cf. equation (2-21)), the radius of this sphere is then given by

$$r_S = \left(\frac{3}{4\pi}\right)^{1/3} \rho(\vec{r}_1)^{-1/3}. \tag{3-4}$$

The radius r_S is sometimes called the *Wigner-Seitz* radius and can be interpreted to a first approximation as the average distance between two electrons in the particular system. Regions of high density are characterized by small values of r_S and vice versa. From standard electrostatics it is known that the potential of a uniformly charged sphere with radius r_S is proportional to $1/r_S$, or, equivalently, to $\rho(\vec{r}_1)^{1/3}$. Hence, we arrive at the following approximate expression for E_X (C_X is a numerical constant),

$$E_X[\rho] \cong C_X \int \rho(\vec{r}_1)^{4/3}\, d\vec{r}_1 . \tag{3-5}$$

What does this mean? We have replaced the non-local and therefore fairly complicated exchange term of Hartree-Fock theory as given in equation (3-3) by a simple approximate expression which depends only on the *local* values of the electron density. Thus, this expression represents a density functional for the exchange energy. As noted above, this formula was originally explicitly derived as an approximation to the HF scheme, without any reference to density functional theory. To improve the quality of this approximation an adjustable, semiempirical parameter α was introduced into the pre-factor C_X which leads to the X_α or *Hartree-Fock-Slater* (HFS) method which enjoyed a significant amount of popularity among physicists, but never had much impact in chemistry,

$$E_{X\alpha}[\rho] = -\frac{9}{8}\left(\frac{3}{\pi}\right)^{1/3} \alpha \int \rho(\vec{r}_1)^{4/3}\, d\vec{r}_1 . \tag{3-6}$$

Typical values of α are between 2/3 and 1. We will later see that the 4/3 power law for the dependence of the exchange interaction on the electron density is also obtained from a completely different approach using the concept of the uniform electron gas, work pioneered by Bloch, 1929, and Dirac, 1930. A detailed discussion of the uniform electron gas and its expressions for exchange and correlation energies awaits the reader in Section 6.4. It is worth mentioning that if the exchange contributions of equation (3-5), with a slightly modified value of C_X, are combined with the Thomas-Fermi energy given by expression (3-2), we end up with an approximation which is known as the *Thomas-Fermi-Dirac* model. This model now includes the kinetic and classical Coulomb contributions as well as the quantum mechanical exchange effects. The important point is that all parts are expressed as pure functionals of the density. Since, just like the original Thomas-Fermi model, the Thomas-Fermi-Dirac extension was also not very successful in chemical applications, we will not discuss it any further.

4 The Hohenberg-Kohn Theorems

Density functional theory as we know it today was born in 1964 when a landmark paper by Hohenberg and Kohn appeared in the Physical Review. The theorems proven in this report represent the major theoretical pillars on which all modern day density functional theories are erected. This chapter introduces these Hohenberg-Kohn theorems and discusses their obvious and maybe not so obvious consequences. We also give an alternative, more modern approach, namely the Levy constraint-search scheme. We go on and discuss the question whether a physically meaningful wave function can be uniquely associated with a certain density. The common denominator in our discussion is the primacy of the application-oriented understanding over the puristic theoretical point of view. Readers who have also an affinity towards the latter and want to learn more about the many theoretical intricacies of the Hohenberg-Kohn theorems are recommended to consult the comprehensive and theoretically sound discussions contained in Parr and Yang, 1989, Kryachko and Ludeña, 1990, Dreizler and Gross, 1995 and Eschrig, 1996.

4.1 The First Hohenberg-Kohn Theorem: Proof of Existence

The first Hohenberg-Kohn theorem provides the proof that our plausibility arguments given at the beginning of the previous chapter are indeed physically justified. To put it differently we are about to show that the electron density in fact uniquely determines the Hamilton operator and thus all properties of the system. The proof originally given by Hohenberg and Kohn in their 1964 paper is disarmingly simple, almost trivial and one may wonder why it took about 40 years after Thomas and Fermi first used the density as a basic variable before their approach was put onto a firm physical foundation. Quoting directly from the Hohenberg/Kohn paper, this first theorem states that '*the external potential* $V_{ext}(\vec{r})$ *is (to within a constant) a unique functional of* $\rho(\vec{r})$ *; since, in turn* $V_{ext}(\vec{r})$ *fixes* $\hat{H}$ *we see that the full many particle ground state is a unique functional of* $\rho(\vec{r})$'. The proof runs as follows and is based on reductio ad absurdum. We start by considering two external potentials V_{ext} and V'_{ext} which differ by more than a constant (since the wave function and hence the charge density is unaltered if a constant is added to the potential, we must require from the outset that the two external potentials differ not only by a constant) but which both give rise to the same electron density $\rho(\vec{r})$ associated with the corresponding non-degenerate ground states of N particles (the limitation to non-degenerate ground states of the original Hohenberg-Kohn argument will later be lifted, see below). These two external potentials are part of two Hamiltonians which only differ in the external potential, $\hat{H} = \hat{T} + \hat{V}_{ee} + \hat{V}_{ext}$ and $\hat{H}' = \hat{T} + \hat{V}_{ee} + \hat{V}'_{ext}$. Obviously, the two Hamiltonians $\hat{H}$ and $\hat{H}'$ belong to two different ground state wave functions, Ψ and Ψ', and corresponding ground state energies, E_0 and E'_0, respectively, with $E_0 \neq E'_0$. However, we assume that both wave functions give rise to the same electron density (this is very well possible, since the prescription of how a density is constructed from a wave function by quadrature, i. e., $\rho(\vec{r}) = N\int\cdots\int|\Psi(\vec{x}_1, \vec{x}_2, \ldots, \vec{x}_N)|^2 ds_1 d\vec{x}_2 \ldots d\vec{x}_N$, is of course not unique). We express this schematically following our notation from Section 1.2 as

$$V_{ext} \Rightarrow \hat{H} \Rightarrow \Psi \Rightarrow \quad \rho(\vec{r}) \quad \Leftarrow \Psi' \Leftarrow \hat{H}' \Leftarrow V'_{ext}.$$

Therefore Ψ and Ψ', respectively, are different, and we can use Ψ' as trial wave function for $\hat{H}$. We must then have by virtue of the variational principle

$$E_0 < \langle \Psi' | \hat{H} | \Psi' \rangle = \langle \Psi' | \hat{H}' | \Psi' \rangle + \langle \Psi' | \hat{H} - \hat{H}' | \Psi' \rangle \tag{4-1}$$

or, because the two Hamilton operators differ only in the external potential

$$E_0 < E'_0 + \langle \Psi' | \hat{T} + \hat{V}_{ee} + \hat{V}_{ext} - \hat{T} - \hat{V}_{ee} - \hat{V}'_{ext} | \Psi' \rangle \tag{4-2}$$

which yields

$$E_0 < E'_0 + \int \rho(\vec{r}) \{V_{ext} - V'_{ext}\} d\vec{r}. \tag{4-3}$$

Interchanging the unprimed with the primed quantities and repeating the above steps of equations (4-1) to (4-3) we arrive at the corresponding equation

$$E'_0 < E_0 - \int \rho(\vec{r}) \{V_{ext} - V'_{ext}\} d\vec{r}. \tag{4-4}$$

After adding equations (4-3) and (4-4), this leaves us with the clear contradiction

$$E_0 + E'_0 < E'_0 + E_0 \text{ or } 0 < 0. \tag{4-5}$$

This concludes the proof that there cannot be two different V_{ext} that yield the same ground state electron density, or, in other words, that the ground state density uniquely specifies the external potential V_{ext}. Using again the terminology of Section 1.2 we can simply add ρ_0 as the property which contains the information about $\{N, Z_A, R_A\}$ and summarize this as

$$\rho_0 \Rightarrow \{N, Z_A, R_A\} \Rightarrow \hat{H} \Rightarrow \Psi_0 \Rightarrow E_0 \text{ (and all other properties).}$$

Since the complete ground state energy is a functional of the ground state electron density so must be its individual components and we can write (where we revert to the subscript 'Ne' to specify the kind of external potential present in our case, which is fully defined by the attraction due to the nuclei)

$$E_0[\rho_0] = T[\rho_0] + E_{ee}[\rho_0] + E_{Ne}[\rho_0]. \tag{4-6}$$

It is convenient at this point to separate this energy expression into those parts that depend on the actual system, i. e., the potential energy due to the nuclei-electron attraction,

$E_{Ne}[\rho_0] = \int \rho_0(\vec{r})V_{Ne}d\vec{r}$, and those which are universal in the sense that their form is independent of N, R_A and Z_A

$$E_0[\rho_0] = \underbrace{\int \rho_0(\vec{r})V_{Ne}d\vec{r}}_{\text{system dependent}} + \underbrace{T[\rho_0] + E_{ee}[\rho_0]}_{\text{universally valid}}. \tag{4-7}$$

Collecting the system independent parts into a new quantity, the *Hohenberg-Kohn functional* $F_{HK}[\rho_0]$, we arrive at

$$E_0[\rho_0] = \int \rho_0(\vec{r})V_{Ne}d\vec{r} + F_{HK}[\rho_0], \tag{4-8}$$

which defines $F_{HK}[\rho_0]$. In other words, if the Hohenberg-Kohn functional is fed with some arbitrary density $\rho(\vec{r})$ it cranks out the expectation value $\langle \Psi | \hat{T} + \hat{V}_{ee} | \Psi \rangle$. This is the sum of the kinetic energy and the electron-electron repulsion operator with the *ground state* wave function Ψ connected with this very density (i. e., Ψ is, among all the many wave functions that yield ρ, the one which delivers the lowest energy),

$$F_{HK}[\rho] = T[\rho] + E_{ee}[\rho] = \left\langle \Psi \middle| \hat{T} + \hat{V}_{ee} \middle| \Psi \right\rangle. \tag{4-9}$$

This, at first glance innocuous-looking functional $F_{HK}[\rho]$ is the holy grail of density functional theory. If it were known exactly we would have solved the Schrödinger equation, not approximately, but exactly. And, since it is a universal functional completely independent of the system at hand, it applies equally well to the hydrogen atom as to gigantic molecules such as, say, DNA! $F_{HK}[\rho]$ contains the functional for the kinetic energy $T[\rho]$ and that for the electron-electron interaction, $E_{ee}[\rho]$. The explicit form of both these functionals lies unfortunately completely in the dark. However, from the latter we can extract at least the classical Coulomb part $J[\rho]$, since that is already well known (recall Section 2.3),

$$E_{ee}[\rho] = \frac{1}{2}\int\int \frac{\rho(\vec{r}_1)\,\rho(\vec{r}_2)}{r_{12}}\,d\vec{r}_1 d\vec{r}_2 + E_{ncl}[\rho] = J[\rho] + E_{ncl}[\rho]. \tag{4-10}$$

$E_{ncl}[\rho]$ is the *non-classical* contribution to the electron-electron interaction containing all the effects of self-interaction correction, exchange and Coulomb correlation described previously. It will come as no surprise that finding explicit expressions for the yet unknown functionals, i. e. $T[\rho]$ and $E_{ncl}[\rho]$, represents the major challenge in density functional theory and a large fraction of this book will be devoted to that problem.

One should note at this point that the ground state density uniquely determines the Hamilton operator, which characterizes all states of the system, ground and excited. Thus, *all properties of all states are formally determined by the ground state density* (even though we would need functionals other than $\int \rho(\vec{r})V_{Ne}d\vec{r} + F_{HK}[\rho]$, which is the functional constructed to deliver E_0 but not properties of electronically excited states). In the next section we will see that the reason why density functional theory is usually termed a ground state only theory is a consequence of the second Hohenberg-Kohn theorem. On the other hand, it is only the ground state density that contains the information about positions and charges of the nuclei allowing the mapping from density to external potential; the density of an excited state cannot be used.

4.2 The Second Hohenberg-Kohn Theorem: Variational Principle

Up to this point we have established that the ground state density is in principle sufficient to obtain all properties of interest. But, how can we be sure that a certain density is really the ground state density that we are looking for? A formal prescription for how this problem should be tackled has been given through the second theorem proven by Hohenberg and Kohn in their 1964 contribution. In plain words, this theorem states that $F_{HK}[\rho]$, the functional that delivers the ground state energy of the system, delivers the lowest energy if and only if the input density is the true ground state density, ρ_0. This is of course nothing else than our old friend, the variational principle which in the present context can be expressed as

$$E_0 \leq E[\tilde{\rho}] = T[\tilde{\rho}] + E_{Ne}[\tilde{\rho}] + E_{ee}[\tilde{\rho}]. \tag{4-11}$$

Stated in still other words this means that for any trial density $\tilde{\rho}(\vec{r})$ – which satisfies the necessary boundary conditions such as $\tilde{\rho}(\vec{r}) \geq 0$, $\int \tilde{\rho}(\vec{r})\, d\vec{r} = N$, and which is associated with some external potential $\tilde{V}_{ext}$ – the energy obtained from the functional given in equation (4-6) represents an upper bound to the true ground state energy E_0. E_0 results if and only if the exact ground state density is inserted into equation (4-8). The proof of the inequality (4-11) is simple since it makes use of the variational principle established for wave functions as detailed in Chapter 1. We recall that any trial density $\tilde{\rho}(\vec{r})$ defines its own Hamiltonian $\tilde{\hat{H}}$ and hence its own wave function $\tilde{\Psi}$. This wave function can now be taken as the trial wave function for the Hamiltonian generated from the true external potential V_{ext}. Thus, we arrive at

$$\langle\tilde{\Psi}|\hat{H}|\tilde{\Psi}\rangle = T[\tilde{\rho}] + V_{ee}[\tilde{\rho}] + \int \tilde{\rho}(\vec{r}) V_{ext} d\vec{r} = E[\tilde{\rho}] \geq E_0[\rho_0] = \langle\Psi_0|\hat{H}|\Psi_0\rangle \tag{4-12}$$

which is the desired result.

Let us summarize what we have shown so far. First, all properties of a system defined by an external potential V_{ext} are determined by the ground state density. In particular the ground state energy associated with a density ρ is available through the functional $\int \rho(\vec{r}) V_{Ne} d\vec{r} + F_{HK}[\rho]$. Second, this functional attains its minimum value with respect to all allowed densities if and only if the input density is the true ground state density, i. e., for $\tilde{\rho}(\vec{r}) \equiv \rho_0(\vec{r})$. Of course, the applicability of this variational recipe is limited to the ground state energy since the property that E_0 is the lowest possible energy of the system is explicitly used (to be precise, it is limited to the lowest lying state within a given symmetry). Hence, we cannot straightforwardly transfer this strategy to the problem of determining energies and properties of electronically excited states (the problem of excited states' properties will be taken on in the following chapter).

Let us pause briefly at this point to scratch at a more formal, theoretical problem. The attentive reader might have noticed that we smuggled in the condition 'and which are associated with some external potential V'_{ext} ' as a restriction for densities to be eligible in the variational procedure. This restriction marks the so-called *V_{ext}-representability* problem of electron densities. It is the problem that among the many densities one may think of, not all are eligible in the context of the Hohenberg-Kohn theorem. Only those should be considered which are associated with an antisymmetric wave function and a Hamilton operator with some kind of external potential (not necessarily restricted to the kind of potentials we have met so far). Intimately connected is the question of how such densities can be recognized. While this is an important problem in some of the more theoretical aspects of density functional theory (for example, it is not known so far, which conditions densities must obey in order to be V_{ext}-representable), it is only of minor relevance from an application's point of view. Most importantly, as we will show in the following section this requirement can be replaced by the much weaker condition that the density must stem from an antisymmetric wave function without the explicit connection to an external potential. Such densities are called *N-representable*. Since virtually all practical applications are in one way or the other related to wave function techniques all densities that occur in these applications trivially satisfy this condition. In any case, in spite of representing an exciting intellectual challenge, V_{ext}- or N-representability problems will not bother us any further, since they belong into the domain of theoretical physics rather than computational chemistry.

4.3 The Constrained-Search Approach

In this section we introduce a different way of looking at the variational search connected to the Hohenberg-Kohn treatment. Recall the variational principle, equation (1-13) as introduced in Chapter 1

$$E_0 = \min_{\Psi \to N} \left\langle \Psi \middle| \hat{T} + \hat{V}_{Ne} + \hat{V}_{ee} \middle| \Psi \right\rangle. \tag{4-13}$$

In words, we search over all allowed, antisymmetric N-electron wave functions and the one that yields the lowest expectation value of the Hamilton operator (i. e. the energy) is the ground state wave function.

In order to connect this variational principle to density functional theory we perform the search defined in equation (4-13) in two separate steps: first, we search over the subset of all the infinitely many antisymmetric wave functions Ψ^X that upon quadrature yield a particular density ρ_X (under the constraint that the density integrates to the correct number of electrons). The result of this search is the wave function Ψ^X_{min} that yields the lowest energy for a given density ρ_X. The second step lifts the constraint of a particular density and extends the search over all densities. We finally identify that density among the many ρ_Γ, $\Gamma = A, B, \ldots, X, \ldots$ as the ground state density, for which the wave function Ψ^Γ_{min} characterized in the first step delivers the lowest energy of all. This way of looking at the minimization problem in density functional theory was introduced by Levy, 1979, and is known as the *Levy constrained-search* formulation and is discussed in detail in Parr and Yang, 1989, or Kryachko and Ludeña, 1990. It can be expressed as

$$E_0 = \min_{\rho \to N} \left(\min_{\Psi \to \rho} \left\langle \Psi \middle| \hat{T} + \hat{V}_{Ne} + \hat{V}_{ee} \middle| \Psi \right\rangle \right) \quad (4\text{-}14)$$

where the inner and outer minimizations correspond to the first and second steps of above, respectively.

The energy due to the external potential is determined simply by the density and is therefore independent of the wave function generating that density. Hence, it is the same for all wave functions integrating to a particular density and we can separate it from the kinetic and electron-electron repulsion contributions

$$E_0 = \min_{\rho \to N} \left(\min_{\Psi \to \rho} \left\langle \Psi \middle| \hat{T} + \hat{V}_{ee} \middle| \Psi \right\rangle + \int \rho(\vec{r}) V_{Ne} d\vec{r} \right) \quad (4\text{-}15)$$

or, introducing the universal functional

$$F[\rho] = \min_{\Psi \to \rho} \left\langle \Psi \middle| \hat{T} + \hat{V}_{ee} \middle| \Psi \right\rangle, \quad (4\text{-}16)$$

this results in

$$E_0 = \min_{\rho \to N} \left(F[\rho] + \int \rho(\vec{r}) V_{Ne} d\vec{r} \right). \quad (4\text{-}17)$$

Given a density, $F[\rho] + \int \rho(\vec{r}) V_{Ne} d\vec{r}$ delivers the corresponding energy and upon minimization, the ground state density and ground state energy are obtained. One should notice that $F[\rho]$ differs from the functional $F_{HK}[\rho]$ given above in equation (4-9) only by the fact that it is defined for all densities that originate from an antisymmetric wave function Ψ. The additional restriction that the density has to be associated with an external potential does not surface in this formulation. Of course, if the input density belongs to the class of V_{ext}-representable densities, as is obviously the case for the ground state density which belongs to the corresponding V_{ext} in the Hamiltonian, the two functionals become identical, $F_{HK}[\rho_0] = F[\rho_0]$. In addition, in the Levy formulation the restriction to non-degenerate ground states of the original Hohenberg-Kohn theorem is lifted. If a ground state density ρ_0 is selected, only one of the wave functions out of a set of functions connected with the same ground state energy (associated with ρ_0) is found in the constrained search.

4.4 Do We Know the Ground State Wave Function in Density Functional Theory?

From a purist theoretical point of view, there is one further important result hidden in the Levy constrained-search strategy: it provides a unique, albeit only formal, route to extract the ground state wave function Ψ_0 from the ground state density ρ_0. This is anything but a trivial problem, since there are many antisymmetric N-electron wave functions that yield the same density via $\rho_0(\vec{r}) = N\int\cdots\int|\Psi(\vec{x}_1,\vec{x}_2,\ldots,\vec{x}_N)|^2 ds_1 d\vec{x}_2\ldots d\vec{x}_N$. Of these, the correct ground state wave function Ψ_0 is the one which yields the lowest energy. Stated in other words, we just have to have a look at all the Ψ's associated with the ground state density ρ_0 and select that one for which $E_{\Psi\rightarrow\rho_0}$ is lowest, which is then Ψ_0. Of course, like so many results presented in this chapter, this one is also absolutely useless in real applications. We have no access to all these wave functions and thus, in real life there is no way whatsoever to identify the correct wave function associated with a particular density. Hence, even though the ground state wave function is in principle accessible once we know the correct ground state density (which, in turn, is provided by minimizing the functional $F[\rho]$), it is fair to say that for all practical purposes, *there is no wave function in density functional theory*. We want to point this out very explicitly, since in the literature there is sometimes a certain laxness about this important fact.[11]

4.5 Discussion

At the conclusion of this chapter let us recapitulate what we know by virtue of the Hohenberg-Kohn theorems and what that they are not able to provide. For good reasons we have introduced the Hohenberg-Kohn theorems as being the bedrock of modern density functional theory. However, at the same time, it cannot be overstressed that these results are not more (but also not less) than mere proofs of existence. All the theorems tell us is that a unique mapping between the ground state density $\rho_0(\vec{r})$ and the ground state energy E_0 exists in principle. However, they do not provide any guidance at all how the functional that delivers the ground state energy should be constructed. It is the kind of result which saves 'hard-core' theoreticians from having sleepless nights, because after Hohenberg and Kohn presented their results it was clear that what people did since Thomas and Fermi, namely employing the electron density as the central variable which contains all the necessary information to describe an atomic or molecular system, is indeed physically sound. However, for those who prefer a pragmatic point of view and are mainly interested in applying density functional theory as a tool to computationally predict the properties of molecules, nothing has visibly changed by the advent of these theorems. The calculations are as hard as before and the Hohenberg-Kohn theorems do not even give a clue as to what kind of approximation should be used for the unknown functionals.

[11] Even though the correct many electron wave function is not available in DFT, we will see in Section 5.3.3 that a related wave function exists, which can often be used for qualitative interpretation.

The second theorem establishes the variational principle. Again, we have to be careful not to overinterpret this result. In any real application of density functional theory we are forced to use an approximation for the functional $F[\rho]$, since the true functional is not available. The variational principle as proven above, however, applies to the exact functional only. This has several unpleasant consequences. First, many conventional wave function based theories, such as the Hartree-Fock or configuration interaction schemes, are strictly variational and the expectation value $E = \langle \tilde{\Psi} | \hat{H} | \tilde{\Psi} \rangle$ is an indicator of the quality of the trial wave function (the lower E is, the better an approximation is $\tilde{\Psi}$ to Ψ_0). In the density functional world, the energy value delivered by a trial functional has absolutely no meaning in that respect. Second, it can well happen that the energies obtained from approximate density functional theory are lower than the exact ones! For example, if we compute the energy of the hydrogen atom with the popular BPW91 functional and a large cc-pV5Z expansion of the one-electron Kohn-Sham functions, the result is $-0.5042\ E_h$, significantly below the exact energy of $-0.5\ E_h$ (explanations of the various acronyms that define the level of calculation will be given in later chapters). The reason for these at first glance unexpected results is that in density functional theory, by using an approximation for the universal functional we in a way use an approximated rather than the exact Hamiltonian, while not paying attention to the wave function (which we do not know anyway). Of course, if we change $\hat{H}$ and use something which is only an approximation to it, the variational principle does not hold anymore. On the other hand, in variational conventional methods we use the exact electronic Hamilton operator from equation (1-4) and compute the energy as an expectation value using more and more sophisticated approximations for the many-particle wave function, exactly the scenario for which the variational principle applies.

Similarly, the constrained-search scheme, even though being very elegant in appearance and strong in formal power, is only of theoretical value and offers no solution to practical considerations. Simply, the program indicated in Section 4.3 cannot be realized – how would we ever be able to search through all wave functions? Since this is obviously impossible, setting up the functional $F[\rho] = \min_{\Psi \to \rho} \langle \Psi | \hat{T} + \hat{V}_{ee} | \Psi \rangle$ is impossible, too. A second point deserves to be mentioned as well. We set out to find a functional that contains an explicit prescription for how to uniquely map an electron density onto an energy, bypassing the complicated N-particle wave function. However, what we ended up with in the constrained search formalism is a definition of the all-decisive functional $F[\rho]$ that explicitly contains the wave function rather than the density – recall equations (4-14) to (4-17). We are not going to pursue these formal aspects of density functional theory any further since it represents a diversion from the real focal point of this book: the role of density functional theory in chemical applications. A competent in-depth discussion with many pointers to the original literature can again be found in the excellent book by Kryachko and Ludeña, 1990.

5 The Kohn-Sham Approach

In this chapter we will show how the Hohenberg-Kohn theorems of the previous chapter can be put to work. As the caption to this chapter indicates, the approach we are discussing has its origin in the second major paper of modern density functional theory, which appeared about a year after the ground breaking contribution by Hohenberg and Kohn. In this report, Kohn and Sham, 1965, suggested an avenue for how the hitherto unknown universal functional of the previous chapter can be approached. At the center of their ingenious idea is the realization that most of the problems with direct density functionals like the Thomas-Fermi method presented in Chapter 3 are connected with the way the kinetic energy is determined. In order to alleviate the situation and realizing that orbital-based approaches such as the Hartree-Fock method perform much better in this respect, Kohn and Sham introduced the concept of a non-interacting reference system built from a set of orbitals (i. e., one electron functions) such that the major part of the kinetic energy can be computed to good accuracy. The remainder is merged with the non-classical contributions to the electron-electron repulsion – which are also unknown, but usually fairly small. By this method, as much information as possible is computed exactly, leaving only a small part of the total energy to be determined by an approximate functional. After introducing the Kohn-Sham scheme, we will discuss some of its major features. In particular we will draw the demarcation line between those properties that apply to 'Kohn-Sham in principle' and what happens to these properties in 'Kohn-Sham in real life'.

5.1 Orbitals and the Non-Interacting Reference System

Let us recall that the Hohenberg-Kohn theorems allow us to construct a rigorous many-body theory using the electron density as the fundamental quantity. We showed in the previous chapter that in this framework the ground state energy of an atomic or molecular system can be written as

$$E_0 = \min_{\rho \to N} \left(F[\rho] + \int \rho(\vec{r}) V_{Ne} d\vec{r} \right) \tag{5-1}$$

where the universal functional $F[\rho]$ contains the individual contributions of the kinetic energy, the classical Coulomb interaction and the non-classical portion due to self-interaction correction, exchange (i. e., antisymmetry), and electron correlation effects,

$$F[\rho(\vec{r})] = T[\rho(\vec{r})] + J[\rho(\vec{r})] + E_{ncl}[\rho(\vec{r})]. \tag{5-2}$$

Of these, only $J[\rho]$ is known, while the explicit forms of the other two contributions remain a mystery. The Thomas-Fermi and Thomas-Fermi-Dirac approximations that we briefly touched upon in Chapter 3 are actually realizations of this very concept. All terms present in these models, i. e., the kinetic energy, the potential due to the nuclei, the classical

Coulomb repulsion, and in the case of the Thomas-Fermi-Dirac model also the exchange contribution are *explicit* functionals of the electron density, making the respective expressions very simple. It turns out, however, that all methods based on the Thomas-Fermi scheme, including the numerous extensions that have been introduced since its original conception, fail miserably when results better than mere qualitative trends are the target. Among the most devastating results was the rigorous proof that within the Thomas-Fermi model no molecular system is stable with respect to its fragments! So, what value for chemistry can a model possibly have in which chemical bonding does not even exist? It quickly became clear that the major reason for the very disappointing performance of the Thomas-Fermi model is the simple functional form for the kinetic energy with its dependence on $\int \rho^{5/3}(\vec{r})d\vec{r}$. Also intuitively we are not surprised that the relationship between the spatial distribution of the electrons as provided by the electron density and their velocities, which are needed for the kinetic energy, is not that trivial. Thus, it seems to be crucial to find a different way to treat the kinetic energy with a better control of the accuracy – and that is exactly what Kohn and Sham set out to do.

To understand how Kohn and Sham tackled this problem, we go back to the discussion of the Hartree-Fock scheme in Chapter 1. There, our wave function was a single Slater determinant Φ_{SD} constructed from N spin orbitals. While the Slater determinant enters the HF method as the approximation to the true N-electron wave function, we showed in Section 1.3 that Φ_{SD} can also be looked upon as the exact wave function of a fictitious system of N *non-interacting* electrons (that is 'electrons' which behave as uncharged fermions and therefore do not interact with each other via Coulomb repulsion), moving in the effective potential V_{HF}. For this type of wave function the kinetic energy can be exactly expressed as

$$T_{HF} = -\frac{1}{2}\sum_{i}^{N}\left\langle \chi_i \middle| \nabla^2 \middle| \chi_i \right\rangle. \tag{5-3}$$

The HF spin orbitals χ_i that appear in this expression are chosen such that the expectation value E_{HF} attains its minimum (under the usual constraint that the χ_i remain orthonormal)

$$E_{HF} = \min_{\Phi_{SD}\to N}\left\langle \Phi_{SD} \middle| \hat{T} + \hat{V}_{Ne} + \hat{V}_{ee} \middle| \Phi_{SD} \right\rangle. \tag{5-4}$$

Of course, all this is not new but only a recapitulation of results from Chapter 1. The important connection to density functional theory is that we now go on to exploit the above kinetic energy expression, which is valid for non-interacting fermions, in order to compute the major fraction of the kinetic energy of our interacting system at hand.

The next step is crucial. We have shown above that the exact wave functions of non-interacting fermions are Slater determinants.[12] Thus, it will be possible to set up a *non-interacting reference system*, with a Hamiltonian in which we have introduced an effective, local potential $V_S(\vec{r})$:

[12] That is, if we are dealing with non-degenerate states. Otherwise the wave function might be a limited linear combination of Slater determinants.

$$\hat{H}_S = -\frac{1}{2}\sum_i^N \nabla_i^2 + \sum_i^N V_S(\vec{r}_i). \tag{5-5}$$

Since this Hamilton operator does not contain any electron-electron interactions it indeed describes a non-interacting system. Accordingly, its ground state wave function is represented by a Slater determinant (switching to Θ_S and φ rather than Φ_{SD} and χ for the determinant and the spin orbitals, respectively, in order to underline that these new quantities are not related to the HF model)

$$\Theta_S = \frac{1}{\sqrt{N!}}\begin{vmatrix} \varphi_1(\vec{x}_1) & \varphi_2(\vec{x}_1) & \cdots & \varphi_N(\vec{x}_1) \\ \varphi_1(\vec{x}_2) & \varphi_2(\vec{x}_2) & & \varphi_N(\vec{x}_2) \\ \vdots & \vdots & & \vdots \\ \varphi_1(\vec{x}_N) & \varphi_2(\vec{x}_N) & \cdots & \varphi_N(\vec{x}_N) \end{vmatrix} \tag{5-6}$$

where the spin orbitals, in complete analogy to equations (1-24) and (1-25), are determined by

$$\hat{f}^{KS}\ \varphi_i = \varepsilon_i \varphi_i, \tag{5-7}$$

with the one-electron Kohn-Sham operator $\hat{f}^{KS}$ defined as

$$\hat{f}^{KS} = -\frac{1}{2}\nabla^2 + V_S(\vec{r}). \tag{5-8}$$

In order to distinguish these orbitals from their Hartree-Fock counterparts, they are usually termed *Kohn-Sham orbitals*, or briefly KS orbitals. The connection of this artificial system to the one we are really interested in is now established by choosing the effective potential V_S such that the density resulting from the summation of the moduli of the squared orbitals $\{\varphi_i\}$ exactly equals the ground state density of our real target system of interacting electrons,

$$\rho_S(\vec{r}) = \sum_i^N \sum_s |\varphi_i(\vec{r}, s)|^2 = \rho_0(\vec{r}). \tag{5-9}$$

5.2 The Kohn-Sham Equations

At this point, we come back to our original problem: finding a better way for the determination of the kinetic energy. The very clever idea of Kohn and Sham was to realize that if we are not able to accurately determine the kinetic energy through an explicit functional, we should be a bit less ambitious and concentrate on computing as much as we can of the true kinetic energy exactly. We then have to deal with the remainder in an approximate manner. Hence, they suggested to use expression (5-3) to obtain the exact kinetic energy of the non-interacting reference system with the same density as the real, interacting one

$$T_S = -\frac{1}{2}\sum_i^N \left\langle \varphi_i \middle| \nabla^2 \middle| \varphi_i \right\rangle . \tag{5-10}$$

Of course, the non-interacting kinetic energy is not equal to the true kinetic energy of the interacting system, even if the systems share the same density, i. e., $T_S \neq T$.[13] Kohn and Sham accounted for that by introducing the following separation of the functional $F[\rho]$

$$F[\rho(\vec{r})] = T_S[\rho(\vec{r})] + J[\rho(\vec{r})] + E_{XC}[\rho(\vec{r})] \tag{5-11}$$

where E_{XC}, the so-called *exchange-correlation energy* is defined through equation (5-11) as

$$E_{XC}[\rho] \equiv (T[\rho] - T_S[\rho]) + (E_{ee}[\rho] - J[\rho]) = T_C[\rho] + E_{ncl}[\rho] . \tag{5-12}$$

The residual part of the true kinetic energy, T_C, which is not covered by T_S, is simply added to the non-classical electrostatic contributions. In other words, the exchange-correlation energy E_{XC} is the functional which contains everything that is unknown, a kind of junkyard where everything is stowed away which we do not know how to handle exactly. Let us also underline that in spite of its name, E_{XC} contains not only the non-classical effects of self-interaction correction, exchange and correlation, which are contributions to the potential energy of the system, but also a portion belonging to the kinetic energy. As indicated by the intimate relation between the orbitals and the density through equation (5-9), T_S is expected to be a functional of ρ. A complementary way of looking at this is to realize that the energy expression of the non-interacting system contains only two components: the kinetic energy and the energy due to the interaction with the external potential. By the Hohenberg-Kohn theorem, the total energy must be a functional of the density. Likewise, the interaction with the external potential is an explicit functional of ρ. Hence, T_S is also necessarily a functional of the charge density. But note that we again do not have a simple expression for T_S where the density enters explicitly – the KS orbitals and not the density ρ appear in equation (5-10).

So far so good, but before we are in business with this concept we need to find a prescription for how we can uniquely determine the orbitals in our non-interacting reference system. In other words, we ask: how can we define V_S such that it really provides us with a Slater determinant which is characterized by exactly the same density as our real system? To solve this problem, we write down the expression for the energy of our interacting, real

[13] Actually, it can be shown that $T_S \leq T$.

system in terms of the separation described by equation (5-11), highlighting the dependence on the orbitals as indicated in equations (5-9) and (5-10):

$$
\begin{aligned}
E[\rho(\vec{r})] &= T_S[\rho] + J[\rho] + E_{XC}[\rho] + E_{Ne}[\rho] \\
&= T_S[\rho] + \frac{1}{2}\iint \frac{\rho(\vec{r}_1)\,\rho(\vec{r}_2)}{r_{12}}\, d\vec{r}_1 d\vec{r}_2 + E_{XC}[\rho] + \int V_{Ne}\rho(\vec{r})d\vec{r} \\
&= -\frac{1}{2}\sum_i^N \left\langle \varphi_i \middle| \nabla^2 \middle| \varphi_i \right\rangle + \frac{1}{2}\sum_i^N \sum_j^N \iint |\varphi_i(\vec{r}_1)|^2 \frac{1}{r_{12}} |\varphi_j(\vec{r}_2)|^2\, d\vec{r}_1 d\vec{r}_2 \\
&\quad + E_{XC}[\rho(\vec{r})] - \sum_i^N \int \sum_A^M \frac{Z_A}{r_{1A}} |\varphi_i(\vec{r}_1)|^2 d\vec{r}_1
\end{aligned}
\tag{5-13}
$$

The only term for which no explicit form can be given, i. e., the big unknown, is of course E_{XC}. Similarly to what we have done within the Hartree-Fock approximation, we now apply the variational principle and ask: what condition must the orbitals $\{\varphi_i\}$ fulfill in order to minimize this energy expression under the usual constraint of $\langle \varphi_i \mid \varphi_j \rangle = \delta_{ij}$? The resulting equations are (for a detailed derivation see Parr and Yang, 1989):

$$
\begin{aligned}
&\left(-\frac{1}{2}\nabla^2 + \left[\int \frac{\rho(\vec{r}_2)}{r_{12}} d\vec{r}_2 + V_{XC}(\vec{r}_1) - \sum_A^M \frac{Z_A}{r_{1A}}\right]\right)\varphi_i \\
&= \left(-\frac{1}{2}\nabla^2 + V_{eff}(\vec{r}_1)\right)\varphi_i = \varepsilon_i \varphi_i \,.
\end{aligned}
\tag{5-14}
$$

If we compare this equation with the one-particle equations from the non-interacting reference system, we see immediately that the expression in square brackets, i. e. V_{eff}, is identical to V_S of equation (5-8) above

$$
V_S(\vec{r}) \equiv V_{eff}(\vec{r}) = \int \frac{\rho(\vec{r}_2)}{r_{12}} d\vec{r}_2 + V_{XC}(\vec{r}_1) - \sum_A^M \frac{Z_A}{r_{1A}} \,. \tag{5-15}
$$

Thus, once we know the various contributions in equation (5-15) we have a grip on the potential V_S which we need to insert into the one-particle equations, which in turn determine the orbitals and hence the ground state density and the ground state energy by employing the energy expression (5-13). It should be noted that V_{eff} already depends on the density (and thus on the orbitals) through the Coulomb term as shown in equation (5-13). Therefore, just like the Hartree-Fock equations (1-24), the Kohn-Sham one-electron equations (5-14) also have to be solved iteratively.

One term in the above equation needs some additional comments, namely V_{XC}, the potential due to the exchange-correlation energy E_{XC}. Since we do not know how this energy should be expressed, we of course also have no clue as to the explicit form of the corresponding potential. Hence, V_{XC} is simply defined as the functional derivative of E_{XC} with respect to ρ, i. e.,

$$V_{XC} \equiv \frac{\delta E_{XC}}{\delta \rho}. \quad (5\text{-}16)$$

It is very important to realize that if the exact forms of E_{XC} and V_{XC} were known (which is unfortunately not the case), the Kohn-Sham strategy would lead to the exact energy, i. e. the correct eigenvalue of the Hamilton operator $\hat{H}$ of the Schrödinger equation. The reader should check for him- or herself that the formalism that we have illustrated in this chapter does not contain any approximation as of yet. Thus, unlike the Hartree-Fock model, where the approximation is introduced right from the start (the wave function is assumed to be a single Slater determinant, which therefore can never deliver the true solution) the *Kohn-Sham approach is in principle exact*! The approximation only enters when we have to decide on an explicit form of the unknown functional for the exchange-correlation energy E_{XC} and the corresponding potential V_{XC}. The central goal of modern density functional theory is therefore to find better and better approximations to these two quantities and we will have a lot more to say about these aspects in the following chapters.

Before we enter a more detailed discussion of various aspects in the Kohn-Sham approach, let us summarize the main features of this procedure:

(i) We define a non-interacting reference system of N particles whose exact ground state is a single Slater determinant Θ_S and whose density ρ_S by construction exactly equals the density of our real, interacting system, ρ_0.

(ii) The orbitals which form this Slater determinant are the solutions of N single particle equations (5-7). This allows the determination of the non-interacting kinetic energy, T_S according to (5-10). The effective potential V_S in the one-electron Hamilton operator must be chosen such that the condition of $\rho_S = \rho_0$ is fulfilled. The next steps aim at finding a way of generating this V_S.

(iii) The energy of the interacting system is separated into the kinetic energy T_S of the non-interacting system, the energy due to the nuclei E_{Ne}, the classical electrostatic electron-electron repulsion energy J and the remainder E_{XC} which consists of the quantum-mechanical contributions to the potential energy (self-interaction correction, exchange and correlation) and the part of the true kinetic energy that is not covered by T_S (equation 5-12).

(iv) This energy expression is subjected to the variational principle with respect to independent variations in the orbitals. The resulting expressions (5-13) show that the effective potential V_S that we need to get the correct orbitals of the non-interacting reference system exactly equals the sum of the potential due to the nuclei, V_{Ne}, the classical Coulomb potential, V_C, and the potential generated by E_{XC}, i. e., V_{XC} (equation 5-15).

(v) Provided that we know the explicit forms of all these potentials, we know V_S and by solving the one-electron equations we obtain the KS orbitals. These define the non-interacting system that shares the same density as our real system. Since in all real applications we do not know the exact V_{XC} we need to introduce an approximation for the exchange-correlation potential.

(vi) The orbitals give us the density via equation (5-9). Inserting this density into the energy expression finally yields the exact ground state density and hence the *exact* ground state energy, again provided we know the exact functionals. In all real applications, however, we have to resort to approximations for the unknown functional E_{XC}.

5.3 Discussion

In the preceding paragraph we have given a detailed survey of the Kohn-Sham approach to density functional theory. Now, we need to discuss some of the relevant properties pertaining to this scheme and how we have to interpret the various quantities it produces. We also will mention some areas connected to Kohn-Sham density functional theory which are still problematic. Before we enter this discussion the reader should be reminded to differentiate carefully between results that apply to the hypothetical situation in which the exact functional E_{XC} and the corresponding potential V_{XC} are known and the real world in which we have to use approximations to these quantities.

5.3.1 The Kohn-Sham Potential is Local

First we point out that the effective potential that occurs in the one-particle equations of the non-interacting reference system $V_S(\vec{r})$ is local in the sense that it is a function of only the spatial variable $\vec{r}$ and is independent on the values of V_S at other points in space, $\vec{r}'$. Due to the equality between $V_S(\vec{r})$ and $V_{eff}(\vec{r})$ demonstrated above, $V_{XC}(\vec{r})$, i. e., the potential responsible for exchange and correlation effects and the difference between T_S and T, must also be local. This is to be contrasted with the non-local exchange contribution that appears in the Hartree-Fock approximation. The result of operating with the Hartree-Fock exchange operator $\hat{K}_j(\vec{x}_1)$ on the orbital $\chi_i(\vec{x}_1)$ depends on the value of χ_i everywhere in the coordinate space, not just $\vec{x}_1$, as we discussed in Section 1.3. We arrive at the fascinating conclusion that the Kohn-Sham equations have a structure that is actually formally less complicated than the Hartree-Fock approximation. Nevertheless, they are exact in principle. We should, however, add that even though the Kohn-Sham potential is a local potential with expressions that are formally less complicated than the corresponding equations in the Hartree-Fock approximation, it will probably have a very complex and non-local dependence on the density.[14] Its value at a particular point in space $V_{eff}(\vec{r})$ will depend on the charge density at all other points in a difficult and for us inaccessible way. The reader should keep in mind that this is necessarily so, because knowledge of the exact exchange-correlation potential is equivalent to exactly solving the Schrödinger equation.

[14] Stated in more mathematical terms: while the exact V_{XC} is local, E_{XC} does not originate from a local kernel.

5.3.2 The Exchange-Correlation Energy in the Kohn-Sham and Hartree-Fock Schemes

We should also clarify at this stage that there are inherent differences between the exchange-correlation energy that appears in the Kohn-Sham formalism and their namesakes, the exchange and correlation energies, as they are defined within the Hartree-Fock picture. Let us state right up front: even though these two quantities are similar in some way, they do not have the same meaning! This is important to realize because the construction of exchange and correlation functionals to be used in the Kohn-Sham scheme are frequently based on the HF-derived definitions of exchange and correlation. Recall from Chapters 1 and 2 that if we employ the hole formalism, the Hartree-Fock exchange energy is given by

$$E_X^{HF} = \frac{1}{2}\iint \frac{\rho_0^{HF}(\vec{r}_1)h_X^{HF}(\vec{r}_1;\vec{r}_2)}{r_{12}} d\vec{r}_1 d\vec{r}_2 \qquad (5\text{-}17)$$

where both electrons are associated with the same spin function. Thus, the exchange energy can be interpreted as the interaction between the HF ground state charge distribution ρ_0^{HF} and the corresponding exchange or Fermi hole h_X^{HF}. The Hartree-Fock correlation energy E_C^{HF} is then defined as the difference between the exact, non-relativistic energy within the Born-Oppenheimer approximation and the Hartree-Fock energy, equation (1-30). We can use an equivalent separation of E_{XC} in the KS scheme and express E_X^{KS} just as its HF counterpart, with the only difference that it is computed from the KS orbitals which are associated with the exact density ρ_0,

$$E_X^{KS} = \frac{1}{2}\iint \frac{\rho_0(\vec{r}_1)h_X^{KS}(\vec{r}_1;\vec{r}_2)}{r_{12}} d\vec{r}_1 d\vec{r}_2 . \qquad (5\text{-}18)$$

As the Hartree-Fock energy is the lowest energy one can possibly get from a single determinant it follows immediately that the correlation energy in the KS scheme using the exact functional must be more negative (larger in an absolute sense) than E_C^{HF}. The decisive difference between E_C^{HF} and E_C^{KS} is, however, that the charge density in the Kohn-Sham approach is by definition the exact density of the real ground state, $\rho_0(\vec{r})$, while the HF orbitals give the HF ground state wave function, whose square certainly does not integrate to the correct ground state density, $\rho_0^{HF}(\vec{r}) \neq \rho_0(\vec{r})$. Hence, in the KS formalism the correlation hole is simply defined as the difference between the total exchange-correlation hole and the exchange only part,

$$h_C^{KS}(\vec{r}_1;\vec{r}_2) = h_{XC}(\vec{r}_1;\vec{r}_2) - h_X^{KS}(\vec{r}_1;\vec{r}_2) \qquad (5\text{-}19)$$

while we have to introduce a term in the HF scheme taking care of the difference between the exact and the Hartree-Fock electron density

$$h_C^{HF}(\vec{r}_1;\vec{r}_2) = (\rho_0(\vec{r}_2) - \rho_0^{HF}(\vec{r}_2)) + h_{XC}(\vec{r}_1;\vec{r}_2) - h_X^{HF}(\vec{r}_1;\vec{r}_2) . \qquad (5\text{-}20)$$

Table 5-1. Contributions to the Hartree-Fock correlation energy [eV] (taken from Baerends and Gritsenko, 1997).

	E_C^{HF}	T_C^{HF}	$E_{Ne,C}^{HF}$	$E_{ee,C}^{HF}$
H_2 at R = 1.401 (R_e)	−1.1	+1.3	−0.5	−1.9
H_2 at R = 5 bohr	−3.9	+8.9	−8.5	−4.4
H_2 at R = 10 bohr	−6.3	+7.9	−8.4	−5.6

Hence, if the HF density is close to the exact density the differences between the exchange and correlation contributions in the Hartree-Fock and Kohn-Sham schemes are small. However, the more $\rho_0^{HF}(\vec{r})$ deviates from $\rho_0(\vec{r})$, the less will the HF exchange and correlation energies parallel their Kohn-Sham counterparts. An obvious consequence of $\rho_0^{HF}(\vec{r}) \neq \rho_0(\vec{r})$ is that the correlation energy in the HF scheme will contain contributions from the nuclear-electron attraction, the classical Coulomb repulsion and the kinetic energy, since these quantities are evaluated using the Hartree-Fock density $\rho_0^{HF}(\vec{r})$, rather than the exact one. Note that the first two contributions to the correlation energy will always be zero in the KS scheme, because the density defined by the KS orbitals equals by construction the ground state density $\rho_0(\vec{r})$. Hence $E_{Ne}[\rho]$ and $J[\rho]$ are computed exactly. In Table 5-1 the consequences of these differences – which are frequently significant, in particular when bonds are stretched – are illustrated for the H_2 molecule at three internuclear distances (where $E_C^{HF} = T_C^{HF} + E_{Ne,C}^{HF} + E_{ee}^{HF}$). Note the large contributions from the electron-nucleus attraction and the kinetic energy, which are caused by the much too diffuse density at elongated H-H distances, as discussed in detail by Baerends and Gritsenko, 1997.

5.3.3 Do the Kohn-Sham Orbitals Mean Anything?

The next point concerns the role of the KS orbitals. Until recently there was a broad consensus that the orbitals satisfying equation (5-14) have no physical significance and that their only connection to the real world is that the sum of their squares add up to the exact density. While this is certainly true in a strict sense, several authors have lately pointed to the interpretative power of the KS orbitals in traditional qualitative molecular orbital schemes, see, Kohn, Becke, and Parr, 1996, Baerends and Gritsenko, 1997, Stowasser and Hoffmann, 1999, and Baerends, 2000. After all, the KS orbitals are not only associated with a one-electron potential which includes all non-classical effects, they are also consistent with the exact ground state density. Actually, the HF orbitals are in a sense much farther away from the real system since they neither reflect correlation effects nor do they yield the exact density. Many authors therefore recommend the KS orbitals as legitimate tools in qualitative MO considerations. On the other hand, one must not confuse the Slater determinant generated from the KS orbitals with the true many-electron wave function! As outlined in the preceding chapter, the exact wave function of the target system is simply not available in density functional theory! Similarly, the eigenvalues ε_i connected to the KS orbitals do not have a strict physical meaning. In Kohn-Sham theory, for example, there is no equivalent of Koopmans' theorem, which could relate orbital energies to ionization energies. There is one exception though: as a direct consequence of the long range behavior of the charge density shown in equation (2-5), the eigenvalue of the highest occupied orbital, ε_{max}, of the KS orbitals equals the negative of the exact ionization energy. Again, however, we have to add a big caveat here: this holds strictly only for ε_{max} resulting from the exact V_{XC}, not for solutions obtained with approximations to the exchange-correlation potential. The exact ionization energy of the hydrogen atom, for example, is 0.5 E_h. None of the approximate exchange-correlation functionals in use today produces a 1s orbital whose absolute energy comes even close. Rather, 1s orbital energies in the order of only –0.23 to –0.28 E_h are obtained, more than 0.2 E_h or 5 eV off the correct value. This disappointing result mirrors deficiencies in the exchange-correlation potentials generated by approximate functionals for E_{XC}, in particular deficiencies of their long range, asymptotic behavior. We will come back to this phenomenon in Section 6.8 of the following chapter. If instead an essentially exact V_{XC} is used, computed from highly sophisticated configuration interaction or similar schemes, the agreement between ε_{max} and –IE improves significantly. That the agreement is not quantitative is due to remaining inaccuracies in the computationally predicted densities (see e. g., Morrison and Zhao, 1995). Finally, we must not forget to mention that the KS orbital energies also play a role in the treatment of excited states in the perturbation theory based treatment of Görling, 1996 or in the framework of time-dependent density functional theory, as we will outline further below. Actually Savin, Umrigar, and Gonze, 1998 showed that there is in many cases a surprisingly good agreement between the ground-state KS eigenvalue differences obtained from the exact KS potential and the corresponding excitation energies. These authors, however, also pointed out that this result only applies if the essentially exact potential is being used. Eigenvalue differences from approximate functionals come not even close.[15]

5.3.4 Is the Kohn-Sham Approach a Single Determinant Method?

One frequently reads the assertion that the Kohn-Sham scheme, just like the Hartree-Fock approximation, is a single determinant approach with all problems and shortcomings connected to this. As a consequence, the Kohn-Sham formalism should, for example, fail to correctly describe the prototype H_2 dissociation just in the same way as demonstrated for the Hartree-Fock case in Section 1.4. However, we have shown above that the Kohn-Sham picture is only a particular rearrangement of the Hohenberg-Kohn theorems and therefore an avenue leading in principle to the exact energy of the electronic Schrödinger equation in all situations, without exception (remember that this holds only for the energy; the exact wave function is not available). In terms of the KS approach the title question rather translates into the problem whether the non-interacting N-electron ground state that shares the same density as the interacting system can be generated by a single Slater determinant built

[15] This is due to the wrong asymptotic decay of approximate functionals, as discussed below in Section 6.8.

from orbitals that are obtained as the N energetically lowest lying orbitals of a simple local Kohn-Sham potential V_S. Such cases are termed *non-interacting pure-state-V_S* representable. For the prototype two-electron closed-shell system H_2, an essentially exact local Kohn-Sham potential V_S can fairly straightforwardly be constructed numerically by 'inverting' equation (5-8). In this case the Kohn-Sham orbitals and the density are connected trivially (apart from an irrelevant phase factor) via

$$\varphi(\vec{r}) = \sqrt{\frac{\rho(\vec{r})}{2}} . \tag{5-21}$$

Nearly exact charge densities are available from top-level conventional quantum chemical calculations, i. e., full configuration interaction using large one electron basis sets. From these, Kohn-Sham orbitals of corresponding quality can be obtained, which lead in an iterative fashion to a very realistic representation of the exact Kohn-Sham potential V_S. Using such an essentially exact Kohn-Sham potential for H_2, Gritsenko and Baerends, 1997, showed that in fact a single Slater determinant is obtained as the Kohn-Sham non-interacting reference system. This is true not only near the equilibrium bond distance, but also when the H-H bond is significantly stretched. Even in such a case with strong non-dynamical electron correlation due to the orbital near-degeneracy discussed in Section 1.5, a single Slater determinant represents the non-interacting Kohn-Sham reference system. Of course, the KS orbitals must increasingly differ from their HF counterparts in order to incorporate the correlation effects and the resulting KS non-interacting wave function is a pretty bad approximation to the true wave function. In other words, the exact H_2 potential curve should be available using a single determinant Kohn-Sham reference system provided that the exact exchange-correlation functional is known. On the other hand, it is an active area of research whether there are also cases where a non-degenerate interacting ground state density cannot be represented by a single Slater determinant (i. e., it is not non-interacting pure-state-V_S representable) and when this is to be expected. For certain internuclear distances the non-degenerate $^1\Sigma_g^+$ ground state of the C_2 molecule has been identified only recently as an example where it is not possible to represent the (essentially) exact interacting density obtained from sophisticated conventional calculations as a single determinant Kohn-Sham solution (Schipper, Gritsenko, and Baerends, 1998a). Rather, an ensemble of a small number of accidentally (i. e., not symmetry dictated) degenerate determinants is required to do so. These densities are called non-interacting *ensemble-V_S* representable. The weights of the individual determinants in the ensemble need to be determined variationally. As a sequel to this study, Schipper, Gritsenko, and Baerends, 1999, provided another, very instructive example in their study of the identity reaction $H_2 + H'_2 \rightarrow HH' + HH'$. In the region around the quadratic saddle point, both, the exact Kohn-Sham solution as well as typical approximate functionals need an ensemble treatment to achieve an accurate barrier height. If instead the standard single determinant Kohn-Sham approach is employed, the barrier is always severely underestimated. A concise but very clear summary of these complications can be found in Baerends, 2000. See also Wang and Schwarz, 1996 for a discussion of these phenomena using the related concept of fractional occupation numbers.

On the other hand, none of the current approximate E_{XC} functionals is able to quantitatively reproduce the subtle details of the non-classical contributions to the energy (such as the left-right correlation in the dissociating H_2), which in fact leads to incorrect dissociation curves in the restricted scheme, very similar to restricted Hartree-Fock. Hence, once more we have an example where it is of utmost importance to clearly distinguish whether we are talking about the Kohn-Sham formalism in principle or about actual implementations of this scheme which necessarily utilize some kind of approximate form for E_{XC}. We will come back to this problem and the relationship to the so-called ensemble densities in Section 5.3.6. We cannot overemphasize how important it is not to let a sloppy way of describing these effects slip in. Again, it is not necessarily the single determinant based regular KS scheme itself which is to blame for the sometimes poor results of approximate Kohn-Sham density functional theory; responsible instead are the deficiencies of the approximations to the exchange-correlation functional. Some of these will occupy us in more detail in the following chapter.

5.3.5 The Unrestricted Kohn-Sham Formalism

The effective potential of the Kohn-Sham equations V_{eff} contains no reference to the spin of the electrons. Hence, for an even number of electrons the KS orbitals necessarily occur in degenerate pairs where the spatial part is shared by an α and a β spin function, akin to the RHF scheme of Chapter 1. Even if we are dealing with a system with an odd number of electrons where the density of the α-spin electrons will differ from the β-spin density, the only, all-decisive variable is still the total density, $\rho(\vec{r}) = \rho_\alpha(\vec{r}) + \rho_\beta(\vec{r})$. No information about the individual spin densities is required. The energy will become a functional of the individual spin densities only if the potential contains parts that are spin dependent, such as an external magnetic field. This is, however, well beyond the scope of this book. The bottom line is that in principle this formalism is suitable for any kind of atom or molecule, be it of closed-shell character or a system with an arbitrary multiplicity. However, as so often this is only the formal point of view which applies to a hypothetical situation, namely that the exact functionals are available. If we think more pragmatically we realize that the current approximate functionals based on the electron density alone do not offer the flexibility to really account for open-shell problems. Therefore, functionals that explicitly depend on the α- and β-spin densities are usually employed in such situations, in analogy to the unrestricted HF approach described earlier. The resulting approximate *spin-density functionals* for exchange and correlation are able to capture more of the essential physics in open-shell species than their spin independent counterparts.

Let us again take the H_2 dissociation problem as a simple but instructive example. Consider the $^1\Sigma_g^+$ ground state density of the H_2 molecule. Obviously the spin density at either nucleus must be zero as dictated by the spatial $D_{\infty h}$ symmetry of this closed-shell system. This requirement of zero spin density is independent of the distance and also applies as the H–H distance grows larger and larger. Thus, even in the limit of a supermolecule of two non-interacting hydrogen atoms, infinitely far apart, the spin density is zero. The energy of this system must of course approach twice the energy of an isolated hydrogen atom. On the other hand it is clear that the ground state of a hydrogen atom is ^2S with one unpaired electron and non-zero spin density. If a spin-density functional is applied to these two, in principle equivalent situations, it will find zero spin density at the nuclei in the supermolecule, but a non-zero spin density in the isolated atom and will assign different energies to both solutions. Consequently any approximate spin-density functional will give the unphysical result that the energy for two isolated hydrogen atoms will not add up to the energy of the supermolecule. Therefore, none of the current approximate functionals is able to recover the correct potential curve for the H_2 molecule if used in the spin restricted form. This situation resembles the problems of the RHF approximation in the H_2 dissociation discussed in Chapter 1 and is schematically shown in Figure 5-1. Conversely, if the corresponding calculations are performed in the *unrestricted Kohn-Sham* (UKS) picture a qualitatively correct potential curve results, including the asymptotic region with the relative energy of the dissociation limit being equal to twice the atomic hydrogen energy. But now, like the UHF solution, the UKS spin densities also break the inversion symmetry as the H–H distance increases, leading to a spin polarization where α-spin density accumulates at one nucleus and β-spin density at the other (Gunnarsson and Lundqvist, 1976, Dunlap, 1987). Thus, the unrestricted density mimics the atomic densities. The symmetry of the charge density is in these cases obviously lower than the symmetry of the molecule and is therefore unphysical. In other words, unrestricted techniques give qualitatively correct energies but wrong densities, whereas spin-restricted methods show the opposite behavior, that is, they give reasonable densities accompanied by incorrect energies. With the approximate functionals in use today it turns out that it is often necessary to allow for the unphysical symmetry breaking in order to achieve satisfactory results (see Salahub, 1987, for an extensive discussion). We postpone a more elaborate discussion of the symmetry problem, including possible remedies, to the next section. At this point we should point out that even though symmetry breaking occurs also within the Kohn-Sham ansatz, this scheme is significantly more robust than HF theory in this respect, as has been shown by, e. g., Bauernschmitt and Ahlrichs, 1996a, and Sherrill, Lee, and Head-Gordon, 1999. This can also be clearly seen in Figure 5-1: the UHF curve appears already at a significantly shorter H–H distance than the UKS one. In other words, the spin-restricted functional allows a reasonably good description of the potential curve for a significantly larger fraction of the dissociation process. As we will see in Chapter 13, this has important consequences for the applicability of restricted functionals for the calculation of saddle points of chemical reactions of formally closed-shell molecules, which are frequently characterized by stretched bonds.

Just as in the unrestricted Hartree-Fock variant, the Slater determinant constructed from the KS orbitals originating from a spin unrestricted exchange-correlation functional is not a spin eigenfunction. Frequently, the resulting $\langle \hat{S}^2 \rangle$ expectation value is used as a probe for the quality of the UKS scheme, similar to what is usually done within UHF. However, we must be careful not to overstress the apparent parallelism between unrestricted Kohn-Sham and Hartree-Fock: in the latter, the Slater determinant is in fact the approximate wave function used. The stronger its spin contamination, the more questionable it certainly gets. In

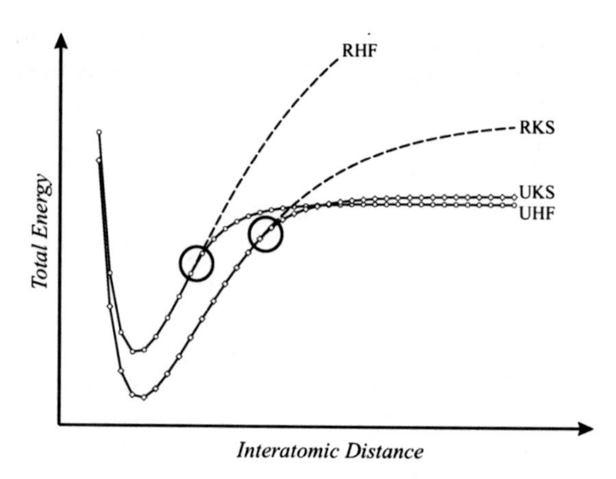

Figure 5-1. H_2 potential curves computed within the restricted and unrestricted Hartree-Fock (RHF and UHF) and Kohn-Sham (RKS and UKS) formalisms.

the former approach, the KS Slater determinant is not the true wave function of the system and the extent to which spin contamination of the KS determinant affects the true wave function is not known. In fact, there are even opinions in the literature that KS determinants for open-shell systems which are not spin-contaminated are actually wrong (see for example Pople, Gill, and Handy, 1995). In any case, it is a noteworthy and comforting fact, even if we do not know how much it is worth, that if unrestricted Hartree-Fock and Kohn-Sham determinants are compared, the deviation of $\langle \hat{S}^2 \rangle$ from the exact value is in most cases considerably less significant for the KS determinant as described by Baker, Scheiner, and Andzelm, 1993 and Laming, Handy, and Amos, 1993. The latter authors suggest that the tendency of unrestricted open-shell Kohn-Sham determinants to have only rather small spin contaminations is due to the local nature of the exchange-correlation functionals as opposed to the non-local Hartree-Fock exchange.

In wave function based methods spin-contaminated unrestricted wave functions are frequently corrected by applying so-called *spin projection and annihilation* techniques. Here, the unwanted contributions to the energy belonging to states of other than the desired $\hat{S}^2$ are removed using several techniques such as spin projection operators or expression of the contaminated wave function in terms of pure spin states and subsequent subtraction of the energies of the unwanted higher spin states. These methods have also been applied to unrestricted Kohn-Sham determinants. However, neither the theoretical soundness of this method nor the quality of the spin-projected energies has been firmly established yet and the general state of affairs in this respect seems to be more critical rather than promising (for a discussion and examples see, e. g., Cramer et al., 1995, Wittbrodt and Schlegel, 1996, Goldstein, Beno, and Houk, 1996 or Rodriguez, Wheeler, and McCusker, 1998).

5.3.6 On Degeneracy, Ensembles and other Oddities

In the preceding section we mentioned the problem of the symmetry broken and hence, unphysical spin densities created in the UKS scheme upon dissociation of the hydrogen molecule. This is just one example of an interesting field where approximate density functional theory faces a plethora of yet unsolved problems, i. e., how to deal with degeneracies due to spin or non-abelian spatial symmetry.[16] For a competent review on this topic, the reader is referred to the beautiful contribution by Savin, 1996, which inspired part of the following exposition. In the following we will enumerate some typical problematic cases. Our main intention is to sharpen the reader's attention in this regard rather than to offer a profound theoretical discussion, let alone general solutions. Due to the complexity of the problem the former forbids itself in the present context, while the latter is still not available in all cases. For a deeper scrutiny we refer the reader to the entry points for the current literature on this subject included in the text.

Whenever symmetry related degeneracies occur, all current approximate Kohn-Sham based density functionals fail in one way or the other. Let us start with a seemingly easy class of systems: atoms. Unless atoms are characterized by completely filled shells, all atomic ground states exhibit spatial or spin degeneracies. Let us take the ^{2}D state of the scandium dication as a simple example. It has a [Ne] $(3s)^2$ $(3p)^6$ $(3d)^1$ configuration, i. e., a singly occupied d-shell. There are a variety of equivalent ways how this occupation can be represented. For example, using real d-orbitals the single electron could reside in any of the five degenerate orbitals. Alternatively one could choose a spherically symmetric ensemble of all five d-orbitals with equal weights of 1/5 or even opt for a complex representation of the d-orbitals. The important point is that each of these representations yield significantly different charge densities. This is shown for the occupation of the five different real d-orbitals in Figure 5-2. Occupying the d_{z^2} orbital yields an energy different from occupying any other d-orbital.

On the other hand, the correct energy of the atomic state must obviously be independent of the particular choice of occupation and be the same in all cases. Hence, any approximate density functional faces the difficult task to deliver the same energy from these different, but yet equivalent atomic densities. The sobering reality is that none of the available functionals is able to master this challenge. This of course provokes the next question, i. e., if there are a number of energies corresponding to the scandium ground state configuration, is there any criterion according to which one can select a particular energy? While we will return to this question in some detail in our later discussion on atomic calculations in Chapter 9, we mention at this point that the usual answer is to opt for the solution which delivers the lowest energy, even if physical symmetry requirements are violated.

Another typical class of examples is given by the dissociation of diatomic molecules as already alluded to above in the case of the H_2 molecule where the correct dissociation behavior was only achieved by allowing for symmetry broken spin densities. This problem

[16] A non-abelian point-group contains irreducible representations of dimension larger than one. Since the degree of degeneracy caused by spatial symmetry equals the dimensionality of the corresponding irreducible representation, this kind of degeneracy is only possible in non-abelian point groups.

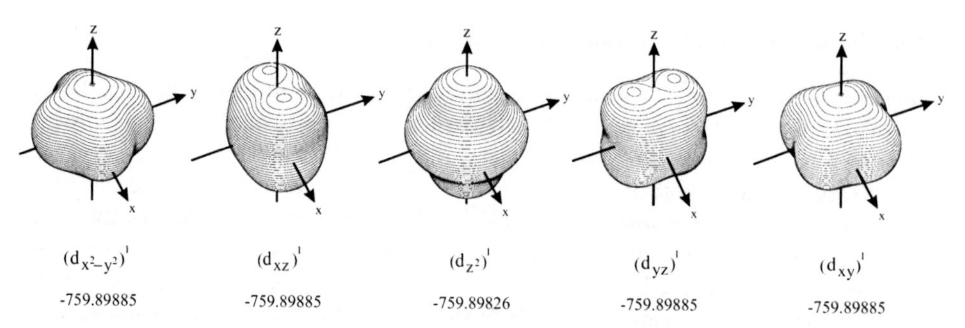

Figure 5-2. Isodensity surfaces (0.001 a.u.) of the d^1-densities generated from integral orbital occupation of the five d-orbitals in Sc^{2+} by one electron in a DFT calculation. The shape of the density resulting from occupation of the d_{z^2}-orbital differs from the other four (which are identical to each other except for their orientation in space) and a slightly different total energy (given in a.u.) is assigned to this particular density.

is, however, more general. Consider the ground state of an arbitrary homonuclear diatomic X_2 which dissociates into two ground state atoms X. At the limit of an infinite X–X distance we have a supermolecule of two non-interacting atoms X. Even though there is no interaction between the two atoms the global wave function has of course still the overall molecular symmetry. It should be obvious from elementary arguments that the energy as well as the charge density of this supermolecule must be obtainable from the corresponding quantities of the isolated atoms. In particular the energy of the supermolecule must be twice the energy of one isolated atom X. This important property is called *size-consistency*.

Let us take the B_2 molecule in its $^3\Sigma_g^-$ ground state as an example. Near the equilibrium distance the dominant configuration is $1\sigma_g^2\ 1\sigma_u^2\ 2\sigma_g^2\ 2\sigma_u^2\ 1\pi_u^2$, i. e., two triplet coupled electrons occupy the lowest bonding π molecular orbital. Hence, the molecular density is cylindrically symmetric, independent of the internuclear distance. This density can easily be represented by a Slater determinant containing two singly occupied π_x and π_y orbitals, which are generated as a linear combination of the corresponding real atomic p-orbitals (L and R stand for left and right, respectively):

$$\pi_x \approx \frac{1}{\sqrt{2}}[p_{x,L} + p_{x,R}] \text{ and } \pi_y \approx \frac{1}{\sqrt{2}}[p_{y,L} + p_{y,R}]. \tag{5-22}$$

At infinite separation, one arrives at two boron atoms each having a donut-like cylindrical density as indicated in Figure 5-3. However, such a density cannot be obtained from real atomic p-orbitals. In other words, the density that results from the supermolecule is simply inaccessible from calculations on the isolated atoms. Whatever we do, we will never generate the correct charge density (and therefore energy) of the dissociated B_2 molecule by calculations of the isolated boron atoms and the requirement of size-consistency is violated. Only if one switches to complex orbitals such as $|p_x \pm ip_y|$, are cylindrical atomic densities possible. But even then, we are still in trouble and face a different problem. Just as

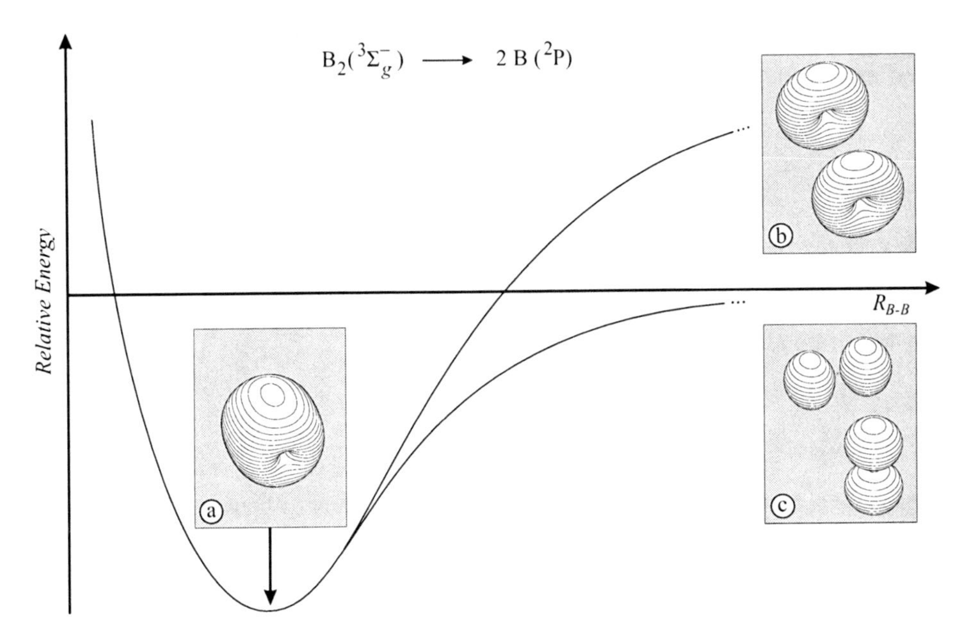

Figure 5-3. The symmetry dilemma in present-day DFT: starting from the cylindrically symmetric molecular π-density (a), the dissociation into atomic fragments can either be computed with correct atomic densities but a wrong energy (b) or a correct energy, but wrong (because symmetry broken) atomic densities (c) (isodensity surfaces at 0.01 a.u. constructed from the p-orbital space; adapted from Savin in *Recent Developments of Modern Density Functional Theory*, Seminario, J. M. (ed.), 1996, with permission from Elsevier Science).

in the preceding example of the scandium density, there are many different, but yet equivalent ways to realize the $1s^2\,2s^2\,2p^1$ occupation of the ^2P atomic ground state of boron. Even if cylindrical atomic densities can be represented by complex orbitals and size consistency is re-established, this atomic density results in a less favorable energy than the energy coming from the use of real atomic p-orbitals shown in the lower half of Figure 5-3. To reach size-consistency with respect to this atomic energy we are forced to break the molecular symmetry in the supermolecule calculation. To be specific, we construct the molecular density by allowing the localization of one open-shell electron in the, say, p_x atomic orbital of the left boron atom and the other electron in the p_y orbital of the right atom. Again we have a situation where the correct energy can only be obtained from an unphysical density.

Are there any remedies in sight within approximate Kohn-Sham density functional theory to get correct energies connected with physically reasonable densities, i. e., without having to use wrong, that is symmetry broken, densities? In many cases the answer is indeed yes. But before we consider the answer further, we should point out that the question only needs to be asked in the context of the approximate functionals: for degenerate states and related problems outlined above, an exact density functional in principle also exists. The real-life solution is to employ the non-interacting *ensemble-V_S* representable densities $\overline{\rho}$ introduced briefly in Section 5.3.4. These densities are not obtained from pure states, but from ground state *ensembles* according to

$$\bar{\rho} = \sum_{i}^{L} w_i \rho_i \ \text{ with } \sum_{i}^{L} w_i = 1 \text{ and } 0 \le w_i \le 1. \tag{5-23}$$

The densities ρ_i are obtained from a set of degenerate KS wave functions and the w_i are the corresponding weights. Without going into details we note that regular density functional theory can be extended to such ensembles. For our problems at hand, we can write down the energy expression as

$$E[\rho_1, \rho_2, \ldots, \rho_L] = \sum_{i}^{L} w_i E[\rho_i]. \tag{5-24}$$

Again, the ρ_i are the equivalent densities obtained from symmetry breaking. Let us clarify this concept by using the examples given above. In the B_2 case, the two equivalent symmetry broken Kohn-Sham Slater determinants are

$$\Theta_1 = \left| \ldots p_{x,L}\, p_{y,R} \right| \text{ and } \Theta_2 = \left| \ldots p_{x,R}\, p_{y,L} \right|. \tag{5-25}$$

These two determinants produce equivalent, but asymmetric densities. In addition, the energies obtained from these densities are the same, i. e. $E[\rho_1] = E[\rho_2]$. If we now insert these two densities in equation (5-24) it is clear that the energy will be invariant to the choice of w_1 and w_2. If we choose $w_1 = w_2 = 1/2$ we will also arrive at the physically correct, i. e. symmetric density. A very similar reasoning can be used for the H_2 dissociation. We again have two equivalent Kohn-Sham spin densities corresponding to

$$\Theta_1 = \left| 1s_L^{\alpha}\, 1s_R^{\beta} \right| \text{ and } \Theta_2 = \left| 1s_L^{\beta}\, 1s_R^{\alpha} \right|, \tag{5-26}$$

which have the same energy. If both spin densities enter with equal weight,

$$\bar{\rho}^{\alpha} = \frac{1}{2}[\rho_L^{\alpha} + \rho_R^{\alpha}] \text{ and } \bar{\rho}^{\beta} = \frac{1}{2}[\rho_R^{\beta} + \rho_L^{\beta}] \tag{5-27}$$

the result is the correct zero spin density,

$$\bar{\rho}^{\alpha} - \bar{\rho}^{\beta} = \frac{1}{2}[\rho_L^{\alpha} - \rho_L^{\beta} + \rho_R^{\alpha} - \rho_R^{\beta}] = 0. \tag{5-28}$$

The strategy to first use broken symmetry solutions and later restore the correct (spin) density by employing ensembles can be applied successfully to solve many degeneracy related problems. However, in practice it is very rarely used because there are hardly any computational schemes that can deal with such situations. In the overwhelming number of cases, symmetry broken solutions are used without bothering about dealing with unphysical spin densities. Finally, we should add that the approach of using broken symmetry and afterwards restoring the correct symmetry by constructing ensembles is not a panacea. There are examples, such as the O_2 dissociation, where even more elaborate strategies reaching beyond the regular Kohn-Sham formalism are required for a solution. For details, see Savin, 1995 and 1996.

5.3.7 Excited States and the Multiplet Problem

We have noted in the derivation of the Hohenberg-Kohn theorems that density functional theory is usually termed a ground state theory. The reason for this is not that the ground state density does not contain the information on the excited states – it actually does! – but because no practical way to extract this information is known so far. However, the properties of excited states and excitation energies in particular are of interest in many respects and a number of strategies how one could approach this problem in the framework of the Kohn-Sham scheme have been put forward. In the following we will concentrate only on those approaches which have found their way into real applications. For a theoretically oriented discussion, see e. g., the important work of Theophilou, 1979 (also reviewed in Parr and Yang, 1989), Gross, Oliveira, and Kohn, 1988a,b or the more recent contributions by Görling, 1996 or Nagy, 1998a,b. In this context we note a very recent report of Görling, 1999 to extend the regular Kohn-Sham schemes with their limitation to ground state properties to excited states. In this method a more general formulation of the Hohenberg-Kohn theorem is presented which allows one to treat ground and excited states on an equal footing. While first exploratory applications to alkali metal atoms show promising performance, routine applications will not be possible for some time. It will be interesting to see whether this or other approaches will have the potential to overcome the excited state problem in density functional theory in a general fashion.

More pragmatically, standard density functional techniques can be used to explore the energetically lowest lying state of each spatial or spin irreducible representation of the system, since they represent in a sense the 'ground state' in that particular symmetry as shown many years ago by Gunnarsson and Lundqvist, 1976. For example, computing the energetic separation between the CH_2 $\tilde{X}^3B_1$ triplet ground state and the lowest lying $\tilde{a}^1A_1$ singlet can be realized within density functional theory by simply setting up the corresponding Kohn-Sham determinants and computing the energy difference between these two (this is the so-called ΔSCF method). Even in cases where the target is a state which is excited even within a certain representation, the pragmatic solution – which has no formal justification! – is to apply the regular ground state scheme, provided the excited state can be written as a single determinant. However, many excited states do not fall into that category and intrinsically need a multi-determinantal description. Instructive examples are provided by the open-shell configurations of atoms which give rise to several terms. For example, the $1s^22s^22p^2$ configuration of the carbon atom leads to three atomic terms, ^3P, ^1D

and ^{1}S. A density functional theory based evaluation of the relative stabilities of these terms is a long-standing problem that holds a prominent place in the theoretical debate, the so-called *multiplet problem*. Let us elaborate on this subject a bit more. If we neglect the effects of spin/orbit coupling, all eligible states of an atom must be simultaneous eigenstates not only of the Hamilton operator $\hat{H}$ but also of the angular momentum operators $\hat{L}^2$ and $\hat{L}_z$, the corresponding spin operators $\hat{S}^2$ and $\hat{S}_z$, and the parity $\hat{\pi}$. Hence, the state of an atom is characterized by the associated quantum numbers, i. e., L, M_L, S, M_S, and π. In density functional theory we are not working with wave functions, which usually are the carriers of these symmetry requirements. Therefore, it must be the exchange-correlation functionals that should in principle contain the dependence on the above quantum numbers. On the other hand, as we will discuss in detail in the following chapter, all current approximate functionals are based on the model of the uniform electron gas and solely depend on the charge or spin densities. They lack any relation to the other quantities relevant for a complete description of an atomic state we have just mentioned. We are therefore facing a considerable conceptual problem inherent to Kohn-Sham density functional theory which transcends the atomic case and applies equally well to other symmetry related problems: how should one describe states which are eigenfunctions of the $\hat{L}^2$ and $\hat{S}^2$ or other operators if we are working merely with an orbital based theory where we have no access to the correct N-electron wave function and its symmetry characteristics? The pragmatic solution adopted is to select the single-determinantal non-interacting Kohn-Sham reference system in such a way that this Slater determinant corresponds to a state of the desired definite values of the conserved quantum numbers. As we will see presently, this creates new problems because of the limitations of Slater determinants.

An early ad-hoc approach to solve the multiplet problem is the *sum method* due to Ziegler, Rauk, and Baerends, 1977, see also von Barth, 1979. Among the central conclusions put forward by these authors is that the energy of a term which is not representable by a *single* Slater determinant but needs a linear combination of determinants to exhibit the correct spatial and spin symmetries, cannot be computed by using the spin densities generated from the corresponding configuration state function, i. e., the proper symmetry adapted linear combination of Slater determinants. Rather, they propose that excited states which cannot be expressed as a single-determinantal wave function should be written as a weighted sum of determinantal energies, according to

$$E = \sum_{j}^{J} C_j E(\Phi_j) \tag{5-29}$$

where the coefficients C_j are fixed by the required symmetry. Let us take a very simple example to illustrate this concept. Consider the helium atom with its ^{1}S ground state characterized by a $(1s)^2$ configuration. If we now transfer one electron from the 1s to the 2s level, i. e. generate a $(1s)^1\,(2s)^1$ occupation, this configuration is connected with two multiplet states, the lowest excited ^{1}S and ^{3}S terms of atomic helium. Recall from basic quantum mechanics that the ^{3}S state consists of three, energetically degenerate components according to values of the quantum number $M_S = 1, 0$, and -1 of the z-component of the total spin, while there is of course only one ^1S component with $M_S = 0$. What are the energies of these excited states? We start out by writing down the corresponding single-determinantal wave functions, but immediately see that one can construct such functions only for the two triplets corresponding to $M_S = \pm 1$:

$$\Phi(^3S, M_S = 1) = \frac{1}{\sqrt{2}} \det\{1s(\vec{r}_1)\ 2s(\vec{r}_2)\} \times (\alpha(s_1)\ \alpha(s_2)) \tag{5-30}$$

$$\Phi(^3S, M_S = -1) = \frac{1}{\sqrt{2}} \det\{1s(\vec{r}_1)\ 2s(\vec{r}_2)\} \times (\beta(s_1)\ \beta(s_2)) \tag{5-31}$$

As orbitals we simply use 1s and 2s atomic functions and our notation should be self-explanatory. Thus, following the assumption that states which are represented by a single determinant can be studied, we could set up a Kohn-Sham determinant corresponding to either equation (5-30) or (5-31) to obtain the energy of the ^3S state in an UKS calculation. The two states with $M_S = 0$, i. e. the ^1S and the remaining component of the triplet (of which we know that it must be energetically degenerate with the $M_S = \pm 1$ components), are more complicated and no longer of a single-determinantal form:

$$\Phi(^3S, M_S = 0) = \frac{1}{2}\left[1s(\vec{r}_1)\ 2s(\vec{r}_2) - 2s(\vec{r}_1)\ 1s(\vec{r}_2)\right] \times \left[\alpha(s_1)\ \beta(s_2) + \beta(s_1)\ \alpha(s_2)\right] \tag{5-32}$$

$$\Phi(^1S, M_S = 0) = \frac{1}{2}\left[1s(\vec{r}_1)\ 2s(\vec{r}_2) + 2s(\vec{r}_1)\ 1s(\vec{r}_2)\right] \times \left[\alpha(s_1)\ \beta(s_2) - \beta(s_1)\ \alpha(s_2)\right]. \tag{5-33}$$

How can we obtain an energy for the excited singlet? The normal prescription is not applicable since there is no single determinant on which a Kohn-Sham calculation could be based. However, the determinant that intuitively comes closest to this state is

$$\Phi(\text{mix}, M_S = 0) = \frac{1}{\sqrt{2}} \det\{1s(\vec{r}_1)\alpha(s_1)\ 2s(\vec{r}_2)\beta(s_2)\}. \tag{5-34}$$

This determinant has the desired $M_S = 0$, but its total spin is not defined. Now comes the trick: we recognize that equation (5-34) is actually a mixture of the functions (5-32) and (5-33) of the $M_S = 0$ states (which can easily be verified by expanding the determinants),

$$\Phi(\text{mix}, M_S = 0) = \frac{1}{\sqrt{2}}[\Phi(^3S, M_S = 0) + \Phi(^1S, M_S = 0)]. \tag{5-35}$$

Now the procedure to get the energy of the singlet is outlined: after reordering equation (5-35) and changing to energies rather than determinants we have

$$E(^1S, M_S = 0) = 2\ E(\text{mix}, M_S = 0) - E(^3S, M_S = 0). \tag{5-36}$$

Hence, we compute the Kohn-Sham energies of the single determinant for the mixed state and of one of the two accessible triplet states. Since all three components of ^3S must have the same energy, we know the energies of both terms on the right hand side of equation (5-36). Through this little detour we finally arrive at the desired result. This scheme can be applied in many situations if we recognize that many (even though not all) multiplet energies can be written as a weighted sum of single determinants Φ_j as in equation (5-29). A partially automated protocol for this technique based on an elegant group theoretical method to obtain the weights of the various determinants Φ_j of mixed symmetry, has been developed by Daul and coworkers as outlined by Daul, 1994 and Daul, Doclo, and Stückl, 1997 and implemented in an auxiliary program to be used together with the *Amsterdam Density Functional* program package (*ADF Single Determinants Fribourg*, *ASF*).

However, note that the sum method has no firm theoretical justification. Not only is the simple assumption that we can characterize excited states through the occupation numbers of the determinant representing the non-interacting reference system questionable, this approach also ignores the fact that the functional for the excited states need not be the same as that for the ground states. Finally, the application of the KS scheme to an unphysical state such as $\Phi(\text{mix}, M_S = 0)$ also carries a question mark. Similarly, the assumption that the same orbitals are used in each of the calculations also adds to the uncertainty of the results. Indeed, there are many examples of inconsistencies of this method. This is most clearly demonstrated by cases where one multiplet energy can be represented in various ways by using different combinations of Slater determinants. Of course, to be consistent the computed energy must be independent of the actually chosen linear combination of determinants. This physical requirement is often not fulfilled and deviations exceeding 0.5 eV for the energy of a given multiplet may occur. In addition, from a technical point of view the method has the disadvantage that several calculations are necessary for obtaining the desired energies and that optimizing the geometry for such states is not straightforward.

An alternative to the sum method, dubbed *spin-restricted open-shell Kohn-Sham* (ROKS), has recently been suggested by Filatov and Shaik, 1998a and 1999. This scheme bears a strong formal similarity to the general spin restricted open-shell version of Hartree-Fock theory. Unlike the UKS based sum method, the non-interacting Kohn-Sham reference wave function uses the same orbitals for α and β electrons and is an eigenfunction of $\hat{S}^2$ and $\hat{S}_z$. Likewise, the ROKS scheme yields one-electron orbitals and non-interacting wave functions that are symmetry adapted. The correct spatial symmetry is introduced via certain relations between the non-interacting wave function and the interacting multiplet energy. The latter corresponds to a symmetry adapted ensemble of Kohn-Sham determinants (which themselves can be viewed as states of mixed symmetry as shown above). A somewhat related scheme applicable for open-shell singlets (*restricted open-shell singlet*, ROSS) has been reported by Gräfenstein, Kraka, and Cremer, 1998. These authors use the relation between an open-shell singlet and the corresponding triplet state and introduce exactly computed exchange integrals to define an energy functional for an open-shell singlet. Taking the $a^1\Delta_g \leftarrow X^3\Sigma_g^-$ excitation energy of the O_2 molecule as an example, where both states originate from the $\ldots 1\pi_g^2$ configuration, ROKS fortuitously reproduces the experimental value of 0.97 eV exactly while the ROSS approach yields a slightly higher excitation energy of 1.18 eV if the BLYP exchange-correlation functional and polarized triple zeta Gaussian basis sets are used, indicating that both schemes perform rather well. Unfortunately, just like the sum method, none of these techniques has been implemented in any standard program package as of yet.

Grimme, 1996 has suggested a different way to bring electronic excitation energies into the realm of density functional theory. His method starts with the configuration interaction scheme restricted to single excitations (CIS), a well established method in wave function based theory to determine excited state energies. The matrix elements of the CIS Hamiltonian are then modified by replacing the Hartree-Fock orbital energies by the corresponding eigenvalues obtained from gradient-corrected Kohn-Sham calculations. In addition, three empirical parameters determined from a representative reference set are included to scale the Coulomb integrals and to introduce an empirical shift of the diagonal CIS matrix elements. Even though this approach also lacks a solid theoretical foundation, computed excitation energies for molecules including fairly large hydrocarbons are within a few tenths of an eV of the experimental data. Grimme's method carries the acronym DFT/SCI for *density functional theory/single excitation configuration interaction*. It has been extended to multi-reference configuration interaction schemes very recently, see Grimme and Waletzke, 1999. It would be interesting to have this method generally available in commonly used quantum chemical programs.

Another, again completely different but apparently very promising approach to the calculation of excitation energies has been developed in the past few years and is based on *time-dependent density functional theory*, TDDFT. From a practical point of view, TDDFT has the important advantage that it can actually be used because it was recently implemented in many quantum chemical programs, such as the 1998 release of *Gaussian* or the current version of *Turbomole*. This technique has a fairly involved theoretical background and we will confine our discussion to a very qualitative level. The reader interested in a more elaborate treatment of the subject is referred to the detailed reviews by, e. g., Casida, 1995, Burke and Gross, 1998 or Petersilka, Gossmann, and Gross, 1998. In a nutshell, this strategy employs the fact that the frequency dependent linear response of a finite system with respect to a time-dependent perturbation has discrete poles at the exact, correlated excitation energies of the unperturbed system. To be more specific, the frequency dependent mean polarizability $\alpha(\omega)$ describes the response of the dipole moment to a time-dependent electric field with frequency $\omega(t)$. It can be shown that the $\alpha(\omega)$ are related to the electronic excitation spectrum according to

$$\alpha(\omega) = \sum_{I} \frac{f_I}{\omega_I^2 - \omega^2}. \tag{5-37}$$

Here ω_I is the excitation energy E_I-E_0 and the sum runs over all excited states I of the system. From equation (5-37) we immediately see that the dynamic mean polarizability $\alpha(\omega)$ diverges for $\omega_I = \omega$, i. e., has poles at the electronic excitation energies ω_I. The residues f_I are the corresponding oscillator strengths. Translated into the Kohn-Sham scheme, the exact linear response can be expressed as the linear density response of a non-interacting system to an effective perturbation.[17] The orbital eigenvalue differences of the ground state KS orbitals enter this formalism as a first approximation to the excitation energies, which are then systematically shifted towards the true excitation energies. Note that in the TDDFT approach only properties of the ground state – namely the ordinary Kohn-Sham orbitals and their corresponding orbital energies obtained in a regular ground state calculation – are involved. Hence, excitation energies are expressed in terms of ground state properties and the problem of whether density functional theory can be applied to excited states is most elegantly circumvented. The TDDFT approach has even been extended from the mere prediction of excitation energies to the computational treatment of excited state surfaces including avoided crossings between states belonging to the same irreducible representation by Casida, Casida, and Salahub, 1998. It is probably fair to say that as of the time of writing TDDFT has the appearance of being the most promising avenue to a satisfactory excited state treatment within approximate density functional theory. An ever increasing number of papers showing the power of this technique has appeared since efficient implementations of TDDFT became generally available in major commercial codes. Errors are usually in the order of a few tenths of an eV, even if difficult situations are considered, such as Rydberg states (Handy and Tozer, 1999) or excited states with substantial double excitation character (Hirata and Head-Gordon, 1999), as we will explore in more detail in Chapter 9.

[17] Note that in all current implementations of TDDFT the so-called *adiabatic approximation* is employed. Here, the time-dependent exchange-correlation potential that occurs in the corresponding time-dependent Kohn-Sham equations and which is rigorously defined as the functional derivative of the exchange-correlation action $A_{XC}[\rho]$ with respect to the time-dependent electron-density is approximated as the functional derivative of the standard, time-independent E_{XC} with respect to the charge density at time t, i. e.,

$$V_{XC}[\rho(\vec{r},t)] = \frac{\partial A_{XC}[\rho]}{\partial \rho(\vec{r},t)} \approx \frac{\partial E_{XC}[\rho]}{\partial \rho_t(\vec{r})} = V_{XC}[\rho_t(\vec{r})]. \tag{5-38}$$

Stated in other words, the zero-frequency limit of A_{XC} is used for treating the finite frequency perturbations. For details see in particular Casida, 1995.

6 The Quest for Approximate Exchange-Correlation Functionals

In the previous chapter we introduced the Kohn-Sham formalism which allows an exact treatment of most of the contributions to the electronic energy of an atomic or molecular system, including the major fraction of the kinetic energy. All remaining – unknown – parts are collectively folded into the exchange-correlation functional $E_{XC}[\rho]$. These include the non-classical portion of the electron-electron interaction along with the correction for the self-interaction and the component of the kinetic energy not covered by the non-interacting reference system. Obviously, the whole endeavor of applying the Kohn-Sham scheme as a tool to get a grip on the Schrödinger equation makes sense only if explicit approximations to this functional are available. The quality of the density functional approach hinges solely on the accuracy of the chosen approximation to E_{XC}. Hence, the quest of finding better and better functionals is at the very heart of density functional theory. In the following we will review the current state of the art regarding approximate functionals for E_{XC}. We start out by showing that unlike in conventional wave function based methods, in density functional theory there is no systematic way towards improved approximate functionals, which in fact represents one of the major drawbacks associated with this approach. Then, we introduce the adiabatic connection, which provides the link between the non-classical potential energy of exchange and correlation and the E_{XC} functional of the Kohn-Sham scheme, with special emphasis on the corresponding hole function. The simple concept of the local density approximation based on the uniform electron gas, which represents the bedrock of almost all current functionals, is discussed. Even though this physical model performs better than anticipated, it is not accurate enough for chemical applications. Hence, ideas about how one can go beyond that approximation have been put forward by many researchers. We will develop the connection from the local density approximation to the more sophisticated generalized gradient approximation up to the nowadays so popular hybrid functionals. These general strategies are realized in many different individual functionals and the most widespread representatives as well as new developments for both classes are presented. We continue with a discussion of the problems due to the self-interaction of the charge density and to the behavior of the corresponding exchange-correlation potentials in the long range asymptotic region. Both aspects are inherent to all approximate exchange-correlation functionals and give rise to unwanted effects. The strengths and weaknesses of the various approaches will be discussed and we conclude this chapter with an assessment as to where future developments might lead.

6.1 Is There a Systematic Strategy?

Before we start looking at possible approximations to E_{XC} we need to address whether there will be some kind of guidance along the way. If we consider conventional, wave function based methods for solving the electronic Schrödinger equation, the quality of the results solely depends on our choice of the approximate wave function. From basic concepts of linear algebra we in fact know the prescription for how the true wave function should be constructed in principle, such as in the full configuration interaction scheme, characterized by an expansion of both the one and the many particle problem in a complete, i. e., essentially infinite basis. Even though this can never be realized because the resulting equations would be much too complicated to be ever solved, this prescription shows us the way how the approximate wave functions can be improved step by step in a systematic manner. Unfortunately a similar beacon guiding us along the way towards our final, albeit unreachable destination does not exist in density functional theory. The origin of this sobering statement is simply that the explicit form of the exact functional is a total mystery to us. Not only is the physics underlying the success of current functionals far from being fully understood, we simply do not have the faintest idea how to arrive at approximations which are closer to the exact functional. All searching for better functionals relies largely on physical or mathematical intuition and has a strong 'trial and error' component. There are, however, a few physical constraints which a reasonable functional has to fulfill. Among those a prominent place is held by the sum rules valid for the exact exchange-correlation holes as outlined in Chapter 2. Of course, the more closely the model hole that emerges from an approximate exchange-correlation functional resembles the true hole, the better this functional will be able to account for the non-classical effects. Other properties include the cusp condition of the correlation hole at zero separation of the two electrons, certain scaling conditions of the exchange and correlation energies and asymptotic properties of the corresponding exchange-correlation potentials, etc.[18] However, one should not expect too much help from such formal boundary conditions, since one of the baffling peculiarities of approximate density functional theory is that functionals which strictly meet these requirements are not necessarily better than others that do not. In fact, some of the most successful approximate functionals violate several of these conditions. It is therefore of immense importance to carefully study the performance of a particular functional with respect to a suitable set of reference data (we will elaborate on these decisive aspects in much detail in the second part of this book). Indeed, the most stringent tests currently available for new functionals are completely empirical and involve the comparison with accurate reference data, such as atomization, ionization and reaction energies, structural data and the like. The most frequently used set of energetic reference data is probably the so-called G2 thermochemical data base which contains more than 50 experimentally well established atomization energies of small molecules containing main group elements originally collected by Curtiss et al., 1991. The ability to reproduce the energetics of this data base or extensions to it has become the de facto standard for measuring the accuracy of a new computational method. The target accuracy is the so-called chemical accuracy which corresponds to an average absolute error of about 0.1 eV or 2 kcal/mol. However, this is a very ambitious goal and up to now only very few and very expensive traditional quantum chemical strategies are able to achieve this kind of accuracy. The performance of approximate

[18] We will not digress on that matter here but rather refer the interested reader to the relevant literature, see Perdew and Burke, 1996, or Adamo, di Matteo, and Barone, 1999.

functionals with regard to the G2 and related reference sets will occupy us in significant detail in Section 9.1. We should also mention in this context that the energy delivered by a particular functional is not the ultimate probe for its quality. The exchange-correlation energy results from the integral over the exchange-correlation potential and in principle the correct energy can be obtained even from an erroneous potential because of a fortuitous error cancellation. Therefore, a more physically motivated test is provided by a point-by-point comparison between the model exchange-correlation potential and accurate potentials derived from high-quality correlated wave function based calculations. However, such accurate potentials are rarely available and this kind of validation is seldom used.

6.2 The Adiabatic Connection

The purpose of this chapter is the illustration of the ways how a good approximation to the exact exchange-correlation functional of Kohn-Sham theory can be found. But before we proceed we need to take a second look at this very quantity and relate E_{XC} with the concept of exchange-correlation holes introduced earlier. The hole functions we discussed in Chapter 2 contained all information about the non-classical contributions to the *potential energy* due to the electron-electron interaction, E_{ncl}. However, we saw in the preceding chapter that E_{XC} as defined in the framework of Kohn-Sham theory also accounts for T_C, the difference between the *kinetic energy* of the real, fully interacting system, T, and the kinetic energy T_S related to the non-interacting reference system,

$$E_{XC}[\rho] = \{T[\rho] - T_S[\rho]\} + E_{ncl}[\rho] = T_C[\rho] + E_{ncl}[\rho]. \tag{6-1}$$

Thus, the information about $T[\rho] - T_S[\rho]$ must be somehow folded into the corresponding hole functions. To do this, imagine that we connect the two systems central for the KS scheme (i. e. the non-interacting reference with no $1/r_{ij}$ electron-electron interaction and the real one where this interaction is operative with full strength) by gradually increasing the *coupling strength parameter* λ from 0 to 1:

$$\hat{H}_\lambda = \hat{T} + V_{ext}^{\lambda} + \lambda \sum_{i}^{N} \sum_{j>i}^{N} \frac{1}{r_{ij}}. \tag{6-2}$$

For each λ the effective external potential V_{ext}^{λ} is adapted such that the density always equals the density of the fully interacting system. Hence $\rho(\vec{r})$ is independent of the actual value of λ. Clearly, for $\lambda = 0$, we recover the Hamilton operator for the non-interacting reference system, and $V_{ext}^{\lambda=0} = V_S$, while $\lambda = 1$ is the regular expression for the real system with $V_{ext}^{\lambda=1} + \sum_{i}^{N} \sum_{j>i}^{N} \frac{1}{r_{ij}} = V_{eff} = \int \frac{\rho(\vec{r}_2)}{r_{12}} d\vec{r}_2 + V_{XC}(\vec{r}_1) - \sum_{A}^{M} \frac{Z_A}{r_{1A}}$. Equation (6-2) describes how these two endpoints are smoothly connected through a continuum of artificial, partially interacting systems. Borrowing from thermodynamics this path is called the *adi-*

abatic connection. In terms of the adiabatic connection the energy of the interacting system $E_{\lambda=1}$ can be expressed as the following integral

$$E_{\lambda=1} - E_{\lambda=0} = \int_0^1 dE_\lambda \text{ , and thus } E_{\lambda=1} = \int_0^1 dE_\lambda + E_{\lambda=0} \tag{6-3}$$

To utilize this relation we now need an explicit expression for dE_λ. To this end we investigate how the total energy E_λ changes upon an infinitesimal change in the coupling strength λ. This energy is the expectation value of the corresponding Hamiltonian

$$d\hat{H}_\lambda = dV_{ext}^\lambda + d\lambda \sum_i^N \sum_{j>i}^N \frac{1}{r_{ij}}, \tag{6-4}$$

and, using the hole formalism, can be expressed as

$$\begin{aligned} dE_\lambda = &\int \rho(\vec{r}) dV_{ext}^\lambda \, d\vec{r} + \frac{1}{2} d\lambda \iint \frac{\rho(\vec{r}_1)\rho(\vec{r}_2)}{r_{12}} d\vec{r}_1 \, d\vec{r}_2 \\ &+ \frac{1}{2} d\lambda \iint \frac{\rho(\vec{r}_1) h_{XC}^\lambda(\vec{r}_1;\vec{r}_2)}{r_{12}} d\vec{r}_1 \, d\vec{r}_2 \end{aligned} \tag{6-5}$$

Inserting equation (6-5) in the integral of equation (6-3) leads to

$$\begin{aligned} E_{\lambda=1} - E_{\lambda=0} = &\int \rho(\vec{r}) \left[V_{ext}^{\lambda=1} - V_{ext}^{\lambda=0} \right] d\vec{r} + \frac{1}{2} \iint \frac{\rho(\vec{r}_1)\rho(\vec{r}_2)}{r_{12}} d\vec{r}_1 d\vec{r}_2 \\ &+ \frac{1}{2} \iint \int_0^1 \frac{\rho(\vec{r}_1) h_{XC}^\lambda(\vec{x}_1;\vec{x}_2)}{r_{12}} d\vec{x}_1 \, d\vec{x}_2 \, d\lambda \end{aligned} \tag{6-6}$$

where we have made use of the λ-independence of the density $\rho(\vec{r})$. Replacing $V_{ext}^{\lambda=0}$ and $V_{ext}^{\lambda=1}$ by V_S and V_{eff}, respectively and using the energy expression for the non-interacting Kohn-Sham system,

$$E_{\lambda=0} = T_S + \int \rho(\vec{r}) V_S d\vec{r} \tag{6-7}$$

and defining the *coupling-strength integrated exchange-correlation hole* $\bar{h}_{XC}$ as

$$\bar{h}_{XC}(\vec{r}_1;\vec{r}_2) \equiv \int_0^1 h_{XC}^\lambda(\vec{r}_1;\vec{r}_2) d\lambda \tag{6-8}$$

we finally arrive at the following equation for the energy of the real, interacting system

$$E_{\lambda=1} = T_S + \int \rho(\vec{r})V_{eff}d\vec{r} + \frac{1}{2}\iint \frac{\rho(\vec{r}_1)\rho(\vec{r}_2)}{r_{12}} d\vec{r}_1 d\vec{r}_2 + \frac{1}{2}\iint \frac{\rho(\vec{r}_1)\bar{h}_{XC}(\vec{r}_1;\vec{r}_2)}{r_{12}} d\vec{r}_1 d\vec{r}_2 \tag{6-9}$$

Thus, the important take-home message is that the exchange-correlation energy of the Kohn-Sham scheme can be expressed through the coupling-strength integrated exchange-correlation hole $\bar{h}_{XC}$. If we know this hole, we know the exchange-correlation energy as demonstrated in the following expression

$$E_{XC} = \frac{1}{2}\iint \frac{\rho(\vec{r}_1)\bar{h}_{XC}(\vec{r}_1;\vec{r}_2)}{r_{12}} d\vec{r}_1 d\vec{r}_2 . \tag{6-10}$$

What does this mean? If we compare equation (6-9) which we just derived with its counterpart which can be deduced from equation (2-19) of Chapter 2 (by adding to the E_{ee} term of the electron-electron interaction described in that expression the kinetic energy and the contribution from the external potential)

$$E = T + \int \rho(\vec{r})V_{eff}d\vec{r} + \frac{1}{2}\iint \frac{\rho(\vec{r}_1)\rho(\vec{r}_2)}{r_{12}} d\vec{r}_1 d\vec{r}_2 + \frac{1}{2}\iint \frac{\rho(\vec{r}_1)h_{XC}(\vec{x}_1;\vec{x}_2)}{r_{12}} d\vec{x}_1 d\vec{x}_2 \tag{6-11}$$

where T is the *true* kinetic energy of the fully interacting system and the last term the non-classical contribution to the electron-electron repulsion, we see that the integration over the coupling-strength elegantly transfers the difference between T and T_S, i. e. the part of the kinetic energy not covered by the non-interacting reference system, into the exchange-correlation hole. In other words, when going from equation (6-11) to equation (6-9) we drastically simplify the expression for the kinetic energy (T) by reducing it to the kinetic energy of the non-interacting reference system (T_S). The price we pay for this is a further complication in the exchange-correlation hole brought about by the additional integration over the coupling strength parameter λ, i. e., we replace h_{XC} by $\bar{h}_{XC}$. Importantly, this integration has no effect on the formal properties of the exchange-correlation hole discussed in Chapter 2, the sum rules, cusp conditions, etc. that apply to h_{XC} apply as well to $\bar{h}_{XC}$. For details on this very fundamental concept see the contributions of Gunnarsson and Lundqvist, 1976, Harris, 1984, Becke, 1988a, 1995, and Jones and Gunnarsson, 1989.

6.3 From Holes to Functionals

As already alluded to above, the analysis of the properties of model hole functions that emerge from approximate exchange-correlation functionals is a major tool for assessing

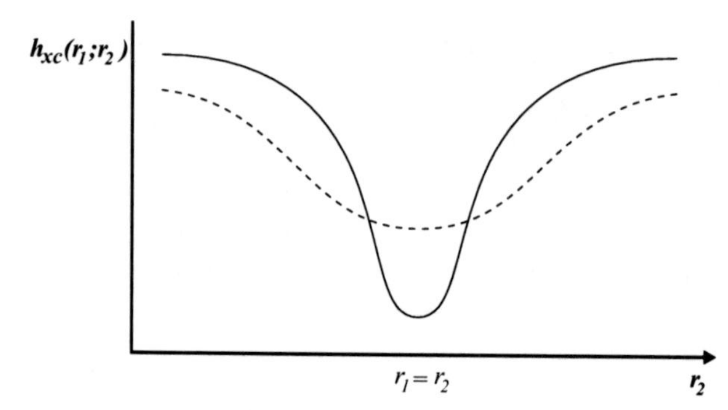

Figure 6-1. Fermi holes of different depths for the on-top density.

the suitability of such functionals. Recall that in Chapter 2 we separated the total exchange-correlation hole into two components, the Fermi and Coulomb holes, with the former being by far the most important contribution to the total hole. The Fermi hole was identified as a non-positive quantity which contains exactly one elementary charge. This is a fairly stringent restriction. For example, for 'normal' positions, where the hole is concentrated around the reference electron it follows that the 'deeper' the hole gets for $\vec{r}_2 \rightarrow \vec{r}_1$ (where it assumes $-\rho(\vec{r}_1)$) the less will it extend into space, i. e., the shorter its range, as schematically shown in Figure 6-1. We should note that the exchange-correlation hole for the special case that $\vec{r}_2 = \vec{r}_1$ is called the *on-top hole*, which has attracted considerable attention lately, see, e. g., Perdew et al., 1997, and Burke, Perdew, and Ernzerhof, 1998.

On the other hand, the Coulomb hole integrates to zero and can be negative as well as positive. Hence this sum rule is of only minor help. There is one additional, very important aspect in this context. The exact hole functions are highly asymmetric entities and it will be very difficult for any approximate hole to recover all the subtle details of its six-dimensional shape (remember that the hole depends on the coordinates of two electrons). However, the expression for the exchange-correlation energy given in equation (6-10) does not rely on the angular details of $\bar{h}_{XC}$ because of the clearly isotropic character of the Coulomb interaction represented by the $1/r_{ij}$ operator and only depends on the distance between any two electrons. Hence, our approximate hole only has to model the *spherically averaged* exact hole about each reference point, which is significantly less complicated (but still complicated enough).

6.4 The Local Density and Local Spin-Density Approximations

In this section we introduce the model system on which virtually all approximate exchange-correlation functionals are based. At the center of this model is the idea of a hypothetical *uniform electron gas*. This is a system in which electrons move on a positive background charge distribution such that the total ensemble is electrically neutral. The number of electrons N as well as the volume V of the gas are considered to approach infinity, while the electron density, i. e., N/V remains finite, $N \to \infty$, $V \to \infty$, $N/V = \rho$ and attains a constant value everywhere. Physically, such a situation resembles the model of an idealized metal consisting of a perfect crystal of valence electrons and positive cores where the cores are smeared out to arrive at a uniform positive background charge. Indeed, the uniform electron gas is a fairly good physical model for simple metals such as sodium. On the other hand, we should note from the start that this model system, which is also known under the label of the homogeneous electron gas, is pretty far from any realistic situation in atoms or molecules, which are usually characterized by rapidly varying densities. The reason why the uniform electron gas has such a prominent place in density functional theory is that it is the only system for which we know the form of the exchange and correlation energy functionals exactly or at least to very high accuracy. We actually already met the exchange functional of this model system in Chapter 3 when we briefly discussed the Dirac exchange functional that appears in the Thomas-Fermi-Dirac method. The idea to use this model for approximating E_{XC} in the Kohn-Sham scheme was already included in the original paper by Kohn and Sham, 1965. Let us tackle the problem now from a slightly different point of view. Central to this model is the assumption that we can write E_{XC} in the following, very simple form

$$E_{XC}^{LDA}[\rho] = \int \rho(\vec{r})\varepsilon_{XC}(\rho(\vec{r}))\, d\vec{r}\,. \tag{6-12}$$

Here, $\varepsilon_{XC}(\rho(\vec{r}))$ is the exchange-correlation energy per particle of a uniform electron gas of density $\rho(\vec{r})$. This energy per particle is weighted with the probability $\rho(\vec{r})$ that there is in fact an electron at this position in space. Writing E_{XC} in this way defines the *local density approximation*, LDA for short. The quantity $\varepsilon_{XC}(\rho(\vec{r}))$ can be further split into exchange and correlation contributions,

$$\varepsilon_{XC}(\rho(\vec{r})) = \varepsilon_X(\rho(\vec{r})) + \varepsilon_C(\rho(\vec{r}))\,. \tag{6-13}$$

The exchange part, ε_X, which represents the exchange energy of an electron in a uniform electron gas of a particular density is, apart from the pre-factor, equal to the form found by Slater in his approximation of the Hartree-Fock exchange (Section 3.3) and was originally derived by Bloch and Dirac in the late 1920's:

$$\varepsilon_X = -\frac{3}{4}\sqrt[3]{\frac{3\,\rho(\vec{r})}{\pi}}\,. \tag{6-14}$$

Inserting equation (6-14) into equation (6-12) retrieves the $\rho^{4/3}$ dependence of the exchange energy indicated in equation (3-5). This exchange functional is frequently called *Slater exchange* and is abbreviated by S. No such explicit expression is known for the correlation part, ε_C. However, highly accurate numerical quantum Monte-Carlo simulations of the homogeneous electron gas are available from the work of Ceperly and Alder, 1980.

On the basis of these results various authors have presented analytical expressions of ε_C based on sophisticated interpolation schemes. The most widely used representations of ε_C are the ones developed by Vosko, Wilk, and Nusair, 1980, while the most recent and probably also most accurate one has been given by Perdew and Wang, 1992. The common short hand notation for the former implementations of the correlation functional is VWN. Hence, instead of the abbreviation LDA, which defines the model of the local density approximation, one frequently finds the acronym SVWN to identify the particular functional. Note that in their paper Vosko, Wilk, and Nusair report several expressions for ε_C. VWN usually implies that the correlation energy density of the homogeneous electron gas has been obtained in the random phase approximation (RPA), while the somewhat less frequently used VWN5 variant (note that this is the one recommended by the authors) denotes the use of the parameterization scheme based upon the results of Ceperly and Alder. Even though these two VWN functionals in most cases perform similarly (Hertwig and Koch, 1997. But see Section 9.4 for examples where VWN and VWN5 perform differently) one should be cautious about which flavor of the VWN functional is actually implemented in the corresponding computer program in order to avoid confusion. Before we go on, we pause for a minute to make a general remark on the nomenclature found in the literature to name a particular functional. While there is no strict rule, most authors now term the functionals as 'XC' where X stands for the exchange part and C for the correlation part as described by the initial letter of the names of the corresponding authors. The letters are augmented by a year, if the same authors developed more than one functional. If the exchange and correlation parts are due to the same authors, the letters are usually given only once.

In the preceding chapter we mentioned that approximate functionals are usually also expressed in an unrestricted version, where not the electron density $\rho(\vec{r})$, but the two spin densities, $\rho_\alpha(\vec{r})$ and $\rho_\beta(\vec{r})$, with $\rho_\alpha(\vec{r}) + \rho_\beta(\vec{r}) = \rho(\vec{r})$ are employed as the central input. Even though from a puristic theoretical point of view the exact functional will not depend on the spin densities (as long as the external potential is spin-independent), approximations to it will benefit from the additional flexibility of having two instead of one variable. In particular, for open-shell situations with an unequal number of α and β electrons, functionals of the two spin densities consistently lead to more accurate results. But also for certain situations with an even number of electrons, such as the H_2 molecule at larger separation, the unrestricted functionals perform significantly better because they allow symmetry breaking. Up to this point the local density approximation was introduced as a functional depending solely on $\rho(\vec{r})$. If we extend the LDA to the unrestricted case, we arrive at the *local spin-density approximation*, or LSD. Formally, the two approximations differ only that instead of equation (6-12) we now write

$$E_{XC}^{LSD}[\rho_\alpha, \rho_\beta] = \int \rho(\vec{r})\varepsilon_{XC}(\rho_\alpha(\vec{r}), \rho_\beta(\vec{r}))\, d\vec{r}\,. \tag{6-15}$$

Just as for the simple, spin compensated situation where $\rho_\alpha(\vec{r}) = \rho_\beta(\vec{r}) = \frac{1}{2}\rho(\vec{r})$, there are related expressions for the exchange and correlation energies per particle of the uniform electron gas characterized by $\rho_\alpha(\vec{r}) \neq \rho_\beta(\vec{r})$, the so-called spin polarized case. The degree of spin polarization is often measured through the spin-polarization parameter

$$\xi = \frac{\rho_\alpha(\vec{r}) - \rho_\beta(\vec{r})}{\rho(\vec{r})}. \qquad (6\text{-}16)$$

ξ attains values from 0 (spin compensated) to 1 (fully spin polarized, i. e., all electrons have only one kind of spin). For details see in particular Appendix E of Parr and Yang, 1989. In the following we do not differentiate between the local and the local spin-density approximation and use the abbreviation LDA for both, unless otherwise noted.

How do we interpret the LDA for the exchange-correlation functional? Let us consider the general case of an open-shell atom or molecule. At a certain position $\vec{r}$ in this system we have the corresponding spin densities $\rho_\alpha(\vec{r})$ and $\rho_\beta(\vec{r})$. In the local spin-density approximation we now take these densities and insert them into equation (6-15) obtaining $E_{XC}(\vec{r})$. Thus, we associate with the densities $\rho_\alpha(\vec{r})$ and $\rho_\beta(\vec{r})$ the exchange and correlation energies and potentials that a homogeneous electron gas of equal, but constant density and the same spin polarization ξ would have. This is now repeated for each point in space and the individual contributions are summed up (integrated) as schematically indicated in Figure 6-2. Obviously, this approximation hinges on the assumption that the exchange-correlation potentials depend only on the *local* values of $\rho_\alpha(\vec{r})$ and $\rho_\beta(\vec{r})$.

This is a very drastic approximation since, after all, the density in our actual system is certainly anything but constant and does not even come close to the situation characteristic of the uniform electron gas. As a consequence, one might wonder whether results obtained with such a crude model will be of any value at all. Somewhat surprisingly then, experience tells us that the local (spin) density approximation is actually not that bad, but rather deliv-

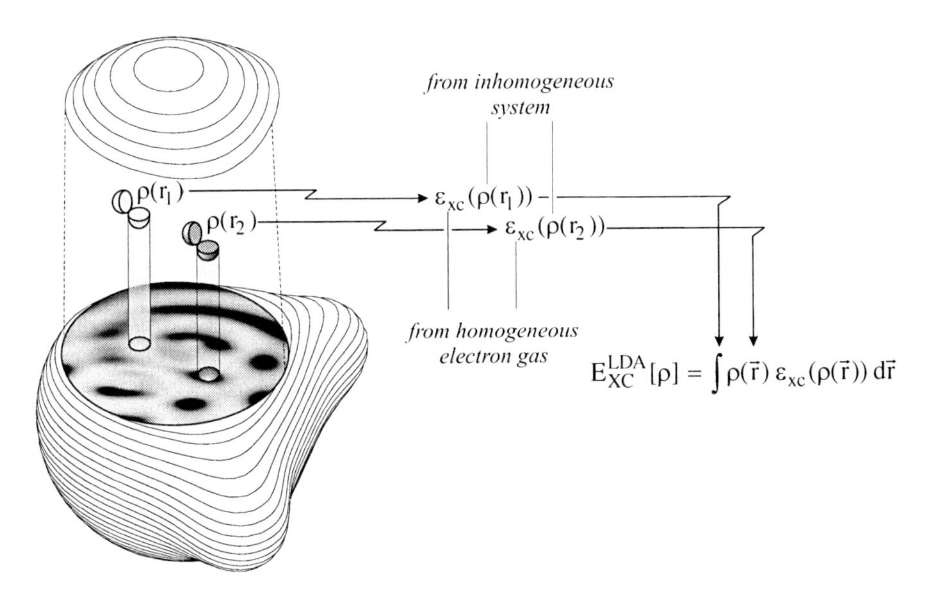

Figure 6-2. The local density approximation.

ers results that are comparable to or even better than the Hartree-Fock approximation. It has proven particularly successful for the determination of molecular properties such as equilibrium structures, harmonic frequencies or charge moments as we will discuss in more detail in later chapters. However, we should not get overexcited about such observations, because a look at energetical details, such as bond energies, immediately shows that for such properties the performance of the LDA is rather poor. If we take the average unsigned deviation from the experimental atomization energies from the G2 data set as an indicator, the LDA deviates by 36 kcal/mol! On the other hand, we need to put this into perspective because the deviation of the HF method is even substantially larger: its error is a hefty 78 kcal/mol, more than twice as large as the error of the local density approximation! While the HF approximation typically underestimates atomization energies, the LDA errs in the opposite direction, giving rise to the notorious overbinding tendency of this approximation, which we will discuss in Chapter 9.

In anticipation of the future discussion on how to improve on the local density approximation, we need to ask ourselves, what are the reasons that the LDA works better than expected from the underlying physical model of the homogeneous electron gas? The clue for an understanding seems to be that the exchange-correlation hole of the uniform electron gas, which is being used as model for the exact hole in the local density approximation, satisfies most of the important relations established for the true hole. Among those are the sum-rules, the behavior for vanishing inter-electronic distance of the exchange part ($h_X(\vec{r}_2 \rightarrow \vec{r}_1; \vec{r}_1) = -\rho(\vec{r}_1)$) and the correlation part (cusp condition), as well as the property of the exchange hole to be negative everywhere. Of course, the LDA model hole and the exact hole differ in many details. The overbinding tendency of the local density approximation can be rationalized in terms of the exchange hole properties. The LDA hole is spherically symmetric and always attached to the reference electron while the exact hole has a pronounced angular structure. In the bonding region between two atoms, the LDA model hole resembles the exact exchange hole, which becomes more isotropic (and symmetric with respect to the reference electron) than in the separated atoms. In an atom, the exact exchange hole is displaced toward the nucleus, whereas the LDA hole remains as it was in the molecular bond: centered on its reference electron. The neglect of this displacement in the LDA causes significant deviations from the exact differential exchange energy upon bond formation, with substantial errors on the atomic asymptote of an atomization process. In other words: the centered LDA exchange hole is a better approximation for the more homogeneous molecular density than for the more inhomogeneous density of atoms. This causes in particular the exchange energy of the molecular system to be too negative, that is, causes a dramatic overbinding (Ernzerhof, Perdew, and Burke, 1997).

Fortunately, only the spherically averaged exchange-correlation hole is of relevance for the exchange-correlation energy, as shown in the preceding section. The agreement between the spherically averaged LDA and exact holes is indeed much better – which is among the reasons why the LDA works at all – and the homogeneous electron gas in fact provides a reasonable first approximation to spherically averaged exchange-correlation holes of real systems. However, we should keep in mind that the LDA hole $h_{XC}^{LDA}(\vec{r}_1; \vec{r}_2)$ will have its highest degree of accuracy for small distances between the reference and the other electron because in the local density approximation we treat the exchange-correlation hole around $\vec{r}_1$ as if the neighborhood were part of a homogenous electron gas of constant density. Clearly, in a real system with considerably varying charge density, this assumption will deteriorate the larger the distance between the reference electron at $\vec{r}_1$ and the other one at $\vec{r}_2$ is.

6.5 The Generalized Gradient Approximation

The only moderate accuracy that the local (spin) density approximation delivers is certainly insufficient for most applications in chemistry. Hence, for the many years in which the LDA was the only approximation available for E_{XC}, density functional theory was mostly employed by solid-state physicists and hardly had any impact in computational chemistry. The situation changed significantly in the early eighties when the first successful extensions to the purely local approximation were developed. The logical first step in that direction was the suggestion of using not only the information about the density $\rho(\vec{r})$ at a particular point $\vec{r}$, but to supplement the density with information about the *gradient* of the charge density, $\nabla\rho(\vec{r})$ in order to account for the non-homogeneity of the true electron density. In other words, we interpret the local density approximation as the first term of a Taylor expansion of the uniform density and expect to obtain better approximations of the exchange-correlation functional by extending the series with the next lowest term. Thus, we arrive at (with σ and σ' indicating α or β spin)

$$E_{XC}^{GEA}[\rho_\alpha,\rho_\beta] = \int \rho\, \varepsilon_{XC}(\rho_\alpha,\rho_\beta)\, d\vec{r} + \sum_{\sigma,\sigma'} \int C_{XC}^{\sigma,\sigma'}(\rho_\alpha,\rho_\beta) \frac{\nabla\rho_\sigma}{\rho_\sigma^{2/3}} \frac{\nabla\rho_{\sigma'}}{\rho_{\sigma'}^{2/3}}\, d\vec{r} + \ldots \tag{6-17}$$

This form of functional is termed the *gradient expansion approximation* (GEA) and it can be shown that it applies to a model system where the density is not uniform but very slowly varying. Unfortunately, and at first glance counterintuitively, if utilized to solve real molecular problems the GEA does not lead to the desired improved accuracy but frequently performs even worse than the simple local density approximation. The reason for this failure is that the exchange-correlation hole associated with a functional such as in equation (6-17) has lost many of the properties which made the LDA hole physically meaningful. For example, the sum rules do not apply any more and the exchange hole is not restricted to be negative for any pair $\vec{r}_1; \vec{r}_2$. Thus, the dependence between the depth of the on-top hole and its extension is lost and the holes as well as the corresponding exchange-correlation energies will be much more erratic. This shows again that it is not so much the model system of the uniform electron gas but much more so the fact that the corresponding exchange-correlation hole system obeys most of the rules of the real system which is responsible for the success of the local density and local spin-density approximations.

In a very elegant (or shall we say brute force) way, this problem was solved by straightforwardly enforcing the restrictions valid for the true holes also for the hole of the beyond-LDA functionals. If there are parts in the GEA exchange holes which violate the requirement of being negative everywhere, just set them to zero. And, in order to have the correct sum rule behavior, well, let us simply truncate the exchange and correlation holes such that $h_X(\vec{r}_1;\vec{r}_2)$ and $h_C(\vec{r}_1;\vec{r}_2)$ contain one and zero electron charges, respectively. Functionals that include the gradients of the charge density and where the hole constraints have been restored in the above manner are collectively known as *generalized gradient approximations* (GGA). These functionals are the workhorses of current density functional theory and can be generically written as

$$E_{XC}^{GGA}[\rho_\alpha,\rho_\beta] = \int f(\rho_\alpha,\rho_\beta,\nabla\rho_\alpha,\nabla\rho_\beta)\, d\vec{r}\,. \tag{6-18}$$

As we will see presently, several suggestions for the explicit dependence of the integrand f on the densities and their gradients exist, including semiempirical functionals which contain parameters that are calibrated against reference values rather than being derived from first principles. In practice, E_{XC}^{GGA} is usually split into its exchange and correlation contributions

$$E_{XC}^{GGA} = E_X^{GGA} + E_C^{GGA} \tag{6-19}$$

and approximations for the two terms are sought individually.

Let us take a closer look at gradient-corrected exchange functionals in order to illustrate the general ideas. In particular, the reader should convince him- or herself that we are dealing with mathematically complex constructs which have been chosen such that the desired boundary conditions which the functionals and corresponding hole functions should satisfy are fulfilled and a satisfactory performance results. One should be aware that it is not the physics but the results obtained from them which dictate the choice of the mathematical constructs. In fact, some of these functionals are not even based on any physical model. In other words, the actual form of E_X^{GGA} and E_C^{GGA} usually does not assist the understanding of the physics these functionals try to describe. This underlines the pragmatic character so typical for approximate density functional theory in general.

We rewrite the exchange part of E_{XC}^{GGA} as

$$E_X^{GGA} = E_X^{LDA} - \sum_\sigma \int F(s_\sigma)\, \rho_\sigma^{4/3}(\vec{r})\, d\vec{r}\,. \tag{6-20}$$

The argument of the function F is the *reduced density gradient* for spin σ

$$s_\sigma(\vec{r}) = \frac{|\nabla\rho_\sigma(\vec{r})|}{\rho_\sigma^{4/3}(\vec{r})}\,. \tag{6-21}$$

s_σ is to be understood as a local inhomogeneity parameter. It assumes large values not only for large gradients, but also in regions of small densities, such as the exponential tails far from the nuclei. Likewise, small values of s_σ occur for small gradients, typical for bonding regions, but also for regions of large density. For example, the combination of large density gradients and large densities close to the nuclei typically leads to values of s_σ in this region which are in between the reduced density gradients in the bonding and tail regions, respectively. Of course, the homogeneous electron gas is characterized by $s_\sigma = 0$ everywhere. Finally, a word on why we divide by the 4/3 power of ρ and not just by ρ itself. This is needed to make s_σ a dimensionless quantity: the dimension of the density is the inverse dimension of volume and hence $[r]^{-3}$. Its gradient has therefore dimensions of $[r]^{-4}$. But this is just the same dimension that $\rho^{4/3}$ has, because of $([r]^{-3})^{4/3} = [r]^{-4}$ and we arrive at the desired dimensionless reduced gradient.

For the function F two main classes of realizations have been put forward (see in particular Adamo, di Matteo, and Barone, 1999). The first one is based on a GGA exchange functional developed by Becke, 1988b. As outlined above, this functional is abbreviated simply as B (sometimes one also finds B88)

$$F^{B} = \frac{\beta s_\sigma^2}{1 + 6\beta s_\sigma \sinh^{-1} s_\sigma}. \tag{6-22}$$

β is an empirical parameter that was determined to 0.0042 by a least-squares fit to the exactly known exchange energies of the rare gas atoms He through Rn. In addition to the sum rules, this functional was designed to recover the exchange energy density asymptotically far from a finite system.

Functionals which are related to this approach include among others the recent FT97 functional of Filatov and Thiel, 1997, the PW91 exchange functional (Perdew, 1991, and Burke, Perdew, and Wang, 1998) and the CAM(A) and CAM(B) functionals developed by Handy and coworkers (Laming, Termath, and Handy, 1993).

The second class of GGA exchange functionals use for F a rational function of the reduced density gradient. Prominent representatives are the early functionals by Becke, 1986 (B86) and Perdew, 1986 (P), the functional by Lacks and Gordon, 1993 (LG) or the recent implementation of Perdew, Burke, and Ernzerhof, 1996 (PBE). As an example, we explicitly write down F of Perdew's 1986 exchange functional, which, just as for the more recent PBE functional, is free of semiempirical parameters:

$$F^{P86} = \left(1 + 1.296\left(\frac{s_\sigma}{(24\pi^2)^{1/3}}\right)^2 + 14\left(\frac{s_\sigma}{(24\pi^2)^{1/3}}\right)^4 + 0.2\left(\frac{s_\sigma}{(24\pi^2)^{1/3}}\right)^6\right)^{1/15}. \tag{6-23}$$

The corresponding gradient-corrected correlation functionals have even more complicated analytical forms and cannot be understood by simple physically motivated reasonings. We therefore refrain from giving their explicit expressions and limit ourselves to a more qualitative discussion of the most popular functionals. Among the most widely used choices is the correlation counterpart of the 1986 Perdew exchange functional, usually termed P or P86. This functional employs an empirical parameter, which was fitted to the correlation energy of the neon atom. A few years later Perdew and Wang, 1991, refined their correlation functional, leading to the parameter free PW91. Another, nowadays even more popular correlation functional is due to Lee, Yang, and Parr, 1988 (LYP). Unlike all the other functionals mentioned so far, LYP is not based on the uniform electron gas but is derived from an expression for the correlation energy of the helium atom based on an accurate, correlated wave function presented in the context of wave function based theory by Colle and Salvetti, 1975. The LYP functional contains one empirical parameter. It differs from the other GGA functionals in that it contains some local components. We should note that all these correlation functionals are based on systems that only include *dynamical*, i. e., short range correlation effects (the uniform electron gas or the helium atom). Non-dynamical effects are not covered by these functionals, a property that we will come back to in the next section.

In principle, each exchange functional could be combined with any of the correlation functionals, but only a few combinations are currently in use. The exchange part is almost exclusively chosen to be Becke's functional which is either combined with Perdew's 1986 correlation functional or the Lee, Yang, Parr one – levels usually abbreviated as BP86 and BLYP, respectively. Sometimes also the PW91 correlation functional is employed, corresponding to BPW91. To be fair, all these flavors of gradient-corrected KS-density functional theory deliver results of similar quality as demonstrated by several studies which assess the performance of these functional. However, in this chapter we will predominantly concentrate on the more formal theoretical aspects of functionals and postpone a detailed view on the actual performance of modern functionals to our discussion in Part B.

We finally note a semantic detail. GGA functionals are frequently termed *non-local* functionals in the literature. This is a somewhat misleading and actually sloppy terminology that should be avoided. In our discussion of the Hartree-Fock scheme in Section 1.3 we introduced the difference between local and non-local operators and showed that the classical Coulomb potential is a local one while the HF exchange contribution represents a typical non-local potential. According to this discussion, all GGA functionals are perfectly local in the mathematical sense: the value of the functional at a point $\vec{r}$ depends only on information about the density $\rho(\vec{r})$, its gradient $\nabla\rho(\vec{r})$, and possibly other information at this very point and is absolutely independent of properties of $\rho(\vec{r}')$ at points $\vec{r}' \neq \vec{r}$. Calling these functionals 'non-local' is only motivated by the fact that these functionals go beyond the 'local' density approximation and of course the observation that knowledge of the gradients is the first step towards accounting for the inhomogeneity of the real density; nevertheless it is sloppy physicists' jargon.

6.6 Hybrid Functionals

We have repeatedly indicated that usually the exchange contributions are significantly larger in absolute numbers than the corresponding correlation effects. Therefore, an accurate expression for the exchange functional in particular is a prerequisite for obtaining meaningful results from density functional theory. However, we have seen in Chapters 1 and 5 that the exchange energy of a Slater determinant can be computed exactly, recall equations (5-17) or (5-18). Thus, why do we bother with complicated, but nevertheless only approximate exchange functionals at all? The straightforward and seemingly most appropriate strategy for arriving at a most accurate exchange-correlation energy seems to be to use the exact exchange energy of equation (5-18) and rely on approximate functionals only for the part missing in the HF picture, i. e., the electron correlation,

$$E_{XC} = E_X^{exact} + E_C^{KS}. \tag{6-24}$$

If applied to atoms this concept indeed delivers promising results. Unfortunately, and at first glance very surprisingly, it does not live up to the expectation at all if applied to molecules and chemical bonding. Against the G2 reference set we noted a mean absolute error of 78 kcal/mol for the Hartree-Fock level (i. e., exact exchange only, where we assume that the HF and KS orbitals are similar). While the inclusion of correlation through an appropriate functional in the spirit of equation (6-24) indeed cuts the error down to 32 kcal/mol, this is nevertheless a disappointing result if we consider that the errors associated with the currently used E_{XC} functionals of the GGA type, where both exchange and correlation are approximated, are in the order of only 5-7 kcal/mol.

What are the reasons for this significant failure of the exact exchange/density functional correlation combination in molecular calculations? For an analysis let us recall our discussion from Chapter 2 about the properties and shapes of the exact hole functions of the H_2 molecule, and in particular have another look at Figure 2-2. We saw that in this simple case the exchange hole corresponds to one half of the density of the σ_g occupied molecular orbital and corresponds to the removal of half an electron from the vicinity of each nucleus. It is completely delocalized and independent from the position of the reference electron. However, the exact full hole is relatively localized, in particular for extended internuclear distances where left-right correlation prevails. To salvage this overall characteristic of the total hole, the exact exchange hole has to be complemented by the correlation hole, which by itself is also delocalized. Thus, both components taken individually are bad representations of the whole, in particular the non-local exchange hole can in no way account for the effects that occur upon bond stretching or in similar situations. As an aside, we note that this is a manifestation of the fact that the separation of E_{XC} in individual exchange and correlation contributions is actually artificial and is only a consequence of the use of a particular reference system, i. e., a single Slater determinant. We re-emphasize that a clear physical meaning can only be attributed to the undivided exchange-correlation energy and hole. If we turn to the approximate holes given by the local density approximation as well as its gradient-corrected extensions we note that they are by construction based on a local model. Thus, these functionals implicitly assume that both, the exchange and the correlation hole are localized holes because all properties are determined by the density and its gradient at one particular point in space. Pictorially speaking, the approximate functionals E_{XC}^{appr} only 'see' their direct neighborhood and are completely 'unaware' of what is going on farther away. But this explains why the simple ansatz of equation (6-24) is bound to fail if applied to molecules. We combine the exact, delocalized exchange hole with a localized model hole for correlation. Because the cancellation between the two individual holes cannot take place (as discussed above and in Section 2.3), the resulting total hole has the wrong characteristics. On the other hand, approximate exchange-correlation holes based on the uniform electron gas are again local and are therefore a better model for the exact hole than equation (6-24). Actually, Becke, 1995, Gritsenko, Schipper, and Baerends, 1997, and Schipper, Gritsenko, and Baerends, 1998b and 1999, as well as others pointed out and verified numerically that current density functionals for exchange with their localized holes effectively reproduce the sum of exact exchange *and* non-dynamical correlation while the corresponding correlation functionals represent only the effects of dynamical electron correlation. By the way, the approximate correlation hole provided by conventional, wave function based techniques indeed has the required long-range characteristics for dealing with non-dynamical correlation, usually accomplished through including energetically low-lying Slater determinants into the wave function. In fact, there are attempts to already include these long-range correlation effects into the 'exact' E_X contribution of equation (6-24) through a multi-configurational SCF (MCSCF) ansatz. Combining this 'exchange plus non-dynamical correlation' portion with a local density functional for dynamical correlation should be more appropriate (see, e. g., the recent reports by Leininger et al., 1997, Borowski et al., 1998, or Gräfenstein and Cremer, 2000). The two major problems with this approach are its significantly increased computational costs and that double-counting of correlation effects cannot be completely excluded.

Rather than pursuing this approach further, we follow a different avenue to exploit exact exchange outlined by Becke, 1993a and 1993b. The theoretical justification of this approach can be extracted from the adiabatic connection sketched in Section 6.2 above. We recall from equation (6-10) that the exchange-correlation energy of the Kohn-Sham scheme is obtained through the coupling-strength integrated exchange-correlation hole. This equation is of course equivalent to the following expression (6-25), where we integrate over the λ-dependent exchange-correlation potential energy, which is nothing else than the non-classical contribution to the electron-electron interaction for different values of λ (note that E_{ncl} corresponds to the pure potential energy contributions, dependent on λ. Only the integration over λ introduces the kinetic energy part into E_{XC})

$$E_{XC} = \int_0^1 E_{ncl}^{\lambda} d\lambda . \tag{6-25}$$

Let us explore first the nature of the integrand E_{ncl}^{λ} for the limiting cases. At $\lambda = 0$ we are dealing with an interaction free system, and the only component which is not included in the classical term is due to the antisymmetry of the fermion wave function. Thus, $E_{ncl}^{\lambda=0}$ is composed of exchange only, there is no correlation whatsoever.[19] Hence, the $\lambda = 0$ limit of the integral in equation (6-25) simply corresponds to the exchange contribution of a Slater determinant, as for example, expressed through equation (5-18). Remember, that $E_{ncl}^{\lambda=0}$ can

[19] Keep in mind that dynamic electron correlation is always connected to the fact that electrons interact as charged species.

be computed exactly, once the KS orbitals are available. On the other hand, for $\lambda = 1$, the non-classical contributions are those of the fully interacting system, containing exchange as well as electron correlation parts. This interacting exchange-correlation energy is not known, but can be approximated – more or less satisfactorily – by any E_{XC} functional. The true exchange-correlation energy is given by the integral of equation (6-25) and we know its value for $\lambda = 0$ exactly and have pretty good approximations for $\lambda = 1$. To exactly evaluate this integral, however, we would need E_{ncl}^{λ} for intermediate values of λ. But this information is not available and we must try to find approximations to this integral. Alternatively, we can analyze this integral from a slightly different point of view. We have seen above that the model holes of the LDA or GGA schemes are reference point centered and relatively localized. Hence, they provide a crude simulation of left-right correlation. As we have discussed at length, this is a desirable feature for describing the hole of the interacting system, which is also localized. On the other hand, at $\lambda = 0$ all there is, is exchange with its delocalized hole and our localized model holes are completely inadequate. Therefore, in terms of holes, to describe the $\lambda = 0$ end of the integration it appears plausible to mix in a certain amount of the pure, exact exchange hole into the overall hole.

Let us be specific. The simplest approximation to solve equation (6-25) is to assume that E_{ncl}^{λ} is a linear function in λ. This leads to

$$E_{XC}^{HH} = \frac{1}{2}E_{XC}^{\lambda=0} + \frac{1}{2}E_{XC}^{\lambda=1}, \tag{6-26}$$

and corresponds to the situation shown schematically in Figure 6-3a. Using the LDA exchange-correlation functional for $E_{ncl}^{\lambda=1}$, equation (6-26) represents the so-called *half-and-half* (HH) combination of 'exact' exchange and density functional exchange-correlation as introduced by Becke, 1993a. In fact this approach showed a promising performance. The absolute average error with respect to the G2 atomization energies amounts to 6.5 kcal/mol, and rivals the value of 5.7 kcal/mol for the gradient-corrected BPW91, if basis-set free, fully numerical results are utilized. The next step taken by Becke, 1993b was to introduce semiempirical coefficients to determine the weights of the various components in this scheme leading to the following extension of equation (6-26):

$$E_{XC}^{B3} = E_{XC}^{LSD} + a(E_{XC}^{\lambda=0} - E_{X}^{LSD}) + bE_{X}^{B} + cE_{C}^{PW91}. \tag{6-27}$$

In this equation we have three parameters. The amount of exact exchange in the functional is determined through a, while b and c control the contributions of exchange and correlation gradient corrections to the local density approximation. As indicated in equation (6-27), Becke utilized his 1988 exchange functional and Perdew and Wang's 1991 correlation functional in his original approach. The three empirical parameters were chosen such that the atomization and ionization energies as well as the proton affinities included in the G2 data base and some total energies were optimally reproduced. This led to a = 0.20, b = 0.72, and c = 0.81. Hence, the amount of exact exchange was reduced relative to the earlier half-and-half scheme, indicative of a large slope of E_{XC}^{λ} at $\lambda = 0$, see Figure

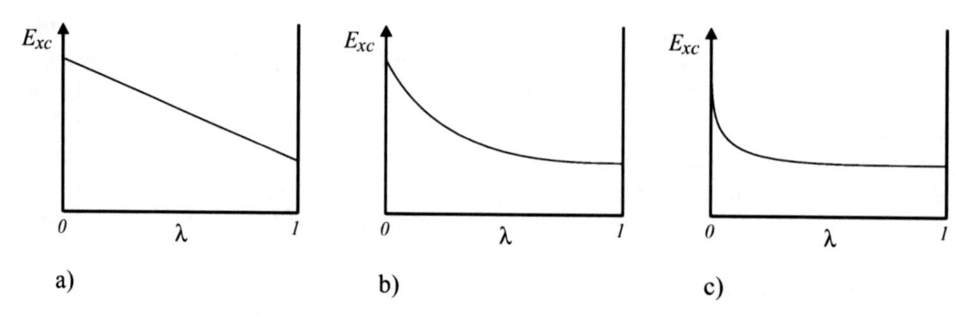

Figure 6-3. λ-dependence of E_{XC}.

6-3b. Most importantly, this three-parameter fit reduced the average absolute error in the G2 atomization energies significantly to only about 2–3 kcal/mol, already very close to the target accuracy of 2 kcal/mol. Of course one should keep in mind that the parameters a, b, and c were fitted to exactly these data and it is a priori completely unclear whether a similarly good performance can also be expected in general. Functionals of this sort, where a certain amount of exact exchange is incorporated are frequently called DFT/HF *hybrid* functionals, because they represent a hybrid between pure density functionals for exchange and exact Hartree-Fock exchange. They are also sometimes referred to as ACM functionals, where the acronym stands for *adiabatic connection method.*

Currently, the most popular hybrid functional is known as B3LYP and was suggested by Stephens et al., 1994. While it is of very similar spirit to the original form proposed, as given in equation (6-27), in B3LYP, the PW91 correlation functional is replaced by the LYP functional. The values of the three parameters were directly taken from Becke's original paper. Thus, the B3LYP exchange-correlation energy expression is (with a, b, and c just as above)

$$E_{XC}^{B3LYP} = (1-a)\,E_{X}^{LSD} + a\,E_{XC}^{\lambda=0} + b\,E_{X}^{B88} + c\,E_{C}^{LYP} + (1-c)\,E_{C}^{LSD}. \quad (6\text{-}28)$$

For the B3LYP functional an unsigned error with respect to the G2 data base of only slightly above 2 kcal/mol was determined.

Since their incarnation in the early nineties these hybrid functionals experienced an unprecedented success (Raghavachari, 2000). In particular the B3LYP functional was an absolute shooting star and soon developed into by far the most popular and most widely used functional. This amazing success was fueled by the surprisingly good performance B3LYP and related functionals demonstrated in many chemical applications, including such difficult areas as open-shell transition-metal chemistry as we will discuss in much detail in later parts of this book.

More recent developments by Becke, 1996a reduced the number of parameters to one

$$E_{XC}^{B1} = E_{XC}^{DFT} + a(E_{X}^{\lambda=0} - E_{XC}^{DFT}) \quad (6\text{-}29)$$

where the amount of exact exchange was empirically determined as a = 0.28 if for E_{XC}^{DFT} a combination of the standard Becke exchange (B) and a new correlation functional due to Becke (B95) was inserted. This functional is commonly referred to as B1B95. Its average absolute error with respect to the G2 atomization energies is only around 2 kcal/mol. However, there are certain technical disadvantages with this functional. Most notably, it depends explicitly on the kinetic energy density in addition to the density and its gradient, which complicates the implementation into standard molecular structure computer programs. This might be the reason, why the new B1B95 has attracted less interest than its three-parameter predecessor, even though inclusion of the kinetic energy density into exchange-correlation functionals seems to have gained a lot of attention lately, as outlined below. In the last part of his series of papers on density-functional thermochemistry, Becke, 1997, introduced a new type of exchange-correlation functional which was based on an elaborate fitting procedure. The exchange-correlation functional was separated into several parts, i. e., exchange, like-spin correlation and unlike-spin correlation and an additional amount of exact, Hartree-Fock exchange, i. e.,

$$E_{XC}^{B97} = E_X^{\alpha\alpha} + E_X^{\beta\beta} + E_C^{\alpha\alpha} + E_C^{\beta\beta} + E_C^{\alpha\beta} + c_X^{HF} E_X^{HF} \tag{6-30}$$

Each component, with the exception of the HF exchange, is expressed in a power series involving the density and the reduced density gradient. These expansions were terminated at second order, since otherwise unphysical, overfitted functionals were obtained. The resulting ten linear coefficients were optimized by a least-square fit to energetical data from the G2 set. Note, that the optimal parameters were determined in a fully numerical (i. e., basis set free), non-self-consistent procedure using LDA densities. If measured against the G2 training set, average absolute and maximum errors of atomization energies of only 1.8 and 5.5 kcal/mol, respectively, were obtained. The amount of exact exchange was determined at 20 % in this B97 functional. As Becke noted, this kind of accuracy is probably as far as one can get with conventional gradient-corrected GGA functionals for exchange and correlation and a certain (but fixed, see below) amount of exact, i. e. $\lambda = 0$, exchange. A year later, Schmider and Becke 1998a, reparameterized the B97 functional with respect to the extended G2 set. The resulting B98 functional retains the good absolute average (1.9 kcal/mol) and maximum errors (9.1 kcal/mol) also for this larger and more demanding training set.[20] Hamprecht et al., 1998, reparameterized the original B97 functional in a self-consistent procedure, i. e., with densities optimized within the same functional and using a TZ2P basis set and termed the resulting B97 flavor B97-1. In the same paper these authors also suggest an extension to the B97 idea. They additionally require in the parameterization scheme that the functional also reproduces nuclear gradients for molecules (i. e., zero if calculations are performed at equilibrium geometries) and, probably even more importantly, that it yields accurate exchange-correlation potentials, a property that will become

[20] Schmider and Becke, 1998a, presented various parameterizations in their paper, which differed in the choice of data included in the fitting. The quoted performance applies to their parameter set 2c, where 148 heats of formation, 42 ionization energies, 25 electron affinities, 8 proton affinities and 10 total energies were included in the training set.

of interest in Section 6-8. Because the least-square procedure now includes much more information, the power series was extended to forth order, increasing the number of parameters to 15. However, unlike in the B97 or B98 schemes, the resulting HCTH functional is a pure GGA functional and contains no exact exchange. In a subsequent paper (Boese et al., 2000), the training set for the parametrization was extended in particular by anions and weakly bound dimers leading to the HCTH/120 and HCTH/147 functionals (the numbers indicate the number of systems used in the training sets. In the original HCTH parametrization a total of 93 systems was used).

Before closing this section let us mention that a fraction of about 20–25 % exact exchange as realized in the above functionals seems to be reasonable also on purely theoretical grounds as shown by Perdew, Ernzerhof, and Burke, 1996, and Burke, Ernzerhof, and Perdew, 1997. These authors proposed parameter-free hybrid functionals of the general form, where the amount of exact exchange has been derived as 25 % from theoretical reasonings through a perturbation theory argument,

$$E_{XC}^{hybrid} = E_{XC}^{GGA} + 0.25(E_X^{HF} - E_X^{GGA}). \tag{6-31}$$

If the PBE exchange-correlation functional is chosen as the GGA component, the PBE1PBE model emerges (some authors prefer to call this functional PBE0). As shown by Adamo and Barone, 1999, PBE1PBE shows promising performance for all important properties, being competitive with the most reliable, empirically parameterized current functionals. However, while about a quarter of exact exchange is reasonable for most regular systems, it should be clear that in general this parameter is certainly not universal but depends on the actual situation. This can be impressively demonstrated using our standard guinea-pig, the H_2 molecule and its molecular ion, H_2^+. Let us consider first the neutral hydrogen molecule. As we have seen in Section 2.3, as the distance of the two nuclei increases, the total exchange-correlation hole gets more and more localized. For infinite distance it is strictly localized and removes exactly one electron from the proton where the reference electron is located while it is zero at the other nucleus. In this situation, the amount of exact exchange in a restricted calculation must go to zero as $r \to \infty$: the correlation holes of approximate functionals are localized. Mixing in any fraction of the delocalized exact exchange hole would therefore lead to an unphysically delocalized total hole (see also, Gritsenko, van Leeuwen, and Baerends, 1996). The situation is completely different for the corresponding radical cation, the one-electron system H_2^+. Here, the exchange-correlation hole obviously contains only the exchange part, which is completely delocalized over the molecule, independent of the internuclear distance. Due to their inherent local character none of the current approximate exchange functionals is capable of correctly representing this situation. In order to describe this delocalized hole a hybrid with 100 % exact exchange would be needed, as discussed by Sodupe et al., 1999. Generally speaking, in situations, where the $\lambda = 0$ limit is represented by degenerate or near-degenerate ground states, the local exchange density functional is a good approximation throughout the whole integration, including $\lambda = 0$. In other words, the λ-dependence of E_{XC}^{λ} is characterized under these circumstances by an extreme slope (approaching $-\infty$) at $\lambda = 0$ and the solution

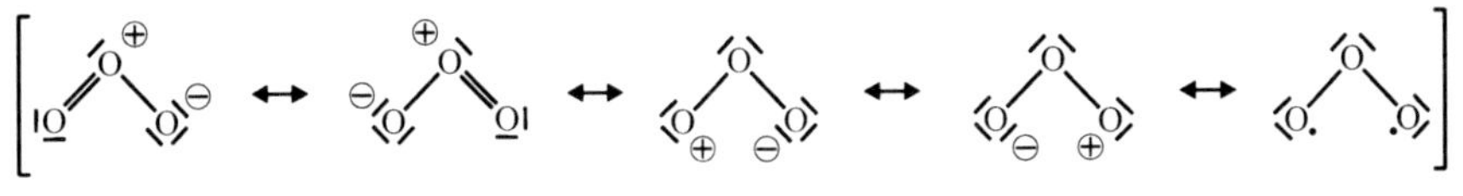

Figure 6-4. Resonance structures of O_3.

Table 6-1. Harmonic frequencies and experimental fundamentals for ozone [cm^{-1}]. Deviations from the experimental result [%] are given in parentheses.

Method	bending (a_1)	antisym. stretch (b_2)	sym. stretch (a_1)
Hartree-Fock	870 (+21.5)	1419 (+30.3)	1541 (+35.8)
MP2	747 (+4.3)	2211 (+203.0)	1170 (+3.1)
B3LYP	750 (+4.7)	1205 (+10.7)	1259 (+10.9)
*m*PW1PW	778 (+8.7)	1296 (+19.0)	1323 (+16.6)
BLYP	688 (−3.9)	991 (−9.0)	1135 (±0)
BP	708 (−1.1)	1054 (−3.2)	1179 (+3.9)
Experiment	716	1089	1135

for $\lambda = 0$ does not contribute to the integral as indicated in Figure 6-3c. A typical example is provided by the ozone molecule which is known to be pathological because of near-degeneracy effects. The ionic and biradical resonance structures indicated in Figure 6-4 both contribute significantly to the overall wave function.

The most sensitive properties in that respect are the vibrational frequencies, in particular the antisymmetric O-O stretching vibration (of b_2 symmetry). Along this vibrational mode the relative weights of the two main contributors of Figure 6-4 to the wave function change. Conventional methods, such as HF or the MP2 approach in particular, where dynamical electron correlation is estimated through second order perturbation theory, fail completely. But also hybrid functionals such as B3LYP or the very recent, one-parameter *m*PW1PW scheme predict harmonic frequencies of O_3 which are in much less harmony with the experimental data than the results obtained from plain GGA protocols, such as BLYP or BP86. Table 6-1 summarizes theoretically predicted harmonic frequencies for ozone from representative computational models employing a flexible cc-pVQZ basis set to expand the KS orbitals.

6.7 Self-Interaction

There is one more problem which is typical for approximate exchange-correlation functionals. Consider the simple case of a one electron system, such as the hydrogen atom. Clearly, the energy will only depend on the kinetic energy and the external potential due to the nucleus. With only one single electron there is absolutely no electron-electron interaction in such a system. This sounds so trivial that the reader might ask what the point is. But consider the energy expression for a one electron system in the Kohn-Sham scheme (which is no different from the general equation (5-14)),

$$E[\rho(\vec{r})] = T_S[\rho] + J[\rho] + E_{XC}[\rho] + E_{Ne}[\rho]. \tag{6-32}$$

The classical electrostatic repulsion term is

$$J[\rho] = \frac{1}{2}\iint \frac{\rho(\vec{r}_1)\,\rho(\vec{r}_2)}{r_{12}}\, d\vec{r}_1 d\vec{r}_2 . \tag{6-33}$$

This term does not exactly vanish for a one electron system since it contains the spurious interaction of the density with itself. Hence, for equation (6-32) to be correct, we must demand that $J[\rho]$ exactly equals minus $E_{XC}[\rho]$ such that the wrong *self-interaction* is cancelled

$$\frac{1}{2}\iint \frac{\rho(\vec{r}_1)\rho(\vec{r}_2)}{r_{12}}\, d\vec{r}_1\, d\vec{r}_2 = -E_{XC}[\rho]. \tag{6-34}$$

And that is where the trouble begins.

We saw in Section 1.3 that by construction the exchange term of the Hartree-Fock model indeed exactly neutralizes the unwanted portion of $J[\rho]$. In particular, for a one electron system equation (6-34) is satisfied and the HF scheme is therefore free of self-interaction errors. On the other hand, in any realization of the Kohn-Sham density functional scheme we have to employ approximations to the exchange-correlation energy which are independent of $J[\rho]$ and we should not expect equation (6-34) to hold. In fact, none of the currently used exchange-correlation functionals is self-interaction free. In Table 6-2 we have summarized the results for the hydrogen atom as obtained with typical exchange-correlation functionals employing a large cc-pV5Z basis set.

We see that the self-interaction error, $J[\rho] + E_{XC}[\rho]$, is in all cases in the order of 10^{-3} E_h or a few hundredths of an eV. In addition, the data in Table 6-2 reiterate some of the facts that we noted before. B3LYP, BP86 and BPW91 yield total energies below the exact result of –0.5 E_h, in an apparent contradiction to the variational principle (see discussion in Sec-

Table 6-2. Energy components [E_h] of various functionals for the hydrogen atom.

Functional	E_{tot}	$J[\rho]$	$E_X[\rho]$	$E_C[\rho]$	$J[\rho] + E_{XC}[\rho]$
SVWN	–0.49639	0.29975	–0.25753	–0.03945	0.00277
BLYP	–0.49789	0.30747	–0.30607	0.0	0.00140
B3LYP	–0.50243	0.30845	–0.30370[a]	–0.00756	–0.00281
BP86	–0.50030	0.30653	–0.30479	–0.00248	–0.00074
BPW91	–0.50422	0.30890	–0.30719	–0.00631	–0.00460
HF	–0.49999	0.31250	–0.31250	0.0	0.0

[a] Includes 0.06169 E_h from exact exchange.

tion 4.5). It is also noteworthy that of the three correlation functionals, only LYP yields the correct result of zero correlation energy for a single electron (i. e., it is self-interaction free), all others deviate non-negligibly from zero.

Of course, this self-correction error is not limited to one electron systems, where it can be identified most easily, but applies to all systems. Perdew and Zunger, 1981, suggested a *self-interaction corrected* (SIC) form of approximate functionals in which they explicitly enforced equation (6-34) by substracting out the unphysical self-interaction terms. Without going into any detail, we just note that the resulting one-electron equations for the SIC orbitals are problematic. Unlike the regular Kohn-Sham scheme, the SIC-KS equations do not share the same potential for all orbitals. Rather, the potential is orbital dependent which introduces a lot of practical complications. As a consequence, there are hardly any implementations of the Perdew-Zunger scheme for self-interaction correction.

Surprisingly, while application of the Perdew-Zunger self-interaction correction improves the results for atoms as expected, this does not necessarily carry over to ground state energies and geometries of molecules, where the self-interaction corrected scheme may even lead to a deterioration of the results as compared to regular approximate Kohn-Sham calculations, as reported by Goedecker and Umrigar, 1997. The reasons for this behavior are, however, not fully understood. Similarly, to what extent the unphysical self-interaction affects the results of density functional calculations in general is not completely clear yet, but it certainly can sometimes have severe consequences. Among the most spectacular examples is the difficulty that approximate exchange-correlation functionals experience when the dissociation of radicals consisting of two identical moieties are studied as pointed out by several authors; e. g., by Merkle, Savin, and Preuss, 1992, and more recently by Bally and Sastry, 1997, and Zhang and Yang, 1998. Sodupe et al., 1999, for example show that Kohn-Sham calculations predict the wrong order of stability for the two low-lying structural isomers of the $(H_2O)_2^+$ dimer, overestimating the stability of the symmetric $H_2O \cdots OH_2$ by some 17 kcal/mol. Even for systems as simple as the one-electron hydrogen molecular ion, H_2^+, the dissociation curve is significantly in error leading to much too small binding energies. The origin of the huge self-interaction error as the H–H bond stretches is the incapability of the intrinsically localized model holes of approximate functionals to describe the delocalized exchange hole of $(H \cdots H)^+$ as $r_{H\text{-}H} \rightarrow \infty$, see also the clear discussion in Perdew and Ernzerhof, 1998. According to Zhang and Yang, 1998, such problems are always to be expected for situations where non-integer number of electrons are involved. The H_2^+ dissociation offers a prototype for this scenario since it leads to a delocalized state according to $H^{+0.5} \cdots H^{+0.5}$ as R_{HH} increases. In general, such cases are to be expected if the ionization energy of one dissociation partner differs by only a small amount from the electron affinity of the other partner. For $H_2^+ \rightarrow H + H^+$ this criterion is perfectly satisfied since the ionization energy of H and the electron affinity of H^+ are identical. These authors go on to speculate, that similar problems should also surface if transition states of chemical reactions with stretched bonds or certain charge-transfer complexes are studied with approximate exchange-correlation functionals. That self-interaction indeed plays a decisive role in the vicinity of transition structures and may heavily affect reaction barriers has been demonstrated, e. g., by Csonka and Johnson, 1998: the barrier for the seemingly simple hydrogen abstraction reaction, $H_2 + H \rightarrow H + H_2$, is raised by approximately 8–9 kcal/mol by inclusion of the self-interaction correction and brings the computed results into much better agreement with the experimental activation barrier, as we will discuss in more detail in Chapter 13. Their paper also includes an instructive and yet concise outline of the Perdew-Zunger procedure.

6.8 Asymptotic Behavior of Exchange-Correlation Potentials

While the behavior of the exchange-correlation potential V_{XC} (recall from equation (5-16) that the exchange-correlation potential V_{XC} is defined as the functional derivative of the exchange-correlation energy E_{XC} with respect to the charge density ρ: $V_{XC} \equiv \frac{\delta E_{XC}}{\delta \rho}$) at distances far from the atom or molecule seems of little importance at first glance, it turned out to be critical for properties which depend not only on the quality of the occupied Kohn-Sham orbitals but also on how well the virtual orbitals are described. Typical examples include atomic electron affinities and properties related to the response of the system to an electromagnetic field, such as polarizabilities, or excitation energies to energetically high-lying states, in particular Rydberg states, computed from the poles of the frequency dependent polarizability (i. e., in the TDDFT scheme). How should the asymptotic V_{XC} look like? We know that for an N-electron system the Coulomb potential $V_C(\vec{r}_1) = \int \frac{\rho(\vec{r}_2)}{r_{12}} d\vec{r}_2$ behaves like N/r when $r \rightarrow \infty$. In order to cancel the unphysical self-interaction in the Coulomb term, the exchange-correlation potential (since correlation effects are much more short-ranged, it usually suffices to analyze only the exchange potential in the asymptotic region) must therefore have a –1/r dependence at large r. The electron far away from the molecule now sees the correct net (N+1-Z) charge (Z being the positive nuclear charge). There are two problems with current popular exchange functionals in this context. First, none of the corresponding potentials has the correct –1/r behavior, they rather all decrease exponentially, i. e., much too fast. As a consequence, these approximate potentials are less attractive than the exact one at large r. The second problem is more subtle and much more difficult to grasp without a detailed theoretical analysis. In the following we will limit ourselves to a pictorial description of the problem without dwelling too much into the physical background. In a celebrated (see Zhang and Yang, 2000) and often quoted paper, Perdew et al., 1982, extended Kohn-Sham density functional theory to fractional electron numbers. In that context they showed that the exchange-correlation potential actually jumps by a constant as the number of electrons passes through an integer. This phenomenon is known as the *derivative discontinuity* in DFT. As a corollary to this it can be shown that none of the currently available approximate functionals, which are all characterized by a continuous potential with respect to variations in the number of electrons is able to model this behavior. This in turn has the rather unexpected consequence that an accurate continuous potential should not vanish asymptotically. Rather, as shown for example by Tozer and Handy, 1998, the asymptotic potential should obey

$$\lim_{r\to\infty} V_{XC}(\vec{r}) = -\frac{1}{r} + I + \varepsilon_{max}. \qquad (6\text{-}35)$$

Here, I and ε_{max} are the lowest ionization energy and the orbital energy of the highest occupied Kohn-Sham orbital, respectively. By the way, these deficiencies of current functionals are the reason for the fact noted in Section 5.3.3 that the exact relationship between these quantities, i. e., that I equals $-\varepsilon_{max}$ is not fulfilled by any approximate exchange-correlation functional. In fact, all approximate functionals give values of $-\varepsilon_{max}$ which are significantly larger (i. e. less negative) than the ionization energy. This reflects the too small attractive character of the corresponding asymptotic potentials.

There have been several attempts to improve the asymptotic behavior of exchange-correlation potentials. For example, van Leeuwen and Baerends, 1994, constructed a potential such that it shows the correct $-1/r$ behavior, however, it still vanishes at infinity and therefore does not take into account the problems connected with the derivative discontinuity. In addition, this LB94 potential cannot be derived as the derivative of an exchange-correlation functional and shows some deficiencies in regions closer to the nucleus. As noted above, Hamprecht et al., 1998, fitted their GGA exchange-correlation HCTH functional to energetics as well as exchange-correlation potentials in order to achieve better characteristics in the asymptotic region. However, their potential also eventually vanishes. For other approaches see also Chermette et al., 1998. While both, LB94 as well as HCTH yielded improved results for low-lying virtual orbitals and related properties as compared to regular exchange-correlation potentials, further improvements were accomplished by an ingeniously simple modification of the HCTH functional. Tozer and Handy, 1998, replaced the potential computed as the functional derivative of the HCTH functional by the asymptotically correct potential from equation (6-35) if the grid point where the potential is being evaluated is 'far' from the molecule.[21] This procedure was termed HCTH(AC) where AC stands for *asymptotically corrected*. The ionization energy and the highest occupied orbital energy needed for this correction were simply taken from regular Kohn-Sham calculations. HCTH(AC) indeed yielded significantly better excitation energies to high-lying Rydberg states and hyperpolarizabilities as we outline in Part B of this book. Before closing this section, we note that hybrid functionals with their exact Hartree-Fock exchange contribution also lead to an amelioration. First, the HF exchange functional obviously shows the correct $-1/r$ decay for large distances. The asymptotic form of the exchange potential in hybrid functionals therefore assumes the improved (but still not correct) form $-a/r$ with $a < 1$ being the amount of exact exchange included (see Casida, 1995). Second, the introduction of some Hartree-Fock exchange leads to a discontinuity in the potential as it goes through an integer particle number. As a consequence, hybrid functionals will – albeit only partially – correct the problems of potentials based on pure density functionals in the asymptotic regime. An extension of the idea of 'asymptotic correction' to hybrid functionals has been presented by Allen and Tozer, 2000. While the correction significantly improves excitation energies to Rydberg States, no overall improvement over the results obtained from HCTH(AC) was achieved.

[21] Tozer and Handy empirically define 'far' in terms of 4.7 times the Bragg-Slater radius of the corresponding atom. The two potentials for 'near' and 'far' are connected through a linear extrapolation.

6.9 Discussion

In the preceding sections we have reviewed the current state of the art in approximate exchange-correlation functionals. The above conclusions support a certain, albeit qualitative hierarchy of functionals of ascending complexity and accuracy as: the local density approximation (LDA) which usually yields good structural properties but frequently fails miserably in binding energies due to overbinding; regular gradient-corrected exchange-correlation functionals (BP86, BLYP, BPW91, PBE and the like) which already provide fairly accurate results, as indicated by absolute average errors of some 5 kcal/mol for atomization energies with respect to the G2 data base; and finally hybrid functionals, which show in many (although not all!) applications the most satisfactory performance. The most prominent example of this class is B3LYP. In Part B of this book we report on the details of how we can break down this hierarchy to individual functionals from each family and how they perform for the prediction of different properties. In this concluding section we will present a glimpse on recent progress how to devise new functionals. While some performance data are also mentioned here, a more detailed portrait on the quality of these and other new developments awaits the reader in Section 9.1.

We already mentioned Becke's recent one-parameter hybrid functional B1B95, which performs better than B3LYP with errors against the G2 set of only slightly above 2 kcal/mol. Of course, the one-parameter scheme is not limited to this particular choice of functionals and any exchange and correlation functional can in principle be used in this protocol. One particularly interesting flavor of such one-parameter hybrids is the modified Perdew-Wang approach (*m*PW1PW) suggested by Adamo and Barone, 1998b. It was designed specifically with non-covalent interactions in mind, but shows a very promising performance across the board.

However, similar accuracy seems to be in reach also for the latest developments in regular GGA functionals without exact exchange, such as demonstrated by Filatov and Thiel's 1997 functional. Still, as we will elaborate on in the application oriented part of this book, none of these functionals is without shortcomings and the hunt for better and more universal functionals is anything but at an end. Novel forms of functionals discussed presently try to explore new forms for exchange-correlation functionals, and we will give a few representative examples. Extending the search for approximate exchange-correlation functionals to schemes that go beyond the GGA by taking second order gradients and the (non-interacting) kinetic energy density into account leads to a new family of functionals, which has been termed *meta-generalized gradient approximation* (meta-GGA) by Perdew et al., 1999. An early example of this kind of functional is provided by the LAP correlation functional due to Proynov, Vela, and Salahub, 1994. This functional involves the Laplacian of the electron density, $\nabla^2\rho(\vec{r})$, for each spin direction and the kinetic energy density as ingredients reflecting inhomogeneity. Also the B95 correlation functional belongs into this category. Filatov and Thiel, 1998, suggest a new functional for E_{XC} which likewise expands the arguments of the usual GGA formulation by including contributions from the Laplacian of the density. Similarly, Schmider and Becke, 1998b, extended their B97 functional by taking into account the Laplacian of the density and the non-interacting kinetic energy density (see also Becke, 1999, for a summary on B97 and extensions to it). Also van Voorhis and Scuseria, 1998, presented a new exchange-correlation functional termed VSXC which depends not only on ρ and $\nabla\rho$, but also on the non-interacting kinetic energy density. Somewhat later Ernzerhof and Scuseria, 1999, Perdew et al., 1999, and Proynov, Chermette and Salahub, 2000, developed alternative formulations of the same motif. In all cases very encouraging results were reported (see, for example, Adamo, Ernzerhof and Scuseria, 2000). These approaches are physically motivated, and without diluting the functionals by any exact exchange, the new functionals reach an accuracy on the G2 test bench which is comparable or even slightly better than that of the actual de facto standard B3LYP. On the other hand, addressing the problem from a very pragmatic point of view, Adamson, Gill, and Pople, 1998, make heavy use of parameterization and thus generate what they call *empirical density functionals*, which also deliver a good overall accuracy. Interestingly, these authors argued that it might not be a proper way to create functionals using large and flexible basis sets (or an infinite basis as used by Becke) during the development phase. Rather than assuming that functionals resulting from such a process will be equally suitable for smaller basis sets, they used a relatively small basis to start with and put more emphasis on the empirical parameterization. Following the lines of the arguments above, it is an appealing idea to assume that the parameterization performed within a small basis expansion set can absorb some deficiencies of the basis limitations itself. Their 'empirical density functional 1' (EDF1 for short) is composed of an adjusted mixture of functional forms for exchange and correlation with X_α, B, LYP as components.

Interestingly, all these latter new functionals achieve their promising performance without exact exchange. Hence, it may well be that future exchange-correlation functionals will get away without any exact exchange mixing and, as van Voorhis and Scuseria, 1998, conclude, that the apparent need for exact exchange mixing is only a relic of the relatively poor quality of the currently used exchange functionals. Nevertheless in concluding this section we have to note that none of the new functionals mentioned above has already gained a significant popularity. Functionals such as BP86, BLYP or B3LYP are still the mainstay in most chemical applications and no serious competitor that could eventually challenge the dominance of these schemes in the short run is actually in sight.

7 The Basic Machinery of Density Functional Programs

The preceding six chapters provided an overview of the theoretical background and current state of the art of modern approximate density functional theory. We now turn to the more practical problem of how the strategies developed so far can be mapped onto computational schemes. To this end, we first introduce the linear-combination-of-atomic-orbitals (LCAO) ansatz, which is the by far most dominant way to make the iterative self-consistent field procedure for solving the one-electron Kohn-Sham equations computationally accessible. This leads immediately to the problem of which kinds of basis sets are suitable in order to expand the Kohn-Sham orbitals in such calculations and according to which criteria one should choose a particular set of basis functions. One of the main questions in this context is, to what extent one can benefit from the vast experience regarding basis sets accumulated in wave function based techniques. Schemes for how the various components appearing in the KS equations are actually determined are discussed with particular emphasis on how the Coulomb energy can be approached. We also give a survey of the techniques employed for the numerical integration of the exchange-correlation potential including grid-free approaches, which circumvent the ubiquitous problems with numerical noise in the grid-based numerical integration. Finally, we will review the development of new algorithms that aim at a linear scaling of the computing time with respect to the size of the molecule which will allow the application of these methods to very large molecules occurring, for example, in biochemistry or material science.

7.1 Introduction of a Basis: The LCAO Ansatz in the Kohn-Sham Equations

Recall the central ingredient of the Kohn-Sham approach to density functional theory, i. e., the one-electron KS equations,

$$\left(-\frac{1}{2}\nabla^2 + \left[\sum_j^N \int \frac{|\varphi_j(\vec{r}_2)|^2}{r_{12}} d\vec{r}_2 + V_{XC}(\vec{r}_1) - \sum_A^M \frac{Z_A}{r_{1A}}\right]\right)\varphi_i = \varepsilon_i \varphi_i . \tag{7-1}$$

The term in square brackets defines the Kohn-Sham one-electron operator and equation (7-1) can be written more compactly as

$$\hat{f}^{KS}\varphi_i = \varepsilon_i \varphi_i . \tag{7-2}$$

Note that the operator $\hat{f}^{KS}$ differs from the Fock operator $\hat{f}$ that we introduced in Section 1.3 in connection with the Hartree-Fock scheme only in the way the exchange and correlation potentials are treated. In the former, the non-classical contributions are expressed via the – in its exact form unknown – exchange-correlation potential V_{XC}, the functional derivative of E_{XC} with respect to the charge density. In the latter, correlation is neglected altogether, while the exchange contribution is given exactly by the action of the exchange operator $\hat{K}$ on a spin orbital χ_i,

$$\hat{K}_j(\vec{x}_1)\,\chi_i(\vec{x}_1) = \int \chi_j^*(\vec{x}_2)\frac{1}{r_{12}}\,\chi_i(\vec{x}_2)\,d\vec{x}_2\,\chi_j(\vec{x}_1)\,. \tag{7-3}$$

The Kohn-Sham equations given above in equations (7-1) or (7-2) represent a complicated system of coupled integro-differential equations (the kinetic energy operator is a differential operator, while the Coulomb contribution is expressed through an integral operator) and we now need to find a computationally efficient way of solving these equations. At the end of this process we obtain as solutions the Kohn-Sham molecular orbitals $\{\varphi_i\}$, which yield the ground state density associated with the particular choice of V_{XC} (should we know the exact V_{XC}, the exact density would result). In principle, a purely numerical approach to solve these equations is possible and a few benchmark calculations for atoms and small molecules using such a technique are available (Becke, 1989). However, numerical procedures are much too demanding for routine applications and other techniques are required. Almost all applications of Kohn-Sham density functional theory therefore make use of the LCAO expansion of the KS molecular orbitals, a scheme introduced by Roothaan, 1951, in the framework of the Hartree-Fock method. In the LCAO approach we introduce a set of L predefined basis functions $\{\eta_\mu\}$ and linearly expand the Kohn-Sham orbitals as

$$\varphi_i = \sum_{\mu=1}^{L} c_{\mu i}\eta_\mu\,. \tag{7-4}$$

If the set $\{\eta_\mu\}$ was complete which would require $L = \infty$, every function φ_i could be expressed exactly via equation (7-4). Of course, in real applications L is finite and it is of crucial importance to choose the $\{\eta_\mu\}$ such that the linear combination of (7-4) provides an approximation of the exact Kohn-Sham orbitals as accurate as possible. It should also be clear that by using a linear combination of predefined basis functions to express the Kohn-Sham orbitals, the originally highly non-linear optimization problem has been simplified into a linear one, with the coefficients $\{c_{\mu i}\}$ being the only variables. When the LCAO method was invented back in the nineteenfifties, the $\{\eta_\mu\}$ were inspired by the exactly known eigenfunctions ('atomic orbitals') of the hydrogen atom, which explains the name. Today, the basis functions are usually chosen according to different, more pragmatic criteria and do not resemble atomic functions anymore, as we will expound presently. For the time being we just assume that we have decided on some set of real basis functions $\{\eta_\mu\}$ (complex basis functions are possible, but to simplify matters we restrict ourselves to real functions). We now insert equation (7-4) into equation (7-2) and obtain in very close analogy to the Hartree-Fock case

$$\hat{f}^{KS}(\vec{r}_1)\sum_{\nu=1}^{L} c_{\nu i}\eta_\nu(\vec{r}_1) = \varepsilon_i \sum_{\nu=1}^{L} c_{\nu i}\eta_\nu(\vec{r}_1)\,. \tag{7-5}$$

If we now multiply this equation from the left with an arbitrary basis function η_μ and integrate over space we get L equations

$$\sum_{\nu=1}^{L} c_{\nu i} \int \eta_\mu(\vec{r}_1)\hat{f}^{KS}(\vec{r}_1)\eta_\nu(\vec{r}_1)d\vec{r}_1 = \varepsilon_i \sum_{\nu=1}^{L} c_{\nu i} \int \eta_\mu(\vec{r}_1)\eta_\nu(\vec{r}_1)d\vec{r}_1 \quad \text{for } 1 \le i \le L \qquad (7\text{-}6)$$

The integrals on both sides of this equation each define a matrix. On the left hand side,

$$F_{\mu\nu}^{KS} = \int \eta_\mu(\vec{r}_1)\hat{f}^{KS}(\vec{r}_1)\eta_\nu(\vec{r}_1)d\vec{r}_1 \qquad (7\text{-}7)$$

is a matrix element of the *Kohn-Sham matrix* while the *overlap matrix* on the right hand side has elements

$$S_{\mu\nu} = \int \eta_\mu(\vec{r}_1)\; \eta_\nu(\vec{r}_1)d\vec{r}_1 \,. \qquad (7\text{-}8)$$

Both matrices are L × L dimensional and, as long as we are dealing with real basis functions, are symmetric, i. e., $M_{\mu\nu} = M_{\nu\mu}$ (in the general case, they are self-adjoint or hermitian, i. e., $M_{\mu\nu} = M_{\nu\mu}^{*}$). Using **S** and **F** and introducing the L × L dimensional matrices **C** containing the expansion vectors

$$\mathbf{C} = \begin{pmatrix} c_{11} & c_{12} & \cdots & c_{1L} \\ c_{21} & c_{22} & \cdots & c_{2L} \\ \vdots & \vdots & & \vdots \\ c_{L1} & c_{L2} & \cdots & c_{LL} \end{pmatrix} \qquad (7\text{-}9)$$

and ε, a diagonal matrix of the orbital energies

$$\boldsymbol{\varepsilon} = \begin{pmatrix} \varepsilon_1 & 0 & \cdots & 0 \\ 0 & \varepsilon_2 & \cdots & 0 \\ \vdots & \vdots & & \vdots \\ 0 & 0 & \cdots & \varepsilon_L \end{pmatrix} \qquad (7\text{-}10)$$

we can rewrite the L equations (7-6) compactly as a matrix equation

$$\mathbf{F}^{KS}\,\mathbf{C} = \mathbf{S}\,\mathbf{C}\,\boldsymbol{\varepsilon}. \qquad (7\text{–}11)$$

Hence, through the LCAO expansion we have translated the non-linear optimization problem, which required a set of difficult to tackle coupled integro-differential equations, into a linear one, which can be expressed in the language of standard linear algebra and can easily be coded into efficient computer programs.

Up to this point the derivation has exactly paralleled the Hartree-Fock case, which only differs in using the corresponding Fock matrix, **F** rather than the Kohn-Sham counterpart, $\mathbf{F}^{\mathbf{KS}}$. By expanding $\hat{f}^{KS}$ into its components, the individual elements of the Kohn-Sham matrix become

$$F_{\mu\nu}^{KS} = \int \eta_\mu(\vec{r}_1)\left(-\frac{1}{2}\nabla^2 - \sum_A^M \frac{Z_A}{r_{1A}} + \int \frac{\rho(\vec{r}_2)}{r_{12}}d\vec{r}_2 + V_{XC}(\vec{r}_1)\right)\eta_\nu(\vec{r}_1)d\vec{r}_1$$

$$= -\frac{1}{2}\int \eta_\mu(\vec{r}_1)\,\nabla^2\,\eta_\nu(\vec{r}_1)d\vec{r}_1 - \int \eta_\mu(\vec{r}_1)\sum_A^M \frac{Z_A}{r_{1A}}\eta_\nu(\vec{r}_1)d\vec{r}_1 \tag{7-12}$$

$$+\int\int \eta_\mu(\vec{r}_1)\frac{\rho(\vec{r}_2)}{r_{12}}\eta_\nu(\vec{r}_1)d\vec{r}_1 d\vec{r}_2 + \int \eta_\mu(\vec{r}_1)\,V_{XC}(\vec{r}_1)\,\eta_\nu(\vec{r}_1)d\vec{r}_1\,.$$

We now need to discuss how these contributions that are required to construct the Kohn-Sham matrix are determined. The first two terms in the parenthesis of equation (7-12) describe the electronic kinetic energy and the electron-nuclear interaction, both of which depend on the coordinate of only one electron. They are often combined into a single integral, i. e.,

$$h_{\mu\nu} = \int \eta_\mu(\vec{r}_1)\left[-\frac{1}{2}\nabla^2 - \sum_A^M \frac{Z_A}{r_{1A}}\right]\eta_\nu(\vec{r}_1)d\vec{r}_1\,. \tag{7-13}$$

Independent of the choice of the $\{\eta_\mu\}$, the one-electron contribution $h_{\mu\nu}$ can be fairly easily computed using well tested algorithms. For the third term we need the charge density ρ which takes the following form in the LCAO scheme

$$\rho(\vec{r}) = \sum_i^N |\varphi_i(\vec{r})^2| = \sum_i^N\sum_\mu^L\sum_\nu^L c_{\mu i}c_{\nu i}\eta_\mu(\vec{r})\eta_\nu(\vec{r})\,. \tag{7-14}$$

The expansion coefficients, which actually contain all relevant information about the charge density, are usually collected in the so-called *density matrix* **P** with elements

$$P_{\mu\nu} = \sum_i^N c_{\mu i}c_{\nu i}\,. \tag{7-15}$$

Thus, we can alternatively express the Coulomb contribution in equation (7-12) solely in terms of the basis functions as the following four-center-two-electron integrals (since the four basis functions η_μ, η_ν, η_λ, η_σ can be attached to a maximum of four different atoms)

$$J_{\mu\nu} = \sum_\lambda^L\sum_\sigma^L P_{\lambda\sigma}\int\int \eta_\mu(\vec{r}_1)\,\eta_\nu(\vec{r}_1)\frac{1}{r_{12}}\eta_\lambda(\vec{r}_2)\,\eta_\sigma(\vec{r}_2)\,d\vec{r}_1 d\vec{r}_2\,. \tag{7-16}$$

Up to this point, exactly the same formulae also apply in the Hartree-Fock case. The difference is only in the exchange-correlation part. In the Kohn-Sham scheme this is represented by the integral,

$$V_{\mu\nu}^{XC} = \int \eta_\mu(\vec{r}_1)\, V_{XC}(\vec{r}_1)\, \eta_\nu(\vec{r}_1) d\vec{r}_1 \,, \tag{7-17}$$

where we have to decide on the explicit form of V_{XC}, whereas the HF exchange integral is given by[22]

$$K_{\mu\nu} = \sum_\lambda^L \sum_\sigma^L P_{\lambda\sigma} \int\int \eta_\mu(\vec{x}_1)\, \eta_\lambda(\vec{x}_1) \frac{1}{r_{12}} \eta_\nu(\vec{x}_2)\, \eta_\sigma(\vec{x}_2)\, d\vec{x}_1 d\vec{x}_2 \,. \tag{7-18}$$

Note again the formal simplicity of equation (7-17) as compared to equation (7-18) in spite of the fact that the former is exact provided the correct V_{XC} is inserted, while the latter is inherently an approximation. The calculation of the formally $L^2/2$ one-electron integrals contained in $h_{\mu\nu}$, equation (7-13) is a fairly simple task compared to the determination of the classical Coulomb and the exchange-correlation contributions. However, before we turn to the question, how to deal with the Coulomb and V_{XC} integrals, we want to discuss what kind of basis functions are nowadays used in equation (7-4) to express the Kohn-Sham orbitals.

7.2 Basis Sets

During the years, a huge collection of basis sets was generated in the context of wave function based approaches to quantum chemistry. Here, the orbitals χ_i which are expressed through the $\{\eta_\mu\}$ are used to construct the approximate *wave function*. It has long been recognized that very large basis sets are needed if high quality wave functions that take also into account electron correlation are the target. In particular, basis functions with complex nodal structures (*polarization functions*, see below) are necessary and in highly correlated calculations the basis set requirements soon lead to computationally very demanding procedures. On the other hand, in the Kohn-Sham scheme the orbitals play an indirect role and are introduced only as a tool to construct the *charge density* according to $\rho(\vec{r}) = \sum_i^N |\varphi_i(\vec{r})|^2$. One should therefore expect that the basis set requirements in Kohn-Sham calculations are less severe than in wave function based ones. Indeed, in most applications this is the case (see, e. g., Bauschlicher et al., 1997, and Martin, 2000).

In the following we will give a very concise overview of the typical kinds of basis sets in use today. In conventional wave function based approaches, such as the Hartree-Fock or

[22] Note that we must use $\vec{x}_i$ and not $\vec{r}_i$ as variable here, because of the spin-dependence of the exchange integral, recall Section 1.3.

configuration-interaction schemes, the set $\{\eta_\mu\}$ is almost universally chosen to consist of so-called cartesian *Gaussian-type-orbitals*, GTO of the general form

$$\eta^{GTO} = N\, x^l y^m z^n \exp[-\alpha r^2]\,. \tag{7-19}$$

N is a normalization factor which ensures that $<\eta_\mu|\eta_\mu> = 1$ (but note that the η_μ are not orthogonal, i. e., $<\eta_\mu|\eta_\nu> \neq 0$ for $\mu \neq \nu$). α represents the orbital exponent which determines how compact (large α) or diffuse (small α) the resulting function is. $L = l + m + n$ is used to classify the GTO as s-functions (L = 0), p-functions (L = 1), d-functions (L = 2), etc. Note, however, that for L > 1 the number of cartesian GTO functions exceeds the number of $(2l+1)$ physical functions of angular momentum l. For example, among the six cartesian functions with L = 2, one is spherically symmetric and is therefore not a d-type, but an s-function. Similarly the ten cartesian L = 3 functions include an unwanted set of three p-type functions.

The preference for GTO basis functions in HF and related methods is motivated by the computational advantages these functions offer, because very efficient algorithms exist for analytically calculating the huge number of four-center-two-electron integrals occurring in the Coulomb and HF-exchange terms. On the other hand, from a physical point of view, *Slater-type-orbitals* (STO) seem to be the natural choice for basis functions. They are simple exponentials that mimic the exact eigenfunctions of the hydrogen atom. Unlike the GTO functions, Slater-type-orbitals exhibit the correct cusp behavior at $r \rightarrow 0$ with a discontinuous derivative (while a GTO has a slope of zero at $r \rightarrow 0$) and the desired exponential decay in the tail regions as $r \rightarrow \infty$ (GTO fall off too rapidly). A typical STO is expressed as

$$\eta^{STO} = N\, r^{n-1} \exp[-\zeta r\;]\, Y_{lm}(\Theta, \phi)\,. \tag{7-20}$$

Here, n corresponds to the principal quantum number, the orbital exponent is termed ζ and Y_{lm} are the usual spherical harmonics that describe the angular part of the function. In fact as a rule of thumb one usually needs about three times as many GTO than STO functions to achieve a certain accuracy. Unfortunately, many-center integrals such as described in equations (7-16) and (7-18) are notoriously difficult to compute with STO basis sets since no analytical techniques are available and one has to resort to numerical methods. This explains why these functions, which were used in the early days of computational quantum chemistry, do not play any role in modern wave function based quantum chemical programs. Rather, in an attempt to have the cake and eat it too, one usually employs the so-called contracted GTO basis sets, in which several primitive Gaussian functions (typically between three and six and only seldom more than ten) as in equation (7-19) are combined in a fixed linear combination to give one *contracted Gaussian function* (CGF),

$$\eta_\tau^{CGF} = \sum_a^A d_{a\tau} \eta_a^{GTO}\,. \tag{7-21}$$

The original motivation for contracting was that the contraction coefficients $d_{s\tau}$ can be chosen in a way that the CGF resembles as much as possible a single STO function. In addition, contracting the primitives is also another trick to reduce the computational burden. For more details on the intricacies of such basis sets, we recommend, e. g., the competent discussion in Feller and Davidson, 1990, or Helgaker and Taylor, 1995.

In density functional theory CGF basis sets also enjoy a strong popularity. They are the natural choice in those programs that offer Kohn-Sham methods as an add-on but were originally designed for wave function theory and hence use CGF to expand the molecular orbitals, such as *Gaussian* or *Turbomole*. Also some of the genuine Kohn-Sham programs use Gaussian basis sets, such as *DGauss* or *DeMon*. However, since in the Kohn-Sham scheme, no exchange integrals as in equation (7-18) appear and, as we will see in a minute, the explicit calculation of the Coulomb integrals (equation (7-16)) can also be circumvented, CGF functions are not the only player in that field and the user may face different types of basis sets depending on which program he or she decides to select. For example, the *Amsterdam Density Functional* code, *ADF*, uses Slater-type orbitals as basis functions. Not only does one get away with less functions than with GTO sets, but also the treatment of non-abelian point group symmetries is easier to implement with exponential functions. This leads to a more convenient input definition and interpretation of the results of high-symmetry molecules and a possible reduction in computing time. Still another kind of basis function is realized within the program *DMol*, which does not use analytical, but *numerical* basis functions (Delley, 1990). Here, the $\{\eta_\mu\}$ are represented numerically on atomic centered grids, with cubic spline interpolations between mesh points. These basis functions are generated by numerically solving the atomic KS equations with the corresponding approximate exchange-correlation functional. Thus, numerical basis sets provide exact energies (within the given functional) for atomic fragments, but obviously necessitate the use of purely numerical (rather than analytical) techniques for solving the integrals over basis functions developed in the preceding section.

We should also mention that basis sets which do not actually comply with the LCAO scheme are employed under certain circumstances in density functional calculations, i. e., *plane waves*. These are the solutions of the Schrödinger equation of a free particle and are simple exponential functions of the general form

$$\eta^{PW} = \exp[i\,\vec{k}\,\vec{r}] \qquad (7\text{-}22)$$

where the vector $\vec{k}$ is related to the momentum $\vec{p}$ of the wave through $\vec{p} = \hbar\vec{k}$. Plane waves are not centered at the nuclei but extend throughout the complete space. They enjoy great popularity in solid state physics for which they are particularly adapted because they implicitly involve the concept of periodic boundary conditions. Unfortunately, the number of plane waves needed to arrive at an acceptable accuracy is usually daunting at best and for this and other reasons applications employing plane wave basis sets are very rare in molecular quantum chemistry. Actually, none of the popular program packages supports this kind of basis set and we will therefore neither discuss plane waves nor recent modifications of this concept, such as the *projector augmented wave method*. Interested readers are directed to the review by Blöchl, Margl, and Schwarz, 1996.

Irrespective of whether we use Gaussian functions, Slater type exponentials or numerical sets, certain categories of functions that can help to characterize the quality of a basis set have become customary in quantum chemistry. The simplest and least accurate expansion of the molecular orbitals utilizes only one basis function (or one contracted function in the case of CGF sets) for each atomic orbital up to and including the valence orbitals. These basis sets are for obvious reasons called *minimal* sets. A typical representative is the STO-3G basis set, in which three primitive GTO functions are combined into one CGF. For carbon, this basis set consists of five functions, one each describing the 1s and 2s atomic orbitals and three functions for the 2p shell (p_x, p_y, and p_z). One should expect no more than only qualitative results from minimal sets and nowadays they are hardly used anymore. The next level of sophistication are the *double-zeta* basis sets. Here, the set of functions is doubled, i. e., there are two functions for each orbital (the generic name 'double-zeta' for such basis sets still points to the beginnings of computational quantum chemistry, when STO functions were in use, where the orbital exponent is called ζ). If we take into account that it is in the valence space where changes in the electronic wave function occur during chemical processes, we can limit the doubled set of functions to the valence orbitals, while the chemically mostly inert core electrons are still treated in a minimal set. This defines the *split-valence* type sets. Typical examples are the 3-21G or 6-31G Gaussian basis sets developed by Pople and coworkers. In most applications, such basis sets are augmented by *polarization functions*, i. e., functions of higher angular momentum than those occupied in the atom, e. g., p-functions for hydrogen or d-functions for the first-row elements. Polarization functions have by definition more angular nodal planes than the occupied atomic orbitals and thus ensure that the orbitals can distort from their original atomic symmetry and better adapt to the molecular environment. Polarized double-zeta or split valence basis sets are the mainstay of routine quantum chemical applications since usually they offer a balanced compromise between accuracy and efficiency. In terms of CGF type basis sets, typical examples are the standard 6-31G(d,p) sets of Hehre, Ditchfield, and Pople, 1972, and Hariharan and Pople, 1973, or the more recent SVP (split-valence polarization) sets of Schäfer, Horn, and Ahlrichs, 1992. Equivalents consisting of two STO functions per atomic orbital or two numerical functions are of comparable importance in their respective domains. In the latter case the doubling of the numerical functions can be achieved, for example, by adding numerically generated atomic orbitals from calculations on doubly or even higher positively charged ions.

It is obvious how these schemes can be extended by increasing the number of functions in the various categories. This results in triple- or quadruple-zeta basis sets which are augmented by several sets of polarization functions including functions of even higher angular momentum. The cc-pVQZ (for *correlation-consistent polarized valence quadruple zeta*) and cc-pV5Z (the 5 stands for quintuple) basis sets mentioned earlier are typical, modern representatives of this approach in terms of Gaussian functions (Dunning, 1989). For example, the large cc-pV5Z contraction consists for the first-row atoms boron to neon of 14s-, 8p-, 4d-, 3f-, 2g-, and 1h-type primitive GTO. These are contracted to 6s- and 5p-type contracted Gaussian functions while the polarization functions are left uncontracted, leading to a final basis set of size 6s, 5p, 4d, 3f, 2g, 1h. This is a valence quintuple set because there is 1 CGF for the 1s core electrons and 5 sets of s- and p-functions representing the corresponding 2s and 2p orbitals of the valence shell. If the so-called spherical harmonic functions are used, where the contaminants in the d- and higher sets are deleted and only the true angular momentum functions are retained, this basis set contains 91 CGF per first-row atom as compared to only 15 in a 6-31G(d,p) basis (by definition the 6-31G(d,p) basis set employs six cartesian d-functions). In a study on the applicability of these correlation-consistent basis sets in Kohn-Sham calculations, Raymond and Wheeler, 1999, concluded that the combination of the B3LYP technique with these sets (cc-pVTZ and better) indeed gives very satisfactory results.

The exponents and contraction coefficients of most Gaussian basis sets have been optimized within the Hartree-Fock or correlated wave function based schemes. A noteworthy exception are the sets of Godbout et al., 1992, which have been explicitly optimized using the LDA approach and represent the standard basis sets provided by the program *DGauss*. In the beginning it was not at all clear whether one could in fact use basis sets that were optimized for representing molecular orbitals in a HF or configuration interaction context to construct the density, as in the Kohn-Sham scheme. However, it fortunately turned out that the results are fairly insensitive with respect to the way the exponents and contraction coefficients have been determined, in particular for the calculation of properties such as energies or equilibrium geometries. Hence, in general it is probably not necessary to use basis sets explicitly optimized for a density functional approach, even though there are a number of special cases where this statement is an oversimplification. Nevertheless, most modern applications of Kohn-Sham density functional theory using Gaussian functions simply employ one of the many standard basis sets, irrespective of their origin in wave function based approaches. In most contemporary program packages the popular sets are provided in an internal basis set library. Should the desired set not be included in that internal library of the program chosen, it can usually be conveniently downloaded even in the appropriate input format from the web-site http://www.emsl.pnl.gov:2080/forms/basisform.html (Feller, Schuchardt, and Jones, 1998).

If the molecules of interest contain elements heavier than, say, krypton, one usually employs a *(relativistic) effective core potential* ((R)ECP), also called *pseudopotential*, to model the energetically deep-lying and chemically mostly inert core electrons, as reviewed recently by Frenking et al., 1996, and Cundari et al., 1996. The potentials are called 'relativistic' if they have been fitted to atomic calculations that explicitly incorporate relativistic effects. The commonly used pseudopotentials have been determined with conventional wave function based methods in mind and it is not a priori clear, whether they are as useful in conjunction with density functional calculations. In particular one might pessimistically expect that an ECP appropriate for a specific exchange-correlation functional should be generated from atomic calculations employing that very functional, causing a Babylonian confusion of tongues. However, the considerable experience accumulated so far shows that this fear is fortunately unfounded and that one can safely use the well-established ECP also in density functional calculations. Specifically, Russo, Martin, and Hay, 1995, as well as van Wüllen, 1996, showed for selected compounds such as transition-metal carbonyls that in comparison with the corresponding all-electron calculations, errors of the same order of magnitude are introduced in density functional and Hartree-Fock calculations employing two different popular flavors of effective core potentials derived from atomic Hartree-Fock computations.

7.3 The Calculation of the Coulomb Term

In this section we discuss the various strategies implemented in common Kohn-Sham programs to compute the classical electrostatic contribution to the electron-electron repulsion

$$J[\rho] = \frac{1}{2}\int\int \rho(\vec{r}_1)\frac{1}{r_{12}}\rho(\vec{r}_2)\, d\vec{r}_1 d\vec{r}_2 \, . \qquad (7\text{-}23)$$

In regular wave function based methods J is determined through the four-center-two-electron integrals $J_{\mu\nu} = \sum_{\lambda}^{L}\sum_{\sigma}^{L} P_{\lambda\sigma}\int\int \eta_\mu(\vec{r}_1)\,\eta_\nu(\vec{r}_1)\frac{1}{r_{12}}\eta_\lambda(\vec{r}_2)\,\eta_\sigma(\vec{r}_2)\, d\vec{r}_1 d\vec{r}_2$ introduced in equation (7-16). Of course, the same approach is valid in density functional calculations. However, the problem inherent to this scheme is the large computational load resulting from the sheer number of integrals. It should be obvious that formally there are some L^4 such integrals which need to be computed and indeed, the handling of the two-electron integrals still constitutes the computational bottle-neck in traditional Hartree-Fock calculations on very large systems. In the HF picture these integrals are not only needed for the classical Coulomb energy but also for the determination of the exchange energy, as indicated by the integrals $K_{\mu\nu}$ shown in equation (7-18). No generally applicable computational alternative determining these contributions without explicit evaluation of the two-electron integrals is known in wave function based methods. In contrast, the exchange (and correlation) contribution in density functional theory is approached via approximate functionals and the evaluation of J is completely uncoupled from the way E_{XC} is treated. Hence, we are not forced to use the $J_{\mu\nu}$ integrals from equation (7-16) but are free to take advantage of more efficient techniques for tackling the classical Coulomb contributions. Many density functional programs indeed use special strategies to compute J which all boil down to simplifying equation (7-16), which we can equivalently write as,

$$J_{\mu\nu} = \int\int \eta_\mu(\vec{r}_1)\,\eta_\nu(\vec{r}_1)\frac{\rho(\vec{r}_2)}{r_{12}}\, d\vec{r}_1 d\vec{r}_2 \, . \qquad (7\text{-}24)$$

We may now expand the density $\rho(\vec{r}_2)$ in terms of an atom-centered auxiliary basis set $\{\omega_\kappa\}$, according to

$$\rho(\vec{r}) \approx \tilde{\rho}(\vec{r}) = \sum_{\kappa}^{K} c_\kappa \omega_\kappa(\vec{r}) \, , \qquad (7\text{-}25)$$

where the tilde indicates that we are dealing with an approximate density (since in practice the auxiliary basis set will never be complete). If we use this approximation, the computational cost for evaluating J is now formally reduced from L^4 to L^2K, since instead of equation (7-16) we now have to solve

$$\tilde{J}_{\mu\nu} = \sum_{\kappa}^{K} c_{\kappa} \iint \frac{\eta_{\mu}(\vec{r}_1)\eta_{\nu}(\vec{r}_1)\omega_{\kappa}(\vec{r}_2)}{r_{12}} d\vec{r}_1 d\vec{r}_2 . \tag{7-26}$$

This technique was originally suggested by Baerends, Ellis, and Ros, 1973, for STO basis functions and later extended by Sambe and Felton, 1975, and others to CGF basis sets. The coefficients c_{κ} are determined either straightforwardly by minimizing the function

$$F = \int \left[\rho(\vec{r}) - \tilde{\rho}(\vec{r})\right]^2 d\vec{r} \tag{7-27}$$

or preferentially by minimizing the Coulomb self-repulsion of the residual density,

$$F' = \iint \frac{\left[\rho(\vec{r}_1) - \tilde{\rho}(\vec{r}_1)\right]\left[\rho(\vec{r}_2) - \tilde{\rho}(\vec{r}_2)\right] d\vec{r}_1 d\vec{r}_2}{r_{12}} . \tag{7-28}$$

The latter approach has the advantage that the exact J is approached strictly from above, however for technical reasons it is only applicable if Gaussian basis functions are employed (Dunlap, Connolly, and Sabin, 1979). Both schemes are of course subject to the constraint that the fitted density is normalized to the total number of electrons, i. e.,

$$\int \tilde{\rho}(\vec{r})\, d\vec{r} = N . \tag{7-29}$$

Among the programs that employ equation (7-26) for evaluating the Coulomb contributions together with equation (7-28) for obtaining the coefficients of the auxiliary basis set are *Turbomole* (where this technique carries the label RI-*J* method, because in the derivation a step occurs which makes use of a mathematical trick called the 'resolution of the identity', see also further below) and *DGauss*, i. e., programs using CGF as basis functions. The STO based program *ADF* takes a slightly different route. The fitted density is here used to construct the Coulomb potential according to

$$V_{Coul}(\vec{r}_1) = \int \frac{\rho(\vec{r}_2)}{r_{12}} d\vec{r}_2 \approx \int \frac{\tilde{\rho}(\vec{r}_2)}{r_{12}} d\vec{r}_2 = \sum_{\kappa}^{K} c_{\kappa} \int \frac{\omega_{\kappa}(\vec{r}_2)}{r_{12}} d\vec{r}_2 = \tilde{V}_{Coul}(\vec{r}_1) . \tag{7-30}$$

The coefficients defining the fitted density are obtained via equation (7-27). To avoid the difficulties of dealing with two-electron integrals in a Slater-type basis, $J_{\mu\nu}$ is evaluated in this context by a numerical integration on a grid as

$$J_{\mu\nu} \approx \int \eta_{\mu}(\vec{r}_1)\, \eta_{\nu}(\vec{r}_1)\, \tilde{V}_{Coul}(\vec{r}_1)\, d\vec{r}_1 \approx \sum_{p}^{P} \eta_{\mu}(\vec{r}_p)\, \eta_{\nu}(\vec{r}_p)\, \tilde{V}_{Coul}(\vec{r}_p)\, W_p . \tag{7-31}$$

Here, the grid contains P points, which are located at positions $\{\vec{r}_p\}$ and the W_p represent the weights in the numerical quadrature scheme. We will encounter a more detailed discussion of such numerical techniques in the following section. Finally, we note that the basis sets used for expanding the density according to equation (7-25) are conveniently chosen to be of the same type as the functions used in the LCAO expansion, i. e., contracted Gaussians in *Turbomole*, *DGauss*, etc. and Slater type functions in *ADF*. Because these functions model the density, which is determined as the sum of squares of the KS orbitals, we can expect as a rule of thumb that the exponents in the auxiliary basis set should cover the range from about twice the smallest to twice the largest exponent. Experience also shows that the number of auxiliary functions K must be some 2-3 times larger than the L basis functions in the LCAO basis set in order to keep the error in the total energy introduced by using only the approximate charge density below a critical threshold of 10^{-4} E_h (ca. 0.06 kcal/mol). A recent, very careful and detailed discussion on the optimization of such auxiliary basis sets has been given by Eichkorn, et al., 1995 and 1997.

A completely different approach is taken in *DMol*. Recall that this program uses numerical basis sets rather than analytical functions. At the center of their implementation is Poisson's equation,

$$\nabla^2 V_{Coul}(\vec{r}) = -4\pi\ \rho(\vec{r}) \tag{7-32}$$

which connects the Coulomb potential with the density. This equation is solved numerically on a grid. In a nutshell, the strategy includes the decomposition of the Coulomb potential into independent single-center contributions which are solved individually. Finally, the bits and pieces are recollected to construct $\tilde{V}_{Coul}(\vec{r})$ on the grid. For details, the reader should consult Delley, 1990, or Becke and Dickson, 1988. This motif of breaking up an integral into atomic contributions will also reappear in the next section where we discuss the numerical integration of the exchange-correlation potential.

The common denominator of all these approaches for evaluating the Coulomb contribution is that no four-center-two-electron integrals such as equation (7-16) are needed. As a consequence, the formal scaling of the computing time is reduced from $O(N^4)$ to $O(N^3)$. But one needs to be very careful to put this conclusion in the right perspective and not to overinterpret the formal scaling properties. It can easily be shown that for both approaches the asymptotic scaling for very large systems reduces to $O(N^2)$. The reason is simply that in extended systems only those integrals will survive where the functions are located on neighboring centers. The majority of the integrals will be close to zero and can safely be neglected because of the vanishing overlap of the basis functions. Johnson, 1995, has termed this the 'N^3 versus N^4 myth'. Nevertheless, the debate whether schemes replacing the exact computation of the Coulomb part by some kind of approximate approach employing fitted densities of potentials are in the long run beneficial or not has certainly not been settled yet. For example, von Arnim and Ahlrichs, 1998, impressively demonstrated that even for large systems with more than 1000 basis functions their implementation of equation (7-26) in the frame of the *Turbomole* program is at least about one order of magnitude faster than the conventional approach of explicitly computing the four-center-two-electron integrals. Well implemented techniques employing fitted densities are probably even competitive with current methods achieving linear scaling (see section 7.6) for cases up to fairly large systems.

7.4 Numerical Quadrature Techniques to Handle the Exchange-Correlation Potential

What we have not discussed so far is how the contribution of the final components of the Kohn-Sham matrix in equation (7-12), i. e., the exchange-correlation part, can be computed. What we need to solve are terms such as

$$V_{\mu\nu}^{XC} = \int \eta_{\mu}(\vec{r}_1)\, V_{XC}(\vec{r}_1)\, \eta_{\nu}(\vec{r}_1) d\vec{r}_1 . \tag{7-33}$$

Unfortunately, the explicit expressions even for the most simple approximations of V_{XC}, such as the LDA, are fairly sophisticated mathematical constructs, as we have seen in the preceding chapter. A general analytical solution of equation (7-33) is therefore out of reach and numerical techniques based on a grid to solve these integrals need to be employed. The art consists now in designing a grid that comprises the most suitable compromise between the desired sufficiently high numerical accuracy and the no less vital requirement of having a computationally efficient scheme. Once we have decided on which mesh to use, $V_{XC}(\vec{r})$ needs to be evaluated at each grid point. From the computational point of view, and if we use a spin-density approach in combination with a GGA functional, this translates into the determination of ρ^{α} and ρ^{β} and their first and second derivatives at each point because a GGA functional contains the gradient of the density and the exchange-correlation potential itself is defined as the functional derivative of the energy with respect to the density.

The straightforward numerical integration of $V_{\mu\nu}^{XC}$ maps equation (7-33) onto

$$V_{\mu\nu}^{XC} \approx \tilde{V}_{\mu\nu}^{XC} = \sum_{p}^{P} \eta_{\mu}(\vec{r}_p)\, V_{XC}(\vec{r}_p)\, \eta_{\nu}(\vec{r}_p)\, W_p . \tag{7-34}$$

In words, the integral of equation (7-33) for the exchange-correlation potential is approximated by a sum of P terms. Each of these is computed as the product of the *numerical* values of the basis functions η_μ and η_ν with the exchange-correlation potential V_{XC} at each point $\vec{r}_p$ on the grid. Each product is further weighted by the factor W_p, whose value depends on the actual numerical technique used.

The first step to build a successful strategy for solving this numerical problem is to find a grid that is best suited for the particular situation. Clearly, the behavior of the exchange-correlation potential is governed by the characteristics of the density $\rho(\vec{r})$, which has cusps at the positions of the nuclei in the molecule. It was therefore quickly apparent that a simple cartesian grid, which does not account for the accumulation of density at the positions of the nuclei, is certainly not the optimal choice. Rather, the implementation of equation (7-34) in most current computer programs follows design principles put forward by Becke, 1988c, where the molecular integration is broken up into separate, but overlapping atomic contributions. In order to limit the number of sub- and superscripts and to keep the exposition as easy and general as possible, we follow the notation generally used and switch to I as the value of the integral and to $F(\vec{r})$ for the integrand, i. e., $I = \int F(\vec{r})\, d\vec{r}$. Then, this decomposition reads

$$I = \sum_{A}^{M} I_A \tag{7-35}$$

where the sum is over the M nuclei and the I_A are the atomic contributions defined as

$$I_A = \int F_A(\vec{r})\, d\vec{r} \,. \tag{7-36}$$

The atomic integrands F_A are chosen such that their sum over all nuclei returns the original function,

$$\sum_{A}^{M} F_A(\vec{r}) = F(\vec{r}) \,. \tag{7-37}$$

The individual $F_A(\vec{r})$ are constructed from the original integrand by the introduction of weight functions $W_A(\vec{r})$ with which $F(\vec{r})$ is multiplied

$$F_A(\vec{r}) = W_A(\vec{r})\, F(\vec{r}) \,. \tag{7-38}$$

$W_A(\vec{r})$ assumes a value close to unity if $\vec{r}$ is close to nucleus A and close to zero near all other nuclei $B \neq A$. It is thus ensured that $F_A(\vec{r})$ is indeed the contribution of $F(\vec{r})$ associated with nucleus A with no (or negligibly small) contributions near the other nuclei and that $F_A(\vec{r})$ has a (near) singularity only at nucleus A. Of course, the weights satisfy

$$\sum_{A}^{M} W_A(\vec{r}) = 1 \,. \tag{7-39}$$

The transition from $W_A(\vec{r}) \approx 1$ to $W_A(\vec{r}) \approx 0$ as the distance from the nucleus A increases needs to be smooth enough such that numerical instabilities are avoided but at the same time also as abrupt as possible such that density peaks from nearby the nuclei are extinguished. The implementation of this concept in the three-dimensional space involves a special choice of coordinates – see Becke, 1988c, for details – but actually leads to a smoothened step function as schematically sketched in Figure 7-1 for the one-dimensional case.

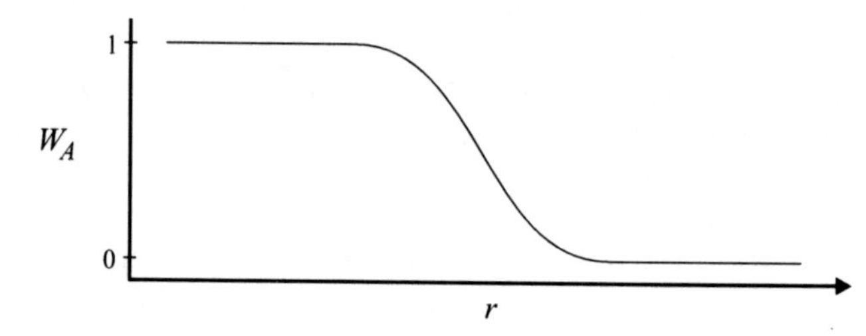

Figure 7-1. Becke partition in one dimension.

The way these weight functions (and in particular their dependence on the distance from the nuclei) are determined involves some mathematical subtleties but is well described in the literature (see Becke, 1988c, Murray, Handy, and Laming, 1993, Treutler and Ahlrichs, 1995) and will not occupy us any further.

Once the atomic contributions F_A are determined, the corresponding integrals I_A are subsequently computed on grids that consist of points on concentric spheres around each atom. Switching to polar coordinates r, θ, and ϕ the radial and angular integrations are separated according to

$$I_A = \int_0^{\infty}\int_0^{\pi}\int_0^{2\pi} F_A(r,\theta,\phi) r^2 \sin\theta \, dr \, d\theta \, d\phi \approx \sum_p^P W_p^{rad} \sum_q^Q W_q^{ang} F_A(r_p, \theta_q, \phi_q). \quad (7\text{-}38)$$

There are P radial and Q angular points, with corresponding radial and angular weights, W^{rad} and W^{ang}, respectively. The total number of grid points per atom is then P × Q. Usually, the angular part is not split up further into separate integrations over the two coordinates θ and ϕ because there are highly efficient algorithms for numerically integrating on the surface of a sphere that clearly outperform the alternative integration over the individual angular coordinates.

The next step is to choose one of the many recipes available for the numerical quadrature of the radial and angular contributions. It is well beyond the scope of this book to offer a detailed discussion of all the pros and cons connected with the various alternatives and we will therefore limit ourselves to a very brief summary of what is currently being used in popular density functional programs. The reader interested in a more comprehensive and in-depth exposition of numerical integration techniques in the context of density functional calculations is encouraged to consult the primary literature quoted. For the integration along the radial coordinate *Gaussian*, for example, employs a Euler-McLaurin scheme originally proposed by Murray, Handy and Laming, 1993. On the other hand, Treutler and Ahlrichs, 1995, prefer a Gauss-Chebyshev type of integration in their implementation of equation (7-38) in the *Turbomole* package. If we turn to the integration of the angular part, i. e., the numerical quadrature on the surface of a sphere, there seems to be a certain consensus that the so-called Lebedev grids offer the best value for money. These are very efficient grids of octahedral symmetry. Usually, Lebedev grids that exactly sum all the spherical harmonics for up to $l = 29$ or $l = 35$ are used in typical density functional programs. This translates into 302 or 434 integration points, respectively, per radial coordinate.

In order to cut down the number of grid points and hence to increase the efficiency of the numerical integration, a technique called *grid pruning* is frequently used. The underlying idea is that as one approaches the nucleus, the electron density loses more and more its angular structure and becomes increasingly spherically symmetric. Hence, for spheres at small distances from the nucleus a progressively smaller amount of angular grid points should suffice. Similar arguments apply if we analyze the situation at large values of r. Here, the actual magnitude of $\rho(\vec{r})$ becomes so small that again we can get away with much less sophisticated angular grids without loosing any significant accuracy. In a pruned grid one exploits these observations and the space around each atom is partitioned into various regions. Within these regions, whose sizes of course depend on the actual atom, Lebedev grids of varying density are used. Close to the nucleus fairly coarse grids are sufficient, while dense ones are employed at intermediate distances and again coarser grids as we move farther away from the nucleus. This technique is used in most modern density functional programs such as *Gaussian* or *Turbomole*. For example, the fairly dense default integration mesh in *Gaussian 98* is a pruned grid containing 75 radial shells and a maximum of 302 angular points per shell. Due to the pruning the actual number of integration points per atom is reduced to only around 7 000, just a third of the regular size of 75 × 302 = 22 650 points.

We now need to address some of the problematic aspects inherent in the numerical quadrature techniques. The origin of all these undesirable features is that none of the numerically approximated quantities are exact. Thus, this numerical noise introduces an additional source of uncertainty into the calculations and one needs to limit the corresponding errors as much as possible. A severe problem is related to the orientation of the angular parts of the atomic grids. Since the angular grids are incomplete, the energies computed by employing these grids are sensitive to rotations of the molecule. In other words, the total energy of a molecule has the very problematic property of not being rotationally invariant. Depending on its orientation in space, small variations in the total energy occur. A solution to this dilemma is the definition of a standard, albeit still arbitrary orientation of the angular grids relative to the Cartesian reference frame. A set of rules characterizing such a standard nuclear orientation can be found in Gill, Johnson, and Pople, 1993. The size of the errors introduced by the lack of rotational invariance is demonstrated in Table 7-1. Here, the molecule HOOF was computed in three different orientations. These calculations employed a pruned, medium sized grid consisting of 50 Euler-McLaurin radial shells and a maximum of 194 angular Lebedev points and a smaller, unpruned grid with 20 radial and 50 angular points (i. e., 1000 points per atom).

While the dependence of the total energy on the orientation is not dramatic, for obvious reasons it is the more severe the coarser the grid is. In addition, the variations will of course increase with the size of the molecule and can therefore become significant for larger molecules. Using a standard orientation solves the problem and the energy determined for orientation I will always be obtained. It should be clear, however, that choosing a particular orientation as the standard is completely arbitrary. One could just as well construct a different set of rules leading to an alternative standard orientation which would be characterized by slightly different total energies. We should also mention that rotational invariance becomes even more important if gradients or harmonic frequencies are being computed. In

Table 7-1. Total energies [E_h] of HOOF in various orientations (taken from Johnson, 1995).

Orientation	(50, 194)-grid	Difference	(20, 50)-grid	Difference
I (Standard)	–247.0686706	0	–247.0669064	0
II	–247.0686463	2.43×10^{-5}	–247.0666277	2.79×10^{-4}
III	–247.0686779	7.30×10^{-6}	–247.0678356	-9.29×10^{-4}

particular harmonic frequencies of low-lying modes can be jeopardized by using techniques that lack rotational invariance.

A second major problem connected to the use of finite grids for the evaluation of the exchange-correlation energy is associated with the determination of derivatives of the energy, such as the gradients used in geometry optimizations. We use

$$E_{XC} \approx \tilde{E}_{XC} = \sum_{t}^{T} W_t f_t \tag{7-39}$$

as the most general expression for the numerical quadrature approximation of E_{XC}, where the weights W_t are the product of the atomic weights W_A and the quadrature weights W_p and W_q. The derivative of $\tilde{E}_{XC}$ with respect to a parameter x is then

$$\frac{d}{dx}\tilde{E}_{XC} = \sum_{t}^{T}\left(f_t \frac{d}{dx} W_t + W_t \frac{d}{dx} f_t \right). \tag{7-40}$$

The first term on the right hand side of equation (7-40), the weight function derivatives, will vanish in the limit of an infinite grid. However, in practical applications we must consider that the atomic weights depend explicitly on the nuclear coordinates and therefore their derivative will not be zero. In particular if coarse grids are used, the contributions of $\frac{d}{dx} W_t$ can be of appreciable magnitude. A different way of looking at this is that the numerical grid for the evaluation of E_{XC} is constructed as the superposition of individual atomic grids. These are not fixed in space but move along with the atoms to which they are attached. Nevertheless, many programs simply neglect these terms. Among the undesirable consequences of this policy is that the calculated gradient does not necessarily vanish exactly at the energy minimum as it should and we are left with the unpleasant situation that at the lowest energy configuration we have a non-zero gradient while the structure with vanishing gradients is not the one with the lowest energy. In some cases if meshes that are too small are employed this may even lead to situations where the optimizer gets so confused that the optimization fails to converge. Fortunately, in most cases this numerical noise is not dramatic if grids of sufficient density are employed as demonstrated by Johnson and Frisch, 1993, and Baker et al., 1994. On the other hand, these errors become more severe if not gradients but higher derivatives are calculated as in the evaluation of harmonic frequencies. For example, using a grid with 50 radial and 194 angular points and a 6-31G(d) basis set, the lowest vibrational frequency of ammonia which corresponds to the 'umbrella' inversion mode amounts to 862 cm^{-1} if the weight derivatives are neglected but to 888 cm^{-1} if they are included. If instead a much smaller grid with 30 and 86 radial and angular points, respectively, is used, the inclusion of the weight derivatives increases the frequency from 879 to 945 cm^{-1}. As an aside, we note that due to the octahedral symmetry of the Lebedev angular grids, the doubly degenerate modes in molecules of D_{3h} or C_{3v} symmetry are not exactly reproduced. For example, in ammonia using the large (50,194)-grid, the two e-symmetric modes split into 1662 and 1685 cm^{-1} and 3505 and 3517 cm^{-1}, respectively.

7.5 Grid-Free Techniques to Handle the Exchange-Correlation Potential

It is clear from the above discussion that the evaluation of the exchange-correlation potential using numerical integration on a finite grid has a disadvantages, mostly due to the 'numerical noise' inherent in this approach. To get rid of these problems it would be desirable to have grid-free implementations to compute E_{XC} and V_{XC}. A first step in that direction has been taken by Zheng and Almlöf, 1993 (see also Almlöf and Zheng, 1997), subsequently taken up by Glaesemann and Gordon, 1998 and 1999. The basic idea is to interpret the density $\rho(\vec{r})$ as a multiplicative operator. Next, the matrix representation **R** of this operator in the basis of the LCAO expansion of the Kohn-Sham orbitals $\{\eta_\mu\}$ is constructed, leading to the matrix elements

$$R_{\mu\nu} = \int \eta_\mu(\vec{r})\, \rho(\vec{r})\, \eta_\nu(\vec{r})\, d\vec{r}\,. \tag{7-41}$$

This step is similar to what we have done in equation (7-7) where we obtained the matrix representation of the Kohn-Sham operator. If we insert expression (7-14) for the charge density in terms of the LCAO functions and make use of the density matrix **P** defined in equation (7-15), we arrive at

$$R_{\mu\nu} = \sum_{\lambda}^{L} \sum_{\sigma}^{L} P_{\lambda\sigma} \int \eta_\mu(\vec{r})\eta_\lambda(\vec{r})\eta_\sigma(\vec{r})\eta_\nu(\vec{r})\, d\vec{r}\,. \tag{7-42}$$

While the computational work for setting up the matrix representation **R** of $\rho(\vec{r})$ scales formally as N^4, this can be cut down to N^3 using again the trick introduced in section 7-3 by expanding the density in terms of an atom centered, orthonormalized auxiliary basis set $\{\omega_k\}$ (recall equation (7-25)). Let us review this simplification under a slightly different perspective. The starting point is again

$$\rho(\vec{r}) \approx \tilde{\rho}(\vec{r}) = \sum_{\kappa}^{K} c_\kappa \omega_\kappa(\vec{r})\,. \tag{7-43}$$

Since we have chosen the $\{\omega_\kappa\}$ to be orthonormal, the expansion coefficients c_κ are related to the density matrix according to

$$c_\kappa = \int \tilde{\rho}(\vec{r})\, \omega_\kappa(\vec{r})\, d\vec{r} = \sum_{\lambda}^{L} \sum_{\sigma}^{L} P_{\lambda\sigma} \int \eta_\lambda(\vec{r})\eta_\sigma(\vec{r})\omega_\kappa(\vec{r})\, d\vec{r} \tag{7-44}$$

(the reader not familiar with these basic techniques of linear algebra should consult, for example, Chapter 1 of Szabo and Ostlund, 1982). Inserting equation (7-44) into equation (7-43) we arrive at the following expression for the density

$$\tilde{\rho}(\vec{r}) = \sum_{\kappa}^{K}\left[\sum_{\lambda}^{L}\sum_{\sigma}^{L} P_{\lambda\sigma} \int \eta_{\lambda}(\vec{r})\eta_{\sigma}(\vec{r})\, \omega_{\kappa}(\vec{r})\, d\vec{r}\right] \omega_{\kappa}(\vec{r}). \tag{7-45}$$

Consequently, the final equation for the now only approximate matrix element $\tilde{R}_{\mu\nu}$ becomes

$$\tilde{R}_{\mu\nu} = \int \eta_{\mu}(\vec{r})\sum_{\kappa}^{K}\left[\sum_{\lambda}^{L}\sum_{\sigma}^{L} P_{\lambda\sigma}\int \eta_{\lambda}(\vec{r})\eta_{\sigma}(\vec{r})\omega_{\kappa}(\vec{r})\, d\vec{r}\right] \omega_{\kappa}(\vec{r})\, \eta_{\nu}(\vec{r})\, d\vec{r} \tag{7-46}$$

or, after slight rearrangement,

$$\tilde{R}_{\mu\nu} = \sum_{\lambda}^{L}\sum_{\sigma}^{L}\sum_{\kappa}^{K} P_{\lambda\sigma}\int \eta_{\mu}(\vec{r})\eta_{\nu}(\vec{r})\omega_{\kappa}(\vec{r})\, d\vec{r} \times \int \omega_{\kappa}(\vec{r})\eta_{\lambda}(\vec{r})\eta_{\sigma}(\vec{r})\, d\vec{r}. \tag{7-47}$$

Hence, we have factored the matrix element $R_{\mu\nu}$ of equation (7-42) which includes expensive four-center integrals into a combination of two three-center integrals through the introduction of the auxiliary basis set with the concomitant reduction in computational costs. This approach is usually known as the *resolution of the identity* (see Kendall and Früchtl, 1997) because if the auxiliary basis set $\{\omega_{\kappa}\}$ were complete, the sum over κ of the corresponding integrals that are formally being inserted will yield unity. (This is the completeness relation of linear algebra, see Szabo and Ostlund, 1982. Note that again we restrict our discussion to real, not complex functions $\{\omega_{\kappa}\}$). We emphasize that this approach constitutes an approximation which is only exact in the limit of a complete auxiliary basis set (see equation (7-43)), which is of course never realized.

After this brief detour we return to the main subject of this section, the implementation of a grid-free KS scheme. Now that we have a matrix representation of the density, we can exploit another well-known fact from linear algebra: a function of a matrix which is expressed in an orthonormal basis can be evaluated by first diagonalizing the matrix, then applying the function on the diagonal elements and finally transforming the matrix back to its original basis. Since the basis functions $\{\eta_{\mu}\}$ are not orthonormal, we need to include as a first step the transformation of **R** to an orthonormal basis, **R'**. Let us illustrate this simple procedure using the LDA exchange functional, where the function is to take the 4/3 power of the density, $f(x) = x^{4/3}$. Without going into any detail about how the required matrix transformations are carried out – be assured that they are possible – the sequence of operations is

$$\begin{pmatrix} R_{11} & R_{12} & \cdots \\ R_{21} & R_{22} & \cdots \\ \cdots & \cdots & R_{LL} \end{pmatrix} \xrightarrow{\text{orthogonalize}} \begin{pmatrix} R'_{11} & R'_{12} & \cdots \\ R'_{21} & R'_{22} & \cdots \\ \cdots & \cdots & R_{LL} \end{pmatrix} \xrightarrow{\text{diagonalize}} \begin{pmatrix} r_1 & 0 & \cdots \\ 0 & r_2 & \cdots \\ \cdots & \cdots & r_L \cdots \end{pmatrix} \xrightarrow{\text{apply } f(x)}$$

$$\begin{pmatrix} r_1^{4/3} & 0 & \cdots \\ 0 & r_2^{4/3} & \cdots \\ \cdots & \cdots & r_L^{4/3} \end{pmatrix} \xrightarrow{\text{undiagonalize}} \begin{pmatrix} F'_{11} & F'_{12} & \cdots \\ F'_{21} & F'_{22} & \cdots \\ \cdots & \cdots & F'_{LL} \end{pmatrix} \xrightarrow{\text{unorthogonalize}} \begin{pmatrix} F_{11} & F_{12} & \cdots \\ F_{21} & F_{22} & \cdots \\ \cdots & \cdots & F_{LL} \end{pmatrix}.$$

The resulting matrix **F** corresponds to the 4/3 power of the original matrix **R**. If more advanced, GGA-type functionals are used rather than the local density approximation, the procedure becomes slightly more complicated due to the more complex forms of the functionals. Here we just briefly sketch the general strategy which is centered around the observation that these functionals can usually be interpreted as a product of operators containing terms proportional to $\rho(\vec{r})^{4/3}$ and to $s = \frac{|\nabla\rho|}{\rho^{4/3}}$, the reduced density gradient. The matrix representation $Q_{\mu\nu}$ of such a product of two functions $f(\vec{r})$ and $g(\vec{r})$ can be approximated if we again introduce an auxiliary basis set $\{\pi_\gamma\}$,

$$\begin{aligned} Q_{\mu\nu} &= \int \eta_\mu(\vec{r})\left[f(\vec{r}) \times g(\vec{r})\right] \eta_\nu(\vec{r})\, d\vec{r} \\ &\approx \sum_\gamma^{G} \int \eta_\mu(\vec{r}) f(\vec{r}) \pi_\gamma(\vec{r})\, d\vec{r} \times \int \pi_\gamma(\vec{r}) g(\vec{r}) \eta_\nu(\vec{r}) d\vec{r}\,. \end{aligned} \tag{7-48}$$

This is nothing else than another resolution of the identity, whose accuracy is determined by the quality of the auxiliary basis set $\{\pi_\gamma\}$. In principle, these techniques could be used to implement any desired functional including the capability to analytically compute energy derivatives as needed in geometry optimizations and many other applications. Their main advantage is that the numerical noise inherent to grid-based numerical integration techniques will be replaced by a smooth, reproducible error due to the incompleteness of the auxiliary basis set. This error is not only independent of the choice of the coordinate system and thus eliminates one significant drawback discussed in the previous section, it furthermore can be controlled by systematically increasing the basis set. However, in particular for gradient-corrected functionals and the evaluation of derivatives, all the experience that has been accumulated so far with these alternatives to the conventional grid-based strategies indicates that very large auxiliary basis sets $\{\omega\}$ and $\{\pi\}$ are needed to obtain reasonable accuracies (see, e. g., Glaesemann and Gordon, 2000). Also, as the basis sets get larger, problems due to linear dependence and other numerical precision problems probably become important. Up to now only a few experimental implementations of grid-free schemes to compute the exchange-correlation contributions in Kohn-Sham theory have been reported. Before closing this section we should mention that Werpetinski and Cook, 1997, describe a different grid-free approach which is, however, limited to exchange-only, local $\rho^{4/3}$ functionals. These authors employ an auxiliary basis set to directly fit the corresponding $\rho^{1/3}$ potential. While their results represent an improvement over conventional grid-based techniques, this approach cannot easily be extended to modern, more complicated functionals and it therefore lacks the generality needed to be a successful contender in this arena.

7.6 Towards Linear Scaling Kohn-Sham Theory

While density functional approaches offer the advantage of obtaining results of better than Hartree-Fock quality for about the price of a HF calculation, their straightforward application to large systems that occur, for example, in biochemistry, catalysis or solution chemistry is still limited because of the high computational costs. There has therefore been an ever growing interest in developing techniques which do not suffer from the problematic scaling properties of conventional algorithms but whose computational efforts grow only linearly with the size of the system.[23]

The contribution which dominates by far the computational effort in current implementations of the Kohn-Sham formalism is the evaluation of the Coulomb contribution J. We noted in Section 7.3 that in the large system limit the determination of J scales quadratically with the size of the system, independent of the choice of either the conventional technique based on four-center-two-electron integrals or strategies employing an auxiliary basis to fit the density. The difficulty with the Coulomb contribution is that on the one hand the electrostatic $1/r$ operator has a singularity at $r = 0$ and is on the other hand also inherently long-ranged. It will be very difficult, if possible at all, to design linear scaling algorithms that properly treat both of these difficulties. All methods attempting to reduce the scaling therefore share a common motif: the Coulomb problem is partitioned into a small, short-range region where J is computed exactly and a large, 'well separated' part, where the interaction is approximated.

One class of strategies is based on the *fast multipole method* (FMM), which was originally designed by Greengard, 1987, to evaluate the Coulomb interaction between point charges. Here, the linear scaling is aimed for by a separation of the physical space around a charged particle. Within the small, so-called *near-field* region the interaction has to be evaluated directly. However, the interaction with the *far-field* region is computed by dividing the physical space into a hierarchical set of cells and approximating the interaction energy between the particles contained in two cells by an intelligent use of multipole expansions. Since the multipole expansions become more accurate as the interaction distance increases, larger and larger cells may be used as the two cells get more distant. Due to this hierarchy of cells, in the limit of very large systems linear scaling is gradually approached, even though its cost also scales with the fourth power of the logarithm of the accuracy required as outlined by Pérez-Jordá and Yang, 1998.

However, unlike point charges, the continuous charge distributions that occur in quantum chemistry have varying extents and the applicability of the multipole approximation is not only limited by the distance but also by the extent or diffuseness of the charge distribution. This additional complexity makes a transfer of the concepts of the fast multipole method to applications in quantum chemistry less straightforward. Therefore it should come as no surprise that several adaptations to extend the applicability of the FMM to the Coulomb problem with continuous charge distributions have been suggested. These lead to

[23] Efforts to tame the unfavorable scaling of electronic structure methods are not limited to density functional theory. For a general summary of the current state of the art see the review by Goedecker, 1999.

schemes such as the *continuous fast multipole method* (CFMM) of White et al., 1994 (see also Johnson et al., 1996), the *Gaussian very fast multipole method* (GvFMM) of Strain, Scuseria, and Frisch, 1996, or the *quantum chemical tree code* (QCTC) of Challacombe, Schwegler, and Almlöf, 1996, to mention just a few. All these techniques achieve close to linear scaling if the system gets large enough. Where the actual break-even point as compared to conventional integration methods occurs, depends on the particular system (for example, compact, three-dimensional molecular systems behave less favorably than two- or one-dimensional chains), basis sets and the desired accuracy.

Rather than splitting the *physical* space into short- and long-range parts as in the above techniques, an alternative is for the Coulomb operator itself to be reformulated and written as a sum of two contributions representing the short- and long-range regimes,

$$\frac{1}{r} \equiv S(r) + L(r) = \frac{f(r)}{r} + \frac{1 - f(r)}{r} \quad \text{with } r = \left|\vec{r} - \vec{r}'\right|. \tag{7-49}$$

The function f(r) divides 1/r into a short-range function S(r) which has a singularity at r = 0 and a non-singular, long-ranged function L(r). The separator function has to be chosen such that f(r) is a rapidly decaying function which approaches unity as $r \rightarrow \infty$. Consequently, 1 – f(r) vanishes at r = 0 and contains all long-ranged components of the 1/r operator, as depicted in Figure 7-2.

S(r) will be treated in real space but needs to be applied only in a small neighborhood around the reference charge distribution. L(r), which represents the bulk of the electrostatic interactions (in terms of their number, not in the contribution to the energy), can be treated using various approximations with favorable scaling properties. For example, in the origi-

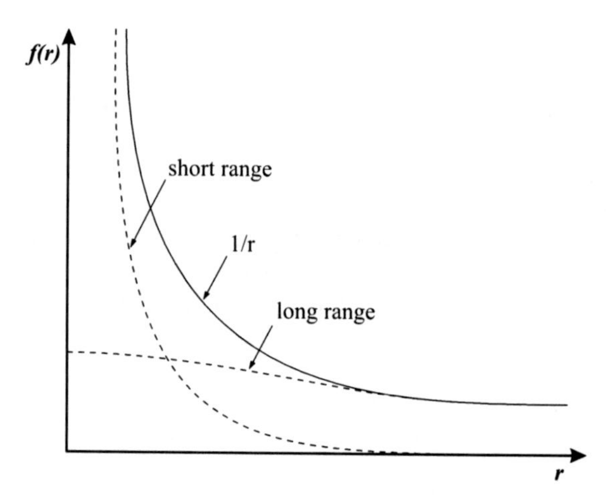

Figure 7-2. Optimal partition of the Coulomb operator (adapted from Lee, Taylor, Dombrowski, and Gill, *Phys. Rev. A*, **55**, 3233 (1997), with permission by the American Physical Society).

nal formulation of this algorithm, christened *KWIK*[24] by Dombroski, Taylor, and Gill, 1996, the function f(r) is chosen to be the error function and the long-range term is approximated by its truncated Fourier series expansion. Among the other possibilities discussed, the most efficient and most drastic one is to neglect the long range part altogether. Obviously this extreme approximation has an enormous effect on the absolute energies. However, if L(r) is sufficiently flat, much smaller consequences are expected for relative energies as has indeed been found by Adamson, Dombroski, and Gill, 1996. To what extent this *Coulomb-attenuated Schrödinger equation* (CASE) approximation and the more sophisticated *Coulomb-attenuated potential* (CAP) extensions to it developed by Gill and Adamson, 1996, will be useful in actual applications has not been established yet.

In summary, the prospect of achieving linear or close to linear scaling in the determination of the Coulomb part in real calculations seems to be rather promising. This is underlined by the optimistic conclusion drawn in a review on linear scaling techniques in DFT by Scuseria, 1999, that 'the "integral bottleneck" that characterized quantum chemistry calculations for many years has clearly been defeated'. Also note that corresponding algorithms have lately been implemented in several popular quantum chemistry codes and are therefore generally available.

The other two possible computational bottlenecks in a Kohn-Sham calculation are the diagonalization of the Kohn-Sham matrix $\mathbf{F}^{\mathbf{KS}}$ and the numerical quadrature of the exchange-correlation functional and potential. In spite of the formal N^3 scaling of the diagonalization, it has been shown to be computationally insignificant unless very large molecular systems are being investigated. Nevertheless, techniques which allow a near-linear scaling in terms of elapsed cpu time and of memory requirement have been reported by various authors (see, e. g., Millam and Scuseria, 1997, Scuseria, 1999). Similarly, novel techniques which also reduce the numerical quadrature to near-linear scaling are starting to emerge (Pérez-Jordá and Yang, 1995, or Stratmann, Scuseria, and Frisch, 1996). Thus, for all three important components of a Kohn-Sham calculation, methods which asymptotically scale linearly with the size of the system are available. In principle, this opens the way to apply these methods to 'realistic' molecules such as enzymes, and non-periodic catalysts etc. However, the mere existence of methods that *asymptotically* scale more favorably than current techniques does not say anything about how large the system has to be in order to make the better scaling an asset. It is the magnitude of the pre-factor that is the decisive property which determines where the cross-over between optimized conventional and linear scaling implementations will occur.

While the linear scaling techniques described so far are targeted at programs using Gaussian functions, related attempts have also been presented for use with Slater functions as described by Fonseca Guerra et al., 1998, in the context of the *ADF* program. We should also mention a completely different strategy for approaching linear scaling, the *divide-and-conquer* method put forward by Yang (see Yang, 1992, and Parr and Yang, 1995). In this scheme, not molecular orbitals as in the conventional Kohn-Sham procedure, but the elec-

[24] The term KWIK is not an acronym but resulted from the final four letters in the central formula of their paper, which contains the string A(**k**,ω)*I*(**k**).

tron density is the central quantity. A large molecule is divided into subsystems in physical space and the densities of these submoieties (which necessarily add up to the total density of the system) are expressed through local basis functions. The individual parts of the energy are evaluated and the energy of the whole system is obtained from the sum of the subsystem contributions. An implementation of this technique within the *DeFT* program has been described by Goh, Gallant, and St-Amant, 1998.

PART B

The Performance of the Model

The past decade has seen considerable efforts in the improvement of density functional methods and present day functionals often yield energy related results approaching so-called chemical accuracy, that is, with errors less than 2 kcal/mol. In fact, in many areas of chemical research it is very difficult, if possible at all, to produce results – either by experimental means or by high quality post Hartree-Fock wave function based computations – which are unquestionably of higher accuracy than those obtained by means of modern functionals. Numerous examples can be found in the chemical literature supporting this statement and there are more success stories for the application of approximate density functional theory every month. On the other hand, the often quoted major drawback of density functional theory is the formal inability to systematically improve the accuracy of quantitative predictions. Apart from some well documented problems for DFT (excited states, highly degenerate systems, weak interactions) it is still impossible to predict DFT errors a priori, only from an intimate knowledge of the fundamental features of the theory. Furthermore, it has frequently been reiterated that if density functional methods definitely fail, and if the reason is not due to the choice of a particular functional, integration grid, or basis set, there is no rigorous procedure to correct the flaws since the underlying reasons of shortcomings in the theory are far from being understood. While not too many years ago this has led many to simply ignore density functional theory despite of its apparent success in many cases, computational chemistry today seems to put this into a more pragmatic perspective. From a practical point of view, i. e., if the focus lies on the efficient solution of a particular chemical problem rather than on a rigorous validation of a theory, the situation for the applicability of DFT methods is not different from the situation of most of the post-Hartree-Fock approaches. If a system is small enough, the whole hierarchy of post-HF methods can be applied in order to substantiate the quality of predictions of DFT. If this is impossible, for like reasons any result from a chosen level of post-HF treatment could be distrusted. The formal advantage of wave function-based theories over approximate density functional theory – namely the potential to improve the correlation treatment step by step up to a point where the interesting observable is converged – is an option only for some small molecular systems. For most chemically interesting problems such a procedure prohibits itself because of the extraordinary scaling of the computational effort with the system size and the slow convergence of correlation effects with basis set size.

Having said this, we like to view modern density functional theory as a very efficient additional tool in the arsenal of computational methods rather than a perfectly different theory, which is used orthogonal to traditional approaches (like valence bond and molecular orbital theory). Just like any newly developed post-HF method,[25] a new functional has to be evaluated by careful calibration to known accurate data, before it should be trusted and applied routinely. Hence, systematic testing of modern density functional methodology for a broad variety of chemically motivated questions is of paramount interest, because the critical evaluation of such results provides the only means to assess the reliability of current DFT methods. This type of research has led already to a substantial body of experience and even for professionals it is not easy to follow up with the vast amount of publications and the rapid development in the field. Therefore, the objective of the second part of this book is to provide a comprehensive survey over systematic benchmarks assessing the accuracy of the different flavors of DFT. We analyze the results stemming from miscellaneous areas of application, ranging from geometry optimizations and vibrational frequencies over to energetic details as well as electric and magnetic properties up to exploration of pathways on potential energy surfaces for chemical reactions. The general strengths and weaknesses of the various functionals presently available are outlined. The selection of examples is of course related to our personal interests and there are many important applications that could not be included because of space limitations. In any case, the main target of this part is to convey a feeling to the reader of how good present day DFT is and, wherever possible, we try to provide an evidence-based hierarchy of density functional models, in order to enable him or her to select the functional and basis set most appropriate for the respective application at hand.

[25] Problems with the recently introduced multiconfiguration perturbation methods might serve as a good example in this respect, see Roos et al., 1996, for a review and further references.

8 Molecular Structures and Vibrational Frequencies

8.1 Molecular Structures

One of the central tasks of computational chemistry is the reliable prediction of molecular structures. It is routine now to carry out geometry optimizations on systems consisting of up to, say, 50 atoms and a lot of experience has been collected with respect to the performance of Hartree-Fock approaches and methods based on Møller-Plesset perturbation theory of low order (see, for example, Hehre et al., 1986). These methods usually perform well within an expected accuracy of ±0.02 Å or better for bond lengths in molecules consisting of main group elements. It has long been recognized that HF theory usually gives bond lengths which are too short, and the description of multiple bonds tends to be problematic due to the neglect of electron correlation. The MP2 approach, conversely, frequently overestimates bond distances but has been a very successful and well-accepted black box treatment of virtually any problem in organic chemistry in the past. However, an entirely different situation has been recognized for transition-metal containing systems, for which an often frustrating performance is observed. For coordinatively saturated closed-shell systems – that is to say, for best cases – deviations exceeding ± 0.1 Å for bond distances involving the metal center are commonplace (see Frenking et al., 1996, for an instructive review). For studies on coordinatively unsaturated open-shell systems the situation for such methods is even worse and is probably condensed best by quoting Taylor: 'Transition-metal chemistry (...) is a graveyard for UHF-based MP methods' (Taylor, 1992).

While automatic geometry optimization schemes employing analytical gradients and modern, effective update algorithms have been around for many years in the Hartree-Fock world, similar strategies for Kohn-Sham density functional theory were first published and implemented in computer programs only at the end of the eighties, see, e. g. Versluis and Ziegler, 1988, Fournier, Andzelm, and Salahub, 1989, Fan and Ziegler, 1991, Andzelm and Wimmer, 1991, and Johnson, Gill, and Pople, 1993. However, as of now structural optimizations using DFT can be performed just as conveniently as in HF schemes.

Transition-metal chemistry in particular was the field where pioneering density functional results have been of unprecedented accuracy for larger systems and impressive to any researcher in the field. Today, it seems that density functional theory has adopted the role of a standard tool for the prediction of molecular structures.

8.1.1 Molecular Structures of Covalently Bound Main Group Elements

Andzelm and Wimmer, 1992, published one of the first comprehensive studies on the performance of approximate density functional theory in which optimized molecular geometries were reported. These authors computed the geometries of several organic species containing the atoms C, N, O, H, and F at the local SVWN level, using a polarized double-zeta basis set optimized for LDA computations. Some trends have been discerned from this study, which provide general 'rules of thumb' for what can be expected from the application of the LDA: While the calculated distances of bonds involving hydrogen atoms were consistently overestimated by up to 0.02 Å, single bonds between heavier atoms were too short by about the same amount. C=C double and aromatic bonds were well described within a few thousandths of an Å, whereas C≡C triple bonds were too long by up to 0.02 Å. Similar trends were observed for single and double bonds between C and O as well as C and N. Single bond lengths were usually underestimated and double bond distances were typically correct or overestimated by up to 0.03 Å, depending on the particular system. Polar C–F bonds agreed nicely with experiment, while a deviation of +0.06 Å was noted for the N-F bond in NF_3. Calculated bond angles were in most cases accurate to within 1° with the exception of NO_2, for which an underestimation of the experimental value by 2.5° has been observed. Equivalent patterns of deviations become obvious from the geometric data published by Johnson, Gill, and Pople, 1993, who used the rather small 6-31G(d) basis set in a comparative study on the performance of six different DFT methods implemented in *Gaussian 92* on a G2 subset.[26] For SVWN optimized structures, a mean absolute deviation of 0.02 Å for bond lengths and 1.9° for bond angles has been observed, which can be compared to results of HF (0.02 Å, 2.0°), MP2 (0.01 Å, 1.8°), and QCISD (0.01 Å, 1.8°) on the same set of molecules. In order to identify effects of the basis set incompleteness on LDA results, Dickson and Becke, 1993, have evaluated geometries of 69 neutral closed-shell species obtained with the basis set free *NUMOL* program at the LDA limit and compared them with results from several other major density functional computer codes along with experimental data where available. The basis set free calculations confirmed the prevailing trends for bond lengths described above but indicated a tendency for an overall contraction of bonds upon improvement of the basis set, very similar to what has been observed for Hartree-Fock optimized geometries. The shortening of bonds upon improvement of the basis set quality led to refined LDA geometries in general, but the systematic overestimation of bonds to hydrogen atoms remained significant. In addition, the description of bond distances between group 1 and group 2 metal dimers such as Na_2 or Mg_2 were found to be problematic for LDA; regardless of whether basis sets are used or not, deviations of up to 0.5 Å occur. As a portrait of trends in deviations depending on bonding situations and basis set qualities we give a few representative examples for hydrocarbons in Table 8-1.

Gradient corrections have been introduced in order to correct for shortcomings of the LDA. Hence, with regard to what we have learned above, one can assume that their application should result in an expansion of those bond lengths that are underestimated at the LDA level, whereas too long bonds should be contracted. To put these simple-minded anticipations into practice, let us consider the bond lengths in the first two columns of Table 8-2.

[26] This set consists of 32 small neutral first-row species and although this is not a very representative testing ground for many chemical problems, it has become a de facto standard used by several other groups. We will refer to it as the 'JGP set' in the following discussions.

Table 8-1. Basis set dependence of SVWN-optimized C-C/C-H bond lengths [Å].

Bond	6-31G(d,p)	6-311++G(d,p)	Limit[a]	Experiment
H-H	— /0.765	— /0.765	— /0.765	— /0.741
H_3C-CH_3	1.513/1.105	1.510/1.101	1.508/1.100	1.526/1.088
H_2C=CH_2	1.330/1.098	1.325/1.094	1.323/1.093	1.339/1.085
HC≡CH	1.212/1.078	1.203/1.073	1.203/1.074	1.203/1.061

[a] Basis set free data from Dickson and Becke, 1993.

Table 8-2. Effect of gradient corrections on computed bond lengths for different bonding situations [Å].

Bond		SVWN[a]	BLYP[a]	SVWN[b]	BLYP[b]	BP86[b]	BPW91[b]	Experiment
H-H	R_{H-H}	0.765	0.748	0.765	0.748	0.752	0.749	0.741
H_3C-CH_3	R_{C-C}	1.513	1.541	1.510	1.542	1.535	1.533	1.526
	R_{C-H}	1.105	1.104	1.101	1.100	1.102	1.100	1.088
H_2C=CH_2	R_{C-C}	1.331	1.341	1.327	1.339	1.337	1.336	1.339
	R_{C-H}	1.098	1.095	1.094	1.092	1.094	1.092	1.085
HC≡CH	R_{C-C}	1.212	1.215	1.203	1.209	1.210	1.209	1.203
	R_{C-H}	1.078	1.073	1.073	1.068	1.072	1.070	1.061

[a] 6-31G(d) basis set; [b] 6-311++G(d,p) basis set.

In the case of H_2, the large overestimation observed for LDA is indeed compensated for at the BLYP level, reducing the deviation from experiment by a significant extent. For the C–H-bonds hardly any bond contraction is seen upon inclusion of gradient corrections. On the other hand, C–C bonds are expanded as expected and, corresponding to the observed error pattern at the SVWN level, single bonds are affected more than triple bonds. However, the underestimation of the C–C single bond visible in the LDA structure is overcompensated at the GGA level, leading to an even larger deviation from experiment, now with an error of reversed sign. For the C=C double bond in turn, the BLYP functional corrects the LDA structure to yield a very good agreement with experiment. Both methods describe the C≡C triple bond as too long, the GGA functionals exaggerate the bond length even a little more than the LDA. Use of the larger 6-311++G(d,p) basis set marginally improves the results for the triple bond in acetylene for all methods by roughly the same amount, whereas C-C single and double bonds remain largely unchanged at the BLYP level. Basis set effects are even less pronounced for bonds involving hydrogen atoms. So far, all three gradient-corrected functionals appear well suited to predict molecular structures of better quality than the LDA. It is noteworthy, however, that the BP86 and BPW91 functionals do not overestimate the length of the C-C single bond as much as the BLYP functional does.

All of the examples discussed above demonstrate that the bonding situation determines the accuracy achievable in LDA or GGA calculations. From a closer inspection of the structural data published in the literature, it becomes obvious that this is indeed generally the case. Furthermore, the merits of the GGA visible in the description of other molecular properties (like binding energies, see following chapter) do not generally lead to improved molecular geometries. In many cases the LDA deficiencies are overcompensated, leading to even larger deviations from experimental data. For example, the proper prediction of the molecular structures of fluorine peroxide, FOOF, and nitrosyl hyperfluorite, FONO, are well known to be vexing problems for standard wave function based methods – results from simple LDA calculations, however, are in good agreement with experiment (Amos, Murray, and Handy, 1993). Although BLYP is certainly among the most prominent GGA functionals and is considered superior to the local density approximation functional SVWN, it produces strikingly worse structures for these species. As another case in point, Altmann, Handy, and Ingamels, 1996, found SVWN results for a set of sulfur-containing compounds by and large closer to experimental values than results from BLYP computations, with differences between the functionals of up to 0.06 Å.[27] Increasing the basis set size from 6-31G(d,p) to TZ2P+f led to a general bond contraction for both functionals (ranging from 0.01 to 0.03 Å) and to an overall improvement with respect to experimental data. For the JGP set of molecules, the BLYP functional yields bond lengths which are on average too long by 0.02 Å and bond angles are generally underestimated by 2°, showing no improvement over Hartree-Fock or SVWN results (Johnson, Gill, and Pople, 1993). A comparison of geometric parameters for the 55 molecules in the G2 set optimized with different functionals revealed that both BLYP and BP86 gave molecular geometries with larger overall deviations from experiment than MP2, the latter GGA performing slightly better than the former (Bauschlicher, 1995). On the other hand, in cases like O_3, S_3, CH_2 and Be_2, which have proved to be notoriously difficult problems for post-HF methods, the gradient-corrected BP86 functional compared favorably with these traditional approaches and SVWN. Only large scale coupled cluster calculations reached a better agreement with experiment (Murray, Handy, and Amos, 1993). The BLYP functional gave slightly worse results, mostly similar to SVWN, in some instances better. Scheiner, Baker, and Andzelm, 1997, conducted an elaborate study on more than 100 molecules consisting of first and second-row elements and were forced to conclude that the performance of the BLYP functional is inferior to that of SVWN. This was particularly so for bonds involving second-row elements. These authors found that both GGA functionals tested, BLYP and BPW91, do in fact provide a noticeable improvement over LDA in the description of bonds to hydrogen atoms, but not for those to second-row elements. Table 8-3 shows that no marked overall improvement results from the use of gradient-corrected functionals when all bond lengths are compared.

Redfern, Blaudeau, and Curtiss, 1997, conducted a comparative study on systems involving third-row atoms. Reported geometries obtained with the BLYP and BPW91

[27] Besides, they noted conspicuous differences of up to 0.07 Å at the LDA level from the use of different programs (*DMol* and *CADPAC*). S-F bonds were longer by 0.03 Å, S-Cl and S-H bonds by 0.04 Å on average if *DMol* was used, while the remaining bonds showed good agreement. This might be indicative of severe shortcomings in the basis sets used (numerical double-zeta and 6-31G(d), respectively), which would underline an increased importance of basis set quality for species containing third-row atoms as noted by several authors.

Table 8-3. Mean absolute deviations from experiment for computed bond lengths [Å]. Taken from Scheiner, Baker, and Andzelm, 1997.

Type	SVWN	BLYP	BPW91	SVWN	BLYP	BPW91
	6-31G(d,p)			TZV2P		
all bonds[a]	0.016	0.021	0.017	0.013	0.016	0.013
first row[b]	0.015	0.017	0.013	0.013	0.013	0.011
bonds to H atoms[c]	0.016	0.014	0.013	0.013	0.010	0.010
second row[d]	0.025	0.042	0.033	0.017	0.033	0.025

[a] Bond distances of all 108 species investigated; [b] bonds involving first row elements and hydrogen atoms; [c] bonds involving at least one H atom; [d] bonds involving at least one second row element.

functionals once more revealed a significantly worse description of bond lengths as compared to MP2 (mean unsigned errors from experiment for bond lengths and angles, respectively: BLYP: 0.05 Å, 1.0°, BPW91: 0.03 Å, 1.0°, MP2: 0.02 Å, 0.4°). Use of the small 6-31G(d) basis set, however, might somewhat obscure the conclusions from this study.

Laming, Termath, and Handy, 1993, showed that there is room for improvement within the B88 exchange functional: after slight modifications in the original functional expression and empirical readjustment of the β-parameter (cf. Section 6.5) with respect to improved geometries and atomization energies for a small set of molecular systems, they favorably tested their new functionals CAM(A)-LYP and CAM(B)-LYP on a reduced G2-set. In combination with a basis set of polarized triple-zeta quality, the mean errors to experimental bond lengths were found significantly reduced compared to BLYP (BLYP: 0.017 Å, CAM(A)-LYP: 0.007 Å, CAM(B)-LYP: 0.009 Å). On the other hand, Adamo and Barone, 1998b, reported only a marginal improvement of computed geometries upon substitution of Becke's exchange functional by the PW functional for exchange: the average deviation for bond lengths in the JPG set is 0.012 Å with the PWPW91 functional compared to 0.014 Å obtained with BPW91 and BLYP (all methods used in combination with the 6-311G(d,p) basis set). Attempts to modify the PW91 exchange functional (leading to the so-called *m*PW functional for exchange as implemented in the program *Gaussian 98*) did not change the performance with respect to geometric parameters of main group species.[28] Also, other newly introduced gradient-corrected functionals did not significantly improve the performance (Neuman and Handy, 1995, Neuman and Handy, 1996, Hamprecht et al., 1998, Adamo and Barone, 1999, Ernzerhof and Scuseria, 1999a). Thus, in conclusion, it appears that for most species the GGA and LDA protocols produce bond lengths of very similar quantitative accuracy with mean deviations from experiment around 0.01 to 0.02 Å for first and second-row species. Bond angles are usually underestimated but generally accurate to within 1° on average. These deviations obey certain trends, which depend on the particular binding situation. Larger deviations have been reported for heavier main

[28] These modifications were not, however, explicitly done with geometric parameters in mind.

group species. It has become obvious in several cases that the BLYP functional is less suited for reliable structure prediction than the BP86 or BPW91 functionals.

In virtually all cases studied, hybrid functionals perform substantially better than LDA or GGA approaches in predicting molecular geometries, and in most comparative studies a 50 % reduction in the mean errors for bond lengths is observed. For example, bond lengths in the G2 set computed using B3LYP and B3P86 show an average absolute deviation from experiment of 0.013 Å and 0.010 Å compared to 0.026 Å and 0.022 Å for the pure GGA functionals BLYP and BP86, respectively. Increasing the basis set from 6-31G(d) to 6-311+G(3df,2p) reduces the mean error further to 0.008 Å for B3LYP (Bauschlicher, 1995). Scheiner, Baker, and Andzelm, 1997, reported the same trends for first and second-row systems using a slightly different hybrid functional implementation. For a set of 20 organic molecules, geometries optimized at the B3LYP/6-31G(d) level were found to be in error by less than 0.005 Å on average for bond lengths, and bond angles were accurate to within a few tenths of a degree. These deviations are of the same order as the uncertainties in the experimental equilibrium structures for most polyatomics (Rauhut and Pulay, 1995). For a set of 13 mostly organic species, the basis set dependence of B3LYP and CCSD(T) structures has been systematically compared and convergence was found to be faster for the DFT method (Martin, El-Yazal, and François, 1995a). For these species, B3LYP was found to give very accurate results with some trends in deviations: the lengths of single bonds were slightly overestimated on average by 0.002 Å, whereas double bonds were too short on average by –0.003 Å and triple bonds by –0.006 Å. Only expensive CCSD(T)/cc-pVQZ calculations gave better structures than the hybrid functional. The uniform error behavior present in the B3LYP geometries led the authors to propose an empirical correction scheme for geometries based on formal bond orders between atoms. This scheme gave marginally improved geometries with a mean absolute error from accurate experimental data in the order of 0.002 Å. This scaling has, however, not been used in subsequent studies.

Improvements over LDA and GGA structures with hybrid functionals are also observed for species containing third-row elements where the B3PW91, for instance, performs better than MP2 (Redfern, Blaudeau, and Curtiss, 1997). Raymond and Wheeler, 1999, reported geometries of a set of challenging species (NO, NO_2, NO_3, O_2, O_3, SO_2, ClO, ClO_2) obtained with the B3LYP functional and used the aug-cc-pVDZ, aug-cc-pVTZ, aug-cc-pVQZ and aug-cc-pV5Z series of basis sets in order to extrapolate to the basis set limit of B3LYP. These authors found a significant reduction in deviations from experimental bond lengths upon improving the basis set from double-zeta to triple-zeta quality, whereas no marked changes resulted from extending the basis to quadruple-zeta and quintuple-zeta. They identified the neutral Cl_2 and the Cl_2^- anion as particularly problematic systems with deviations from experiment of 0.03 Å and 0.09 Å, respectively, in the extrapolated basis set limit. Average deviations ranging from 0.031 Å (aug-cc-pVDZ) to 0.009 Å (extrapolated limit) were reported for all other systems and many computed results fell within the experimental uncertainty. An overall accuracy of 1.0° (aug-cc-pVDZ) and 0.8° (aug-cc-pVQZ) was observed for bond angles. Further particularly challenging test cases for density functionals are XONO and XNO_2 with X = F, Cl, and Br. For the geometries of these species, the hybrid functionals provide a crucial improvement over GGA functionals and the LDA, although they do not reach the performance of large scale coupled cluster calculations (Lee, Bauschlicher, and Jayatilaka, 1997). Further examples for the performance of density functional theory regarding geometry prediction of open-shell species have recently been reviewed by Ventura, 1997.

At first sight, the good performance of hybrid functionals like B3LYP might be rather surprising, since Becke did not use geometric data as input for the optimization of those parameters on which most of the modern hybrid functionals rely. However, taking the typical deviations of the constituent functional ingredients, namely Hartree-Fock and BLYP, into account, it becomes more obvious why the ubiquitous B3LYP functional behaves so well: bond lengths evaluated with the former method are usually too short, and we just learned that BLYP generally gives bonds that are too long. Consequently, a composite of both should profit from error cancellations. While this certainly is only a superficial rationalization for the very satisfactory performance of the hybrid approach, this simple view is nevertheless often sufficient to extrapolate the behavior of such functionals if the performance of the constituent HF and GGA methods is known. In any case, the generally observed high quality of structures optimized by the B3LYP functional has led several authors to suggest that such geometries (and zero-point energy corrections from harmonic frequency calculations, see below) should be used instead of MP2 geometries within the framework of highly accurate extrapolation schemes like G2 and CBS. And indeed, first applications of procedures altered in this way revealed not only an improved computational efficiency but also slightly reduced average errors (Bauschlicher and Partridge, 1995, Mebel, Morokuma, and Lin, 1995, Montgomery et al., 1999). With respect to the general quality of structural predictions it is apparent that the admixture of exact HF exchange seems to influence the results more than the particular choice of local or non-local parts of exchange and correlation functionals within a particular hybrid functional. While a blend of 50 % exact exchange in older procedures (Becke, 1993a) does not lead to significantly better geometries compared to the pure GGA, virtually all functionals including a fraction of 20-25 % HF exchange yield very similar results, all of high quality. As a summary, Table 8-4 offers error statistics that have been obtained from the application of various density functional methods to structures of main group species: the hybrid functionals containing three empirically fitted mixing parameters (B3LYP and B3PW91) perform essentially identically to more recently developed hybrids containing only one mixing parameter determined on theoretical reasoning (B1LYP, B1PW91, see Adamo and Barone, 1997, PBE1PBE, see Ernzerhof and Scuseria, 1999, and Adamo and Barone, 1999, or B98, see Bienati, Adamo, and Barone, 1999). Despite the use of various functionals for exchange and correlation, or modifications thereof (*m*PW, Adamo and Barone, 1998b), the performance of the methods is significantly improved upon admixing exact exchange in the order of some 20 %.

Table 8-4. Compilation of mean absolute deviations for bond lengths [Å] / bond angles [degrees] for small main group molecules from different sources.

32 1st row species, 6-31G(d) basis, Johnson, Gill, and Pople, 1993					
HF	0.020 / 2.0		SVWN	0.021 / 1.9	
MP2	0.014 / 1.8		BLYP	0.020 / 2.3	
QCISD	0.013 / 1.8				
33 1st row species, TZ2P basis, Laming, Termath, and Handy, 1993					
SVWN	0.090 / 1.9		CAM(A)LYP	0.007 / 1.7	
BLYP	0.013 / 1.7		CAM(B)LYP	0.009 / 1.5	
13 species, Martin, El-Yazal, and François, 1995a					
CCSD(T)/cc-pVDZ	0.018 / 2.2		B3LYP/cc-pVDZ	0.009 / 1.7	
CCSD(T)/cc-pVTZ	0.014 / 0.6		B3LYP/cc-pVTZ	0.004 / 0.3	
CCSD(T)/cc-pVQZ	0.002 / 0.4		B3LYP/cc-pVQZ	0.004 / 0.3	
20 organic molecules, Rauhut and Pulay, 1995					
BLYP/6-31G(d)	0.012 / 0.6		B3LYP/6-31G(d)	0.003 / 0.5	
108 1st and 2nd row species, Scheiner, Baker, and Andzelm, 1997					
	6-31G(d,p)	DZVP	TZVP	TZ2P	UCC[a]
HF	0.021				
MP2	0.014				
SVWN	0.016	0.016	0.014	0.013	0.013
BLYP	0.021	0.024	0.020	0.016	0.016
BPW91	0.017	0.019	0.016	0.013	0.012
ACM	0.011	0.011	0.009	0.010	0.009
40 species cont. 3rd row elements, 6-31G(d) basis, Redfern, Blaudeau and Curtiss, 1997					
MP2	0.022 / 0.4		B3LYP	0.030 / 0.5	
BLYP	0.048 / 1.0		B3PW91	0.020 / 0.5	
BPW91	0.020 / 0.5				
32 1st row species, 6-311G(d,p) basis, Adamo and Barone, 1997, 1998, 1999					
BLYP	0.014		B3LYP	0.004	
BPW91	0.014		B1LYP	0.005	
PWPW91	0.012		B3PW91	0.008	
*m*PWPW91	0.012		B1PW91	0.005	
PBEPBE	0.012		*m*PW3PW91	0.008	
BHLYP	0.015		*m*PW1PW91	0.010	
			PBE1PBE	0.010	
40 1st and 2nd row species, TZ2P basis, Hamprecht *et al.*, 1998					
BLYP	0.019 / 0.4		B3LYP	0.008 / 0.2	
HTCH	0.013 / 0.7				

[a] uncontracted aug-cc-pVTZ basis.

8.1.2 Molecular Structures of Transition-Metal Complexes

Accurate experimental information on equilibrium geometries for transition-metal complexes is much more limited than for main group molecules. Since reliable data exists for several carbonyl complexes these systems have served in the past to test the performance of various theoretical methods, including density functional theory. The commonly accepted working hypothesis describing trends in electronic structure of these complexes is the Dewar-Chatt-Duncanson model, according to which the binding between metal and ligands is governed by the interplay of donor and acceptor contributions. A balanced description of these effects has been a long-standing challenge for computational methods, in particular for complexes of the first transition-metal row. For this class of complexes, the Hartree-Fock model generally overestimates metal-ligand (M-L) bond lengths, typically by 0.1 to 0.3 Å, whereas the performance of MP2 strongly depends on the electronic situation at the metal center: good agreement with experimental data is commonly found for geometries of 4d and 5d species provided that relativistic effects are properly accounted for. For elements of the first transition-metal row, however, MP2 tends to underestimate M-L bond lengths significantly. For the LDA, a general trend to underestimate the lengths of M-L bonds is well documented. Nevertheless, it performs favorably compared to HF and frequently gives smaller deviations from experiment than MP2. The systematic underestimation of M-L bonds involving the metal centers by the LDA is compensated to a large extent by the application of gradient corrections – a typical lengthening of bonds involving metal atoms is in the order of 0.05 Å, which compares to 0.01-0.02 Å for main group elements. However, owing to the variable quality of GGA geometries for main group species outlined above, the situation remains unconvincing for the structures of the ligands where significant deviations occur. As for species consisting of main group elements, the best overall description of structural properties within the DFT framework is once more found for hybrid functionals including some 20 % of exact exchange. Let us consider these statements in more detail for the chromium hexacarbonyl complex $Cr(CO)_6$, a well studied compound for which a highly accurate experimental structure obtained from neutron diffraction is available. Table 8-5 contains a representative collection of geometric data computed at different levels of theory.

Clearly visible are common trends for HF and MP2 structures: bond distances are overestimated by the former and underestimated by the latter method. The CCSD(T) optimization yields an improved Cr-C bond but, interestingly, the worst description of the C-O bond among all methods shown. This flaw is certainly a consequence of the lack of higher angular momentum functions in the basis set used for the ligand atoms – however, the computational demands of this level of theory are already respectable even considering present computational standards. In view of the results of standard wave function based methods the performance of the SVWN functional is not bad, although the Cr-C distance is far too short. For this bond, the BP86 gradient-corrected functional yields a very good agreement with experiment, but a worse C-O bond length results. BLYP overestimates both the Cr-C as well as the C-O bond distance, but there might be some room for improvement by using more flexible basis sets. The two hybrid methods give a balanced improvement over the LDA structure for both bond types and the agreement with experiment is good.

Table 8-5. Computed bond lengths [Å] for the $Cr(CO)_6$ complex in O_h symmetry. Experimental values: $R_{Cr\text{-}C}$ = 1.918 Å, $R_{C\text{-}O}$ = 1.141 Å (see ref. 70 in Jonas and Thiel 1995).

Bond	HF	MP2	CCSD(T)	SVWN	BP86	BLYP	B3P86	B3LYP
	2.010[a]	1.862[a]	1.939[c]	1.865[d]	1.911[d]	1.942[f]	1.901[d]	1.927[d]
$R_{Cr\text{-}C}$	2.017[b]	1.874[b]		1.866[e]	1.910[e]	1.937[g]		1.929[g]
	1.970[c]				1.908[f]			1.921[h]
	1.111[a]	1.154[a]	1.178[c]	1.145[d]	1.156[d]	1.157[f]	1.141[d]	1.142[d]
$R_{C\text{-}O}$	1.111[b]	1.154[b]		1.145[e]	1.153[e]	1.164[g]		1.150[g]
	1.118[c]				1.154[f]			1.155[h]

[a] Doubly polarized triple-zeta basis on C and O, ECP/triple-zeta basis on Cr (Jonas and Thiel, 1995); [b] doubly polarized triple-zeta basis on C and O, Wachters basis on Cr (Jonas and Thiel, 1995); [c] Wachters basis on Cr, triple-zeta basis on C and O (Barnes, Liu, and Lindh, 1993); [d] 6-311+G(d) basis as implemented in *Gaussian* (viz., modified Wachters basis on Cr) (Spears, 1997); [e] triple-zeta STO on Cr, polarized double-zeta STO on C and O (Ziegler, 1995); [f] double numerical basis as implemented in *DMol* (Delley, 1994); [g] Wachters basis on Cr, 6-31G(d) basis on C and O (Hamprecht *et al.*, 1998); [h] extended Wachters basis on Cr, polarized double-zeta basis (D95*) on C and O (Koch and Hertwig, 1998).

Table 8-6 displays M-C and C-O bond lengths for the hexacarbonyls of Cr, Mo, and W determined at different levels of theory together with experimental data. First we note the general expansion of M-C bonds by the GGA treatment improving the LDA geometry for the Cr and Mo complexes, but not for $W(CO)_6$. For this species, a nearly perfect agreement with experiment for W-C and C-O bond lengths is seen with SVWN, while BP86 overestimates both distances. However, neither LDA nor BP86 calculations reflect the experimental trends in M-C bond lengths, but the metal-CO bond lengths increase steadily from Cr to W. Better agreement is obtained if relativistic effects are included within the GGA treatment, either by means of perturbation theory (BP86+QR) or by use of relativistic effective

Table 8-6. Bond lengths for neutral hexacarbonyl complexes of Cr, Mo, and W in O_h symmetry [Å].

	$Cr(CO)_6$		$Mo(CO)_6$		$W(CO)_6$	
Method	$R_{M\text{-}C}$	$R_{C\text{-}O}$	$R_{M\text{-}C}$	$R_{C\text{-}O}$	$R_{M\text{-}C}$	$R_{C\text{-}O}$
SVWN[a]	1.866	1.145	2.035	1.144	2.060	1.144
BP86[a]	1.910	1.153	2.077	1.152	2.116	1.154
BP86+QR[a]	1.910	1.153	2.076	1.153	2.049	1.155
BP86/ECP[b]	1.908	1.154	2.065	1.153	2.075	1.154
B3LYP[c]	1.921	1.155	2.068	1.155	2.078	1.156
MP2/ECP[b]	1.862	1.154	2.031	1.152	2.047	1.153
Experiment	1.914	1.141	2.063	1.145	2.058	1.148

[a] Triple-zeta STO on the metal, polarized double-zeta STO on C and O (Ziegler, 1995); [b] doubly polarized triple-zeta basis on C and O, ECP/triple-zeta basis on the metal (Jonas and Thiel, 1995); [c] polarized double-zeta basis on C and O, ECP/triple-zeta basis on the metal (Koch and Hertwig, 1998).

core potentials (BP86/ECP, see Frenking et al., 1996). Both approaches give essentially the same accuracy (see also van Wüllen, 1996) and correctly reproduce the experimentally observed trend in bond lengths. It is obvious from Table 8-6 that the presence of relativistic effects results in metal-ligand bond contractions in the order of 0.06 Å for $W(CO)_6$, which is just about the usual amount by which LDA underestimates the length of such bonds. Hence, the excellent SVWN estimate for the W-C bond length seems to be predominantly caused by error cancellation, whereas the C-O bond is indeed very well described at this level. Finally, the MP2 results of Jonas and Thiel reflect also for these species what is now common knowledge for such systems (see Frenking and Wagener, 1998, and references cited therein): the simplest post-HF method, i. e. MP2, cannot deal with 3d transition-metal compounds, while reasonable geometries are usually obtained for coordinatively saturated closed-shell complexes of 4d and 5d elements.

The trends derived from these example cases are representative for most of the geometric data published on coordinative transition-metal compounds. In the past decade the BP86 functional has become the preferred computational workhorse for the handling of transition-metal complexes. Accurate geometries were obtained with this functional for the neutral carbonyl complexes $Fe(CO)_5$, $Ru(CO)_5$, $Os(CO)_5$, $Ni(CO)_4$, $Pd(CO)_4$, and $Pt(CO)_4$ (Jonas and Thiel, 1995). Where the results could be compared to experimental data, deviations for metal ligand bonds did not exceed 0.01 Å. For structures of ionic hexacarbonyl complexes, a larger maximum deviation of 0.06 Å has been noted, but the direct comparison with experiment is somewhat hampered by the presence of crystal packing or counter-ion effects in X-ray structures (Jonas and Thiel, 1996). Related findings have been reported for the structures of $V(CO)_6^-$ (Spears, 1997) and a variety of other complexes (for typical examples, see Ziegler, 1995, Rosa et al., 1996, Bérces, 1996, or Ehlers et al., 1997). Experimental trends in bond lengths are well reproduced for different kinds of coordinatively bonded complexes despite the presence of non-negligible errors in absolute values, e. g., for C-O bond lengths.

By and large, the BP86 functional has been shown to be a valuable tool for the assessment of transition-metal coordination chemistry. Unfortunately, other functionals have not been tested as extensively as BP86 but a recent review shows that the B3LYP functional gives essentially equally good structures for this class of compounds (Frenking and Wagener, 1998). The molecular structures of the complexes $Fe(CO)_5$, $Fe_2(CO)_9$, and $Fe_3(CO)_{12}$ have been studied using the BP86 and B3LYP functionals in combination with basis sets of double-zeta and polarized double-zeta quality, and the results were carefully compared to available experimental and existing theoretical data in a vividly written report (Jang et al., 1998). The experimental geometries are reproduced well within 0.02 Å and 0.4°, which is just about the order of magnitude by which different experiments deviate from each other. Notable qualitative differences between both theoretical methods occurred for the description of axial and equatorial M-L bond lengths in $Fe(CO)_5$, but in view of the experimental contradictions found for this issue it is hard to state which functional performs better. However, B3LYP compared favorably with CCSD(T) results for this compound. In a study on the actinide complexes UF_6, NpF_6, and PuF_6, the bond lengths computed with HF, SVWN, BLYP, and B3LYP were compared with experimental structures obtained by electron diffraction techniques (Hay and Martin, 1998). Employing a newly developed relativistic ECP/ basis set combination for the metal atoms and the 6-31G(d) basis on F, the observed deviations in bond lengths were moderate with the best performance found for SVWN, the worst for BLYP. At the conventional HF level the bond lengths were underestimated by 0.02 Å while the density functionals produced distances which were too long with errors of 0.01 Å (SVWN), 0.04 Å (BLYP), and 0.02 Å (B3LYP).

In a comparative study on optimized geometries of 25 transition-metal complexes, qualitative differences for SVWN and BP86 structures have been noted (Bray et al., 1996). A general expansion of bond lengths ranging from 0.02 to 0.09 Å was found when going from SVWN to the gradient-corrected BP86 functional. This trend is consistently present in all types of bonds investigated. However, a better agreement with experiment is found for BP86 structures of those complexes, which predominantly contain coordinative metal-carbon bonds (0.01 Å vs. 0.05 Å mean deviation for BP86 and SVWN, respectively), whereas SVWN gives better results for complexes containing covalent bonds between metal and non-carbon atoms (0.07 Å vs. 0.03 Å mean deviation for BP86 and SVWN, respectively). From what we have learned in Section 8.1.1, this behavior is not totally unexpected and it can be rationalized after reinspection of the published data. Weak bonds (e. g., coordinative bonds like M-CO, M-CN or M-NO) are systematically underestimated in length in LDA structures. Stronger M-Cl and M-OH bonds agree very well in length with experiment whereas double bonds like M=O come out too long from LDA optimizations. Inclusion of gradient corrections affords a systematic elongation of bonds, which leads to an improved description of weak bonds and exaggerated bond distances for stronger bonds. Thus, akin to the deviation patterns noted earlier for bonds between main group atoms, the particular binding situation determines the accuracy resulting from the LDA and GGA treatment.

8.2 Vibrational Frequencies

Vibrational spectroscopy is of utmost importance in many areas of chemical research and the application of electronic structure methods for the calculation of harmonic frequencies has been of great value for the interpretation of complex experimental spectra. Numerous unusual molecules have been identified by comparison of computed and observed frequencies. Another standard use of harmonic frequencies in first principles computations is the derivation of thermochemical and kinetic data by statistical thermodynamics for which the frequencies are an important ingredient (see, e. g., Hehre et al. 1986). The theoretical evaluation of harmonic vibrational frequencies is efficiently done in modern programs by evaluation of analytic second derivatives of the total energy with respect to cartesian coordinates (see, e. g., Johnson and Frisch, 1994, for the corresponding DFT implementation and Stratman et al., 1997, for further developments). Alternatively, if the second derivatives are not available analytically, they are obtained by numerical differentiation of analytic first derivatives (i. e., by evaluating gradient differences obtained after finite displacements of atomic coordinates). In the past two decades, most of these calculations have been carried out at the Hartree-Fock level in combination with small to medium sized basis sets and the results have been systematically compared to experimental data. At this level, the calculated frequencies are commonly overestimated quite systematically by ca. 10 %, which can be traced back to the missing electron correlation, basis set deficiencies and the neglect of anharmonicity. As the observed deviations are in most cases uniform, a simple empirical scaling of the computed frequencies or diagonal force constants allows for a substantial improvement of the results in most instances. The scale factors are usually transferable within the same class of compounds. A force field determined in this way can predict vibrational frequencies to within 10 to 20 cm^{-1} accuracy for systems which are well-behaved in the HF approximation, such as for simple organic molecules. For systems demanding a higher degree of electron correlation, however, the Hartree-Fock method fails to give even qualitatively correct answers and errors are generally non-systematic. This category of species is found among transition-metal compounds, systems containing multiple bonds, and open-shell species. Faulty geometric parameters have been recognized as a key problem in this regard (for an in-depth discussion and potential remedies see Allen and Császár, 1993, and references cited therein). If we would restrict ourselves to conventional wave function based methods, the accurate prediction of vibrational frequencies for transition-metal complexes is only possible by means of sophisticated wave function based theory (large scale CI or coupled-cluster approaches) but the large number of electrons intrinsically renders its application prohibitively expensive, in particular for systems with low point-group symmetry. Consequently, the computationally efficient treatment of electron correlation and the availability of analytical first and second derivatives – which are not at hand for most highly correlated post-HF approaches – spurred on the interest in approximate density functionals in this important field of application. The systematic and accurate assessment of the performance of DFT for force fields of transition-metal complexes is somewhat hampered by the smaller number of experimentally well characterized systems, but the available studies are very encouraging.

8.2.1 Vibrational Frequencies of Main Group Compounds

From several early studies it has become obvious that harmonic frequencies computed at the LDA level are generally as close to experiment as those obtained from MP2 theory (Fan and Ziegler, 1992, Bérces and Ziegler, 1992, Murray et al., 1992, Handy et al., 1992). The particular performance of frequency evaluation employing the simple SVWN functional has been investigated in some detail by Andzelm, and Wimmer, 1992, for a set of small molecules consisting of C, N, O, H, and F atoms. C=C and C≡C bond stretching frequencies were found to be very close to experimentally derived harmonic frequencies, whereas C–C single bond stretches were overestimated by about 1 %. The C–H bond stretching frequencies were typically too low by about 2 %. These deviations are fully in line with the discussion of bond lengths above: for those bond types which are overestimated in length, frequencies which are too low result, and vice versa. In addition, a general underestimation of low frequency bending and torsional modes has been noted, which directly contributes to a general underestimation of zero-point vibrational energies. LDA underestimates these energies to about the same extent as MP2 overestimates them.

Given the sometimes erratic behavior of optimized molecular structures found in the JGP set, a remarkably similar overall performance for SVWN and BLYP with respect to predicted harmonic vibrational frequencies was noted by Johnson, Gill, and Pople, 1993. Both methods compared favorably with (unscaled) HF and MP2 results and the mean absolute deviations from available experimental harmonic frequencies were reported as SVWN: 75 cm^{-1}, BLYP: 73 cm^{-1}, HF: 168 cm^{-1}, MP2: 99 cm^{-1}, QCISD: 42 cm^{-1}. Hertwig and Koch, 1995, have systematically studied vibrational frequencies for main group homonuclear diatomics and found the BP86 functional to perform slightly better than BLYP, but both GGA schemes describe experimental data remarkably better than MP2 or CISD (mean absolute errors for the species Li_2 to Cl_2 employing the 6-311G(d) basis set are HF: 218 cm^{-1}, MP2: 138 cm^{-1}, CISD: 104 cm^{-1}, BP86: 39 cm^{-1}, BLYP: 48 cm^{-1}). The superior performance of GGA functionals over both the simple LDA or the hybrid functionals in the difficult case of the ozone system – an exacting testing ground for post-HF methods, on which many fail to even give qualitatively correct answers – has already been mentioned in Section 6.6. BLYP also gave improved frequencies compared to SVWN for the demanding FOOF and FONO systems despite larger deviations found in bond lengths (Amos, Murray, and Handy, 1993). Several other case studies documented the high quality of harmonic frequencies predicted at the GGA level (see, e. g., Florian and Johnson, 1994, Hutter, Lüthi, and Diederich, 1994, Florian and Johnson, 1995, Wheeless, Zhou, and Liu, 1995, and Michalska et al., 1996). But just as importantly, the evaluation of frequencies at the LDA level can lead to chemically meaningful answers for large scale cases where any higher level of theory cannot be applied within the limits of available computing resources (for an example see, Hill, Freeman, and Delley, 1999).

Zhou, Wheeless, and Liu, 1996, have systematically investigated the reliability of results from six different density functional methods employing the small 6-31G(d) basis set. These authors computed harmonic frequencies for typical organic molecules such as ethylene, formaldehyde, glyoxal, acrolein, and butadiene, as well as some deuterated derivatives for which experimental data is available. The results indicate that frequencies obtained by the three hybrid methods B3LYP, B3P86, and BHLYP somewhat overestimate observed fundamentals (151 data points, mean absolute errors: 51 cm^{-1}, 56 cm^{-1}, and 109 cm^{-1}, respectively) but that B3LYP and B3P86 results are closer to available experimental harmonic values (16 data points, errors: 21 cm^{-1}, 22 cm^{-1}, and 96 cm^{-1}). The SVWN, BLYP, and BP86 functionals, in turn, deviate more strongly (errors: 85 cm^{-1}, 73 cm^{-1}, and 79 cm^{-1}, respectively) but a better agreement is seen for the direct comparison with experimental fundamentals (errors: 28 cm^{-1}, 16 cm^{-1}, 19 cm^{-1}). At first sight the excellent accord between harmonics obtained with the LDA and the two GGA functionals and experimental fundamentals is surprising in light of the only mediocre quality of the corresponding structural predictions. The good performance of the GGA functionals in the determination of directly observed fundamental frequencies implies that the correlation between equilibrium structures and frequencies is not as strong for current DFT methods as it is for conventional wave function based methods. This can, however, be attributed to a cancellation of errors: the BLYP functional systematically overestimates bond lengths, which leads to a general underestimation of computed force constants. This bias is compensated for by the systematic overestimation somewhat inappropriately introduced by the direct comparison of computed harmonic and observed fundamental frequencies (fundamentals are usually smaller than their harmonic counterparts due to anharmonicity effects). This view seems justified by the fact that C–H stretching frequencies have been identified as the main source of anharmonicity and if these vibrations were excluded, BLYP and B3LYP gave very similar results. However, neither the computation of anharmonic vibrational energy levels nor the experimental determination of harmonic frequencies is routinely practical for polyatomic molecules. Hence, it seems that the B3LYP and B3P86 methods give results which are theoretically more sound but that GGA frequencies might be a pragmatic way for the interpretation of directly observed experimental frequencies without the need to account for anharmonicity effects. A similar conclusion has been put forward by Finley and Stephens, 1995, after studying the vibrational frequencies of a set of 11 small first-row compounds. Interestingly, these authors also noted a significant improvement of computed lower frequencies upon improvement of the basis set quality. For more complicated situations, however, the application of gradient-corrected functionals may lead to large errors in predicted frequencies and the overall performance of hybrid functionals is found superior – in these instances, the failures are caused by large errors in the GGA structures (Lee, Bauschlicher, and Jayatilka, 1997). A variety of case studies verifies the high quality of vibrational spectra computed with hybrid methods, which usually outperform LDA and gradient-corrected methods and give results close to sophisticated post-HF methods and experiments (see, e. g., Barone, Orlandini, and Adamo, 1994a, Martin, El-Yazal, and François, 1995b, Kozlowski, Rauhut, and Pulay, 1995, Martin, El-Yazal, and François, 1996, Kesyczynski, Goodman, and Kwiatkowski, 1997, Kwiatkowski and Leszczynski, 1997, Stepanian et al., 1998a, 1998b, 1999, Devlin and Stephens, 1999, or Bienati, Adamo, and Barone, 1999). Interestingly, the kinetic energy density dependent VSXC functional yields frequencies of similar quality than B3LYP without the use of Hartree-Fock exchange (Jaramillo and Scuseria, 1999).

After it has become clear that DFT methods are in general well-behaved in predicting vibrational frequencies and that deviations from experimental results occur quite systematically, some consideration has been given to the development of generic scaling factors. Florian and Johnson, 1994, were probably the first to show that the systematic deviations apparent in density functional calculations on formamide could benefit from scaling. Based on investigations on a set of 20 small molecules with over 300 experimental fundamentals, Rauhut and Pulay, 1995, published scaling factors for the BLYP/6-31G(d) and B3LYP/6-31G(d) levels of theory (0.990, and 0.963, respectively). The fortuitous agreement of BLYP frequencies with anharmonic experimental frequencies leads to a scaling factor close to unity and thus the mean deviation from experiment only changes from 30 cm^{-1} to 26 cm^{-1} after scaling. However, the mean deviation for B3LYP improves rather significantly from 74 cm^{-1} to 19 cm^{-1} upon scaling. These authors have also shown that the accuracy of computed frequencies can be further improved by application of their scaled quantum mechanical (SQM) force field procedure. In this model the molecular force field is expressed in a set of standardized valence internal coordinates which are then sorted into groups sharing common scaling factors. Such factors – eleven in number for this specific set of molecules – are derived for each group separately by a least squares fit procedure to experimental frequencies. This type of scaling yields reduced errors for both hybrid methods tested (B3LYP, and B3PW86) and, in particular, frequencies in the fingerprint region profited strongly. Application of this procedure has led, for example, to a reassignment of some fundamentals for the infrared spectrum of aniline. The disadvantage of this approach is of course the need to transform the computed force constants into a set of nonredundant valence coordinates, the manual derivation of which is a tedious affair and error-prone for larger molecules.

In a seminal paper, Scott and Radom, 1996, have investigated the performance of a variety of modern functionals (BLYP, BP86, B3LYP, B3P86, and B3PW91) in combination with the 6-31G(d) basis set for predicting vibrational frequencies and zero-point vibrational energies for a large suite of test molecules. By fitting computed data to a basis of 1066 individual experimental vibrations they developed a set of uniform scaling factors relating the computed harmonic frequencies to experimental fundamentals. Table 8-7 shows the resulting scaling factors along with some data allowing an assessment of the performance for each of the methods after scaling on the test set of 122 molecules.

All DFT methods perform better than HF and MP2; the hybrid techniques are even better than the costly QCISD. Both GGA functionals show scaling factors close to unity which means that they can be used without scaling, but they do not perform quite as well as

Table 8-7. Frequency scaling factors, rms deviation, proportion outside a 10 % error range and listings of problematic cases [cm^{-1}] for several methods employing the 6-31G(d) basis set. Taken from Scott and Radom, 1996.

Method	f[a]	RMS[b]	10 %[c]	Problematic Cases (Deviations larger than 100 cm^{-1})
HF	0.8953	50	10	233(O_2), 221(O_3, F_2), 180($^1A_1$-CH_2), 164(F_2O), 139(N_2), 120(N_2F_2), 115(HOF, NF_3), 103($NClF_2$)
MP2	0.9434	63	10	660(O_3), 304(NO_2), 277(N_2), 225(O_2), 150(HF), 149($^1A_1$-CH_2), 142(HC_2H), 136(HC_4H), 131(ClNS), 120(ClC_2H), 117(H_2), 115($^3B_2$-CH_2), 111(C_2N_2), 101(FCN)
QCISD	0.9537	37	6	202($^1A_1$-CH_2), 129(HF), 117(C_2H_2), 101(O_3)
BLYP	0.9945	45	10	224($^1A_1$-CH_2), 189(H_2), 165(HF), 116(OH), 113(SO_3), 112($^3B_1$-CH_2), 111(SO_2), 109(C_2H_2)
BP86	0.9914	41	6	229($^1A_1$-CH_2), 142(H_2), 115(HF), 114($^3B_2$-CH_2), 106(F_2)
B3LYP	0.9614	34	6	204($^1A_1$-CH_2), 132(HF), 125(F_2), 121(H_2), 110(O_3)
B3P86	0.9558	38	4	204($^1A_1$-CH_2), 146(F_2), 139(O_3)
B3PW91	0.9573	34	4	204($^1A_1$-CH_2), 140(F_2), 137(O_3)

[a] Scale factor; [b] root mean square error after scaling in cm^{-1}; [c] percentage of frequencies that fall outside by more than 10 % of the experimentally observed fundamentals.

methods including exact exchange. The hybrid functionals show scaling factors similar to MP2 and QCISD whereas their accuracy is superior to all traditional methods with respect to the criteria presented in Table 8-2. B3PW91 performs best, closely followed by the popular B3LYP functional.

In addition to the uniform scaling factors given above, these authors proposed separate scaling factors for zero-point vibrational energies, for low frequency vibrations, and for correcting thermal contributions to enthalpies and entropies. The evaluation shows significant differences between uniform and separately optimized scaling factors and the authors recommend that the latter should be used in order to improve the theoretical predictions. For the BLYP functional, additional tests with a larger 6-311G(df,p) basis set and different integration grid sizes indicated only minor influences on the deviations from experimental data. Uniform scaling parameters for vibrational frequencies and zero-point vibrational energies have been independently developed by others (Bauschlicher and Partridge, 1995, for B3LYP, Wong, 1996, for SVWN, BVWN, BLYP, B3LYP, and B3P86, Jaramillo and Scuseria, 1999, for VSXC and B3LYP), and it is pleasing to note that, where the investigations overlap, the results are in good mutual agreement.

8.2.2 Vibrational Frequencies of Transition-Metal Complexes

The pioneering comprehensive study of monometal carbonyls by Jonas and Thiel, 1995, was probably the first devoted to the computation of vibrational spectra for such compounds using density functional theory. These authors systematically compared experimental spectra and calculated vibrational data for several neutral tetra-, penta-, and hexacarbonyl complexes, $M(CO)_n$, n = 4-6). They used HF, MP2, BP86, and BLYP and found the BP86 functional in combination with a doubly polarized triple-zeta basis on the ligands and a relativistic ECP/triple-zeta basis on the metal to be very well suited for predictive purposes. Most of the DFT results reported were in very good agreement with experiment, whereas HF structures and frequencies were found to be completely inadequate. Results from MP2 calculations were satisfactory only for third and, to a lesser extent, second-row transition-metal complexes, whereas significant deviations (partly exceeding 100 cm^{-1}) occurred for first-row transition-metal complexes. In agreement with the aspects considered above, the BP86 functional tends to underestimate the C–O stretching modes rather uniformly by some 20 to 40 cm^{-1} while M–C stretching modes were accurate to within 20 cm^{-1}. A subsequent study on several metal- carbonyl hydrides corroborated the good quality of BP86 results, which were found to be superior to both, HF and MP2 approaches (Jonas and Thiel, 1996). The computed harmonic M–C stretching frequencies were again slightly lower than experimental values, whereas M–H bond stretches were overestimated by up to 50 cm^{-1} for third-row complexes and accurate to within 10 cm^{-1} for the heavier hydrides. These studies were subsequently extended to include charged carbonyl complexes, where a highly uniform small deviation between computed harmonics and observed fundamental frequencies for a given type of vibration was found (Jonas and Thiel, 1998). Computed raw frequencies could even be improved by the application of constant shift factors (28 cm^{-1} for C–O and –13 cm^{-1} for M–C stretching modes) to the computed harmonics.[29] The B3LYP functional gave slightly larger deviations than BP86 in related work on several isoelectronic hexacarbonyl complexes, if the computed harmonic frequencies were compared to directly observed fundamentals (Szilagyi and Frenking, 1997). In the study on $Fe(CO)_5$, $Fe_2(CO)_9$, and $Fe_3(CO)_{12}$ by Jang et al., 1998, the results could be compared to experiment, and an improved performance was found for B3LYP after scaling the harmonic frequencies by a factor of 0.97, a scale factor which is quite similar to the values for main group compounds described above. BP86 results were found to be in good agreement without scaling. A study on the actinide complexes UF_6, NpF_6, and PuF_6 employing HF, SVWN, BLYP and B3LYP reported harmonic frequencies which were compared to experimental gas-phase and matrix results (Hay and Martin, 1998). The computed and experimental harmonic frequencies differed on average by 50 cm^{-1}(HF), 19 cm^{-1} (SVWN), 33 cm^{-1} (BLYP), and 21 cm^{-1} (B3LYP) and a correlation with deviations for computed bond lengths was recognized.

In summary, systematic comparisons between computed and experimental frequencies have proven the applicability of DFT for the evaluation of vibrational spectra on coordinatively saturated closed-shell transition-metal complexes. A much lower computational cost in combination with a high overall accuracy compared to correlated traditional approaches renders the well tested BP86 functional a highly valuable research tool with excellent predictive power for this field of research. This assessment is corroborated by an ongoing series of studies by Andrews and coworkers, who use density functional theory as a standard tool to augment and interpret experimentally measured infrared data, even on small coordinatively unsaturated, open-shell species (which have as yet received not as much attention as the standard closed-shell complexes). A detailed discussion of the huge amount of published data is well beyond the scope of this book and the reader is rather referred to representative pieces of work, which provide pointers to further publications. These studies include the infrared spectra of $Cu(CO)^{+}_{1-4}$, $Cu(CO)_{1-3}$, and $Cu(CO)^{-}_{1-3}$ (Zhou and Andrews, 1999a, see literature cited for $Ni(CO)^{-}_{x}$ and $Co(CO)^{-}_{x}$) as well as related ruthenium and osmium complexes (Zhou and Andrews, 1999b, see literature cited for iron complexes), nitrides and N_2 complexes of rhodium (Citra and Andrews, 1999, and references cited therein for related work on Fe, Co, and Ni), nitric oxide complexes (Kushto and Andrews, 1999), oxygen complexes of chromium, molybdenum, and tungsten (Zhou and Andrews, 1999c), scandium (Bauschlicher, 1999) and yttrium and lanthanum (Andrews et al., 1999).

[29] As discussed for the two complexes $[V(CO)_6]^-$ and $Cr(CO)_6$ these constant shifts obtained for BP86 are just about the order of magnitude of anharmonic effects present in such compounds; see Spears, 1997.

9 Relative Energies and Thermochemistry

Information about the energetic properties of molecules is at the heart of every quantum chemical investigation. In this chapter we will take a close look at the performance of current approximate density functionals when it comes to the determination of such important energetic properties as atomization energies, ionization energies (IE) and electron affinities (EA). In particular the first quantity is of great value since it provides an idea of the ballpark figure we are talking about when we are interested in computing the thermochemistry of chemical reactions. The energies needed to remove (IE) or add an electron (EA) to an atom or molecule are obviously of significant interest in their own right. For example, the interpretation of photoelectron spectroscopy experiments is greatly facilitated if accurate ionization energies are available. Furthermore, since the computation of all these different quantities poses severe and often different demands on the method chosen, the accuracy with which a functional delivers such energies is a probe for its versatility. Somewhat counterintuitively at first sight, the reliable calculation of energetic information for atoms, in particular for transition-metals, is especially difficult and not without ambiguity. We therefore devote a complete section to this problem. Finally, we take up the discussion on excitation energies from Section 5.3.7 and give an overview of the current state of the art in the determination of electronically excited states and the corresponding transition energies using density functional theory. We conclude with a few remarks about the ability of approximate density functional theory to reproduce singlet-triplet gaps in carbenes and related species.

9.1 Atomization Energies

Chemical reactions involve the cleavage and formation of bonds within molecules. The calculation and prediction of thermochemical data has long been a vivid field for quantum chemistry (see, e. g., Irikura and Frurip, 1998). For example, whenever experimental data is not available and empirical estimates fail, some type of quantum chemistry usually becomes involved to obtain the missing information, but 'computational thermochemistry' is of great relevance also in many other areas. In practice, there is always the need to reach some compromise between accuracy and computational effort. Hartree-Fock theory provides an exact treatment of exchange and scales well with the molecular size, but it suffers from severe deficiencies in describing chemical bonding due to the neglect of correlation energy contributions. Except for isodesmic (or related) reactions (see, e. g., Hehre et al., 1986) it cannot be used for thermochemical predictions. The introduction of dynamic and static correlation effects by means of post-HF wave function based methods improves the situation to a desired accuracy, but severely suffers from the notoriously unfavorable scaling with molecular size. Notwithstanding the rather positive appraisal of the local density approximation for the evaluation of molecular geometries and vibrational frequencies in the preceding chapter, binding energies obtained with this method are generally very inac-

Table 9-1. Deviations between computed atomization energies and experiment for the JGP test set employing the 6-31G(d) basis set [kcal/mol]. Taken from Johnson, Gill and Pople, 1993.

	HF	MP2	QCISD	SVWN	SLYP	BVWN	BLYP
mean abs. dev.	86	22	29	36 (40)[a]	38	4 (4)[a]	6
mean dev.	–86	–22	–29	36 (40)[a]	38	0 (4)[a]	1

[a] Basis set free results taken from Becke, 1992.

curate. The literature is full of results giving testimony to the inability of this method to deliver even qualitatively meaningful answers for problems related to chemical energetics. This problem was recognized more than two decades ago and was actually the main stimulus for the development of gradient corrections and, later, hybrid functionals.

Computed atomization energies, i. e., of the (hypothetical) reactions in which a molecule is broken up into its constituent ground state atoms, are often very error-prone since their evaluation necessitates the breaking of each bond in a molecule. While we usually have a closed-shell molecule on the left hand side of the reaction, the ground state atoms defining the right hand side are open-shell with varying multiplicity. Hence large differential correlation effects are typical for these reactions. Therefore, the calculation of atomization energies is a stringent test for any computational strategy and the deviations from experiments seen in such studies can probably be considered as upper bounds. As an instructive example, Table 9-1 shows some error statistics for atomization energies obtained with different methods in combination with the rather small 6-31G(d) basis set for the JGP set of 32 first and second-row species.

It is apparent that the Hartree-Fock level is characterized by an enormous average deviation from experiment, but standard post-HF methods for including correlation effects such as MP2 and QCISD also err to an extent that renders their results completely useless for this kind of thermochemistry. We should not, however, be overly disturbed by these errors since the use of small basis sets such as 6-31G(d) is a definite 'no-no' for correlated wave function based quantum chemical methods if problems like atomization energies are to be addressed. It suffices to point out the general trend that these methods systematically underestimate the atomization energies due to an incomplete recovery of correlation effects, a reliable assessment of which requires sufficiently large and flexibly polarized basis sets.[30] The errors are systematic because correlation effects are always stronger in molecular systems than in their fragments (most correlation effects are roughly proportional to the number of spin-paired electrons). An insufficient recovery of electron correlation leads to a lack of stabilization of the parent molecular systems, which causes the underbinding tendency. The LDA, in turn, shows a notorious overbinding for every single molecule in the test set except Li_2, and deviations from experimental atomization energies as large as 220 % occur

[30] The G2 extrapolation scheme – which is a prescription to extrapolate the quality of QCISD(T)/ 6-311++G(3df,2p) calculations – actually reaches chemical accuracy for the G2 test set of species, but only with an empirical correction, depending on the number of electron pairs in a molecule in order to better account for these effects.

Table 9-2. Signed deviations [kcal/mol] between computed atomization energies (employing a 6-31G(d) basis) and experiment. Taken from Johnson, Gill, and Pople, 1993.

Molecule	SVWN	SLYP	BVWN	BLYP	Molecule	SVWN	SLYP	BVWN	BLYP
CH	7	6	3	0	F_2	47	55	11	18
$CH_2(^3B_1)$	21	19	2	–2	O_2	57	68	10	19
CH_3	31	28	3	–2	N_2	32	39	–1	6
CH_4	44	40	4	–3	CO	37	46	–5	1
C_2H_2	50	55	–9	–6	CN	37	45	3	9
C_2H_4	70	71	–3	–4	NO	44	53	5	13
C_2H_6	86	85	–1	–6	CO_2	82	99	–3	11

for particular species like F_2 (see below). Although better than the HF approximation, the LDA is certainly not a useful thermochemical tool with a mean absolute deviation of 36 kcal/mol and it has largely been abandoned for this kind of studies. In an attempt to amend the situation by inclusion of a gradient-corrected correlation functional (SLYP) one ends up with even larger errors, which is irksome at first sight. A spectacular improvement, though, results from the application of gradient corrections to exchange (but not to correlation): the BVWN functional affords a reduced mean absolute deviation of 4 kcal/mol. Inspection of the last two columns in Table 9-1 again shows that inclusion of gradient corrections to correlation (to yield the BLYP functional) slightly decreases the overall accuracy. The comparison with basis set free results for the same set of species taken from a paper published by Becke, 1992a, reveals no significant influence of the basis set size on the overall performance of the SVWN or the BVWN functional.[31] Becke has shown in subsequent work that neither the choice of the PW91 parameterization of the uniform electron gas (instead of VWN) nor the addition of PW91 gradient corrections for correlation significantly changes the overall picture (Becke, 1992b). Also in this latter study, a slightly larger mean absolute deviation occurred for the G2 set of molecules upon inclusion of gradient corrections to correlation as compared to exchange-only corrections.

A closer look at the original data published by Johnson, Gill, and Pople, 1993, reveals that use of the gradient-corrected LYP correlation functional instead of VWN increases atomization energies for non-hydride species quite significantly, while those containing hydrogen atoms are reduced. This rather systematic trend is portrayed in Table 9-2 for a few example cases.

Use of the LYP correlation functional apparently reduces the overbinding proportional to the number of hydrogen atoms by about 1 kcal/mol per H from CH to CH_4. On the contrary, for the nonhydride diatomics listed, the atomization energies increase by 4–5 kcal/mol per atom upon substitution of VWN by the LYP functional. For species like C_2H_2,

[31] The largest deviations occur for different species, however. The BVWN results published by Becke were obtained in a post-LDA manner at LDA optimized geometries as opposed to the data published by Johnson, Gill, and Pople, 1993, which were computed selfconsistently at geometries corresponding to the respective level of theory. Hence, it is difficult to unambiguously pin down the origin of these differences.

C_2H_4, and C_2H_6, in which both types of bonds are broken, the particular stoichiometry determines which functional performs better. The very same trends are observed for other correlation functionals (see Becke, 1992b). Such behavior is clearly unsatisfying and an indication that an only incomplete error cancellation is operative. Consequently, quite large errors can occur for unfortunate cases (ranging from −12 kcal/mol for H_2O to +19 kcal/mol in the case of NO for the BLYP functional). While BP86 compares favorably with other GGA functionals for the evaluation of molecular structures and vibrational frequencies, it is defeated by BLYP when it comes to determining atomization energies. In combination with the 6-31G(d) basis set, the BLYP functional yields atomization energies for the G2 set with a mean absolute deviation 2 kcal/mol smaller than BP86 and a significantly smaller maximum deviation (Bauschlicher, 1995a). Using a large 6-311+G(3df,2p) basis, the situation becomes even worse for BP86: the mean unsigned error is almost 5 kcal/mol smaller at the BLYP level and the difference in maximum errors is nearly 10 kcal/mol in favor of the latter.[32] The newer PW91 correlation functional performs equally well or marginally better than LYP when used in combination with the Becke exchange term. Both show very similar mean absolute deviations for over 100 atomization energies evaluated in the comprehensive study of Scheiner, Baker, and Andzelm, 1997 (cf. Table 9-3).

Although observed maximum deviations are sometimes substantial and evidently far from chemical accuracy, the fact that the overall errors for gradient-corrected functionals are more than five times smaller than those of the traditional wave function based methods shows nonetheless their general suitability for thermochemical studies at a modest level of computational effort. The BLYP functional in combination with small basis sets would lend itself particularly well to thermochemical studies on extended systems, where the computational demands of larger bases or correlated post-HF methods are prohibitive. However, for medium sized hydrocarbons a large underestimation of atomization energies has been observed for the BLYP functional. The BP86 functional overestimated the same atomization energies twice as much as BLYP underestimated them, so here BPW91 seems to be the GGA functional of choice with only very moderate deviations (Curtiss et al., 1997). All in all, the introduction of gradient corrections to exchange is the key to improved thermochemical data, whereas the influence of corrections to the correlation term is rather modest (inclusion of the latter, however, has important consequences for the accuracy of computed ionization energies, see below). The importance of gradient corrections to exchange is not completely unexpected considering the fact that exchange, which is the dominant component of the exchange-correlation energy in Kohn-Sham theory, is largely in error in the LDA. Gunnarson and Jones, 1985, have argued convincingly that much of the exchange error inherent to the LDA stems from an improper incorporation of angular characteristics and nodal structures of the Kohn-Sham orbitals. Differential exchange effects are generally overestimated and errors in atomization energies are largest for molecular systems, in which substantial changes in the orbital nodal structure occur – e. g., upon bond formation from atoms resulting in occupied antibonding orbitals as in O_2. Similar argu-

[32] Remarkably, the BLYP functional approaches or sometimes surpasses the accuracy of hybrid functionals if small basis sets are used (see Table 9-5).

ments can be used to rationalize, for example, the huge overbinding of F_2 by the LDA (borrowing from an illuminating paper by Ernzerhof, Perdew, and Burke, 1997). The experimental dissociation energy of the fluorine molecule into two ground state atoms (F_2 $(^1\Sigma_g^+) \rightarrow 2F\ (^2P_u)$) amounts to 37 kcal/mol. Hartree-Fock and the local density approximation both give ridiculously wrong results: with the 6-31G(d) basis set the former yields –34 kcal/mol (the fluorine molecule is unbound at the HF level with respect to the two constituting F atoms) while the latter gives a binding energy of 84 kcal/mol, overshooting the experimental D_0 by more than a factor of two. Since F_2 formally contains a single bond, these large errors are somewhat irritating. It is, however, well known that the lone pairs are strongly interacting in molecular F_2, which in fact is the origin of the problems for both methods. HF fails because it does not account for correlation energy and the electron pairs repel each other too strongly. If correlation is included through second or fourth order Møller-Plesset perturbation theory, very realistic binding energies of 35 and 30 kcal/mol, respectively, result. To LDA, on the contrary, overlapping lone pairs are nothing but a higher electron density. This method overestimates the exchange stabilization brought about by these orbital interactions, which leads to an overestimation of the molecular binding energy. Incorporation of explicit density gradient dependencies into the exchange terms repairs the shortcomings to a large extent (BVWN gives 47 kcal/mol), but still, GGA functionals do not quite reach chemical accuracy. It is clear that the gradient-corrected functionals represent a major improvement over the local density approximation and deliver average errors which sometimes get close to our target accuracy of 2 kcal/mol. Likewise, the data indicate that even though the various GGA functionals differ significantly in their mathematical appearance, they all perform quite similarly. However, we also note that the maximum deviations are significant and that we are still a long way from a density functional approach that is able to generally provide chemical accuracy.

In view of this situation, Becke has taken the next logical step and improved the GGA performance by admixture of exact exchange as we have already discussed to some extent in Chapter 6. His first approach, the half-and-half scheme (Becke, 1993a), did not include gradient corrections and was not much of an improvement over GGA functionals in terms of thermochemical accuracy. However, a subsequently suggested parameterized version, which included gradient corrections to exchange and correlation, gave impressively reduced mean errors for atomization energies of the G2 set (Becke 1993b). This was the forerunner of the now widely used B3LYP hybrid (Stephens et al., 1994) which today is the most popular density functional and is implemented into most major computer codes. For details on these functionals, the reader should leaf back to Section 6.6.

Bauschlicher and Partridge, 1995, tested the B3LYP functional in combination with different basis sets on the G2 set of molecules. In combination with the 6-31G(d) basis, it yields an accuracy comparable to that of the pure BLYP functional (5.2 kcal/mol average error). This only mediocre performance improves significantly to a mean absolute error of only 2.2 kcal/mol, if the larger 6-311+G(3df,2p) basis is used, regardless of whether the geometries were obtained at this level or with the much more affordable 6-31G(d) basis set. Also, use of the aug-cc-pVTZ basis gave an improved average error of 2.3 kcal/mol, almost reaching the desired goal of chemical accuracy. However, in this study the atomization

Table 9-3. Mean absolute deviations (MAD) from experiment [kcal/mol] for 44 atomization energies and number of results that deviate by less than 5, between 5 and 10, and over 10 kcal/mol from experiment. Taken from Martell, Goddard, and Eriksson, 1997.

Basis set	MAD	< 5	[5..10]	> 10	MAD	< 5	[5..10]	> 10
	BLYP				B3LYP			
6-31G(d,p)	7.6	21	12	11	5.6	26	15	3
6-311G(d,p)	6.8	23	11	10	6.8	23	14	7
cc-pVDZ	7.3	20	13	11	8.5	13	20	11
cc-pVTZ	7.2	20	12	12	3.1	36	5	3
	BP86				B3P86			
6-31G(d,p)	12.7	9	9	26	6.7	21	12	11
6-311G(d,p)	10.4	15	7	22	6.2	24	12	8
cc-pVDZ	9.9	13	10	21	5.4	25	15	4
cc-pVTZ	14.2	5	8	31	6.9	21	11	12
	BPW91				B3PW91			
6-31G(d,p)	7.0	22	11	11	5.6	24	17	3
6-311G(d,p)	6.1	24	11	9	6.9	21	17	6
cc-pVDZ	7.2	20	12	12	8.4	13	19	11
cc-pVTZ	7.8	19	11	14	3.8	36	6	2

energy for SO_2 proved to be very problematic and the results for this molecule were found to be extremely sensitive to the basis set quality. Martell, Goddard and Eriksson, 1997, studied the performance of the three commonly used GGA functionals, namely BP86, BLYP, and BPW91 together with the corresponding hybrid functionals B3P86, B3LYP, and B3PW91 on a set of 44 small first and second-row molecules. They used four different basis sets (6-31G(d,p), 6-311G(d,p), cc-pVDZ, and cc-pVTZ) in order to assess the reliability of predicted atomization energies with these moderately sized bases.[33] Comparing the results for different basis sets and functionals it is important to firstly realize that no particular method provides results superior to all others. The authors noted a general underestimation of atomization energies for the two hybrid functionals B3LYP and B3PW91, which contrasts with the overestimation found for the three pure GGA protocols and, to a smaller extent, for the B3P86 hybrid. Looking at the mean absolute errors for all method/basis set combinations documented in Table 9-3, the B3LYP functional gives the highest accuracy, closely followed by B3PW91. Larger errors occur for the B3P86 hybrid, which only marginally surpasses the BPW91 and BLYP gradient-corrected functionals in terms of accuracy. The worst performance is found for the BP86 functional, for which deviations from experiment below 5 kcal/mol are the exception. The error pattern compiled in Table

[33] Molecular geometries, which were not reported in this study, have been obtained using the 6-31G(d,p) and cc-pVDZ basis sets. Thus, the 6-311G(d,p) and cc-pVTZ results refer to single point energy calculations only.

Table 9-4. Average shifts in atomization energies upon basis set enlargement [kcal/mol]. Based on data taken from Martell, Goddard, and Eriksson, 1997.

	BLYP	B3LYP	BP86	B3P86	BPW91	B3PW91
6-31G(d,p) → 6-311G(d,p)	–2.6	–2.6	–2.6	–2.0	–2.5	–2.4
cc-pVDZ → cc-pVTZ	4.8	5.5	5.2	5.9	5.0	5.9

9-3 reveals that the B3LYP and B3PW91 hybrid functionals give quite reliable atomization energies in combination with the cc-pVTZ basis set. The most problematic systems in this study were SO_2, ClO_2 and CCl, which also pose severe difficulties for traditional quantum chemical methods.

A disturbing trend in the basis set dependence is seen from the mean unsigned errors listed in Table 9-3. Reduced errors occur for the pure GGA functionals and B3P86 when improving the quality of the Pople-type basis from 6-31G(d,p) to 6-311G(d,p) – exactly what one would expect for any well-behaved quantum chemical method. Yet the opposite trend emerges for the correlation consistent basis sets when going from the polarized double-zeta cc-pVDZ to the polarized triple-zeta cc-pVTZ basis set. Better results are obtained with the smaller basis set. Only the B3LYP and B3PW91 results show the expected behavior, these two functionals deliver the smallest mean absolute errors of all methods if combined with the large cc-pVTZ basis set. These baffling findings can be rationalized by inspection of the data summarized in Table 9-4.

In spite of large differences in mean errors obtained with the various methods tested, substitution of the 6-31G(d,p) basis by the larger 6-311G(d,p) set yields a systematic shift to reduced atomization energies on average by 2.5 kcal/mol, irrespective of the method used. Conversely, use of the larger cc-pVTZ instead of the cc-pVDZ basis set brings about an increase in atomization energies, on average the order of +5.5 kcal/mol. Apparently, the 6-311G(d,p) basis yields a better description of isolated atoms, whereas the improved correlation consistent basis stabilizes molecular systems quite significantly with respect to the atoms. These shifts explain why GGA functionals, which usually overestimate atomization energies, perform better with the larger Pople-type 6-311G(d,p) and the smaller cc-pVDZ correlation consistent basis set. The clearly visible reason is once again error cancellation.

Another study by Scheiner, Baker, and Andzelm, 1997, has quite extensively addressed the evaluation of atomization energies with respect to different functionals and basis sets of varying quality. In this work it has been observed that polarized double- and triple-zeta basis sets which have been explicitly optimized at the LDA level (denoted $DZVP_{LDA}$ and $TZVP_{LDA}$) are better suited for LDA and GGA calculations than for hybrid functionals. Use of a standard TZV2P basis for BLYP and BPW91 led to ca. 1 kcal/mol larger errors compared to results obtained with the $TZVP_{LDA}$ basis set, whereas just the opposite has been observed for the B3LYP hybrid functional. Furthermore, the latter functional still showed a remarkable drop by 2.5 kcal/mol in mean absolute deviations if a large uncontracted aug-cc-pVTZ basis set was used – in contrast, only marginal improvements (below 0.3 kcal/mol) were seen for BLYP and BPW91. Apparently, the basis set requirements for converged results are higher for hybrid methods than for GGA functionals. Furthermore, it seems that hybrid methods get along much better with standard basis sets taken from the wave function ab initio world than GGA functionals do.

Redfern, Blaudeau, and Curtiss, 1997, have tested the BLYP, B3LYP, BPW91, and B3PW91 functionals with respect to the accuracy of atomization energies computed for a set of 19 molecules containing third-row, non-transition-metal elements. They used a rather large 6-311+G(3df,2p) basis set for single point energy calculations on top of MP2/6-31G(d) geometries (which might not give the highest accuracy possible for the density functional treatment). Among the functionals tested, the B3PW91 approach afforded the lowest average unsigned and maximum error (2.1 and 5.7 kcal/mol, respectively) which compared nicely to much more costly G2 calculations (1.2 and 5.2 kcal/mol). The B3LYP functional gave slightly worse energetics (3.3 and 6.2 kcal/mol), whereas the pure GGA functionals led to larger mean errors and substantial maximum deviations (BLYP: 5.3 and 24.2 kcal/mol; BPW91: 4.5 and 24.7 kcal/mol).

Before we end this discussion let us take up the thread from Section 6-9 and present some results pertaining to the large number of new functionals that have emerged in the past few years. The literature contains a variety of attempts to further improve the accuracy of density functional methods, which essentially follow two distinct lines, namely (a) the fitting of adjustable functional parameters to some kind of experimental data and (b) the fulfillment of theoretically derived and physically meaningful conditions. Examples, for instance, belonging to the first category are the CAM(A) and CAM(B) exchange functionals reported by Laming, Termath, and Handy, 1993. Two different fitting procedures to experimental data have been applied and for a G2 subset the resulting functionals showed a non-uniform performance. If combined with the LYP correlation functional, the CAM(A)-LYP functional gave significantly improved geometries as compared to BLYP but at the same time much worse mean errors for atomization energies. In contrast, the CAM(B)-LYP functional showed the reverse behavior with worse geometric parameters than CAM(A)-LYP but smaller errors for atomization energies (CAM(A)-LYP: 21.9 kcal/mol, CAM(B)-LYP: 6.5 kcal/mol, BLYP: 9.5 kcal/mol). Such a situation is of course by no means satisfying. On the one hand, the results show that there is definitely room for improvement within the particular formulations of GGA functionals by means of fitting procedures and CAM(B)-LYP might indeed appear as a useful improvement over the BLYP functional. On the other hand, one could expect from theoretical reasoning that an improved description of molecular binding also leads naturally to a better performance in structure prediction, which obviously is not the case. It is hence apparent that reparameterization does not necessarily improve the fundamental physics but rather exerts some shift on the outcome of error compensation effects. Therefore, one can rightly argue that this might lead to a non-systematic performance for molecules or properties not included in the fitting set, which would render the quality of such corrections very difficult to judge a priori. Obviously, only extensive testing can identify the particular advantages and potential caveats of such functionals.

The idea of modifying existing functionals by fitting particular terms to accurate experimental data has been tempting to others, too. Stewart and Gill, 1995, have reparameterized a simplified LYP correlation functional formalism, and tested its performance for atomization energies in combination with Becke's 1988 (B) exchange term. This new simple functional, referred to as Becke-Wigner (BW), has been tested with the 6-31G(d) and 6-311+G(3df,2p) basis sets to evaluate the atomization energies of the G2 set. The resulting mean absolute errors were slightly in favor of BLYP (6-31G(d): 5.5 and 4.5, 6-311+G(3df,2p): 4.9 and 4.7 kcal/mol, for BW and BLYP, respectively). However, the computed data for individual molecules were quite different. If combined with the rather modestly sized 6-31+G(d) basis set, the empirical density functional EDF1 by Adamson, Gill, and Pople, 1998, yields reasonably good results with mean absolute errors for atomization energies of 3.2 kcal/mol compared to 4.4 and 5.9 kcal/mol for BLYP and B3LYP in this basis, respectively. No improvement has been found upon exact exchange admixture. It will be interesting to see a further assessment of the accuracy of this functional in future applications. Neumann and Handy, 1995, implemented the Becke-Roussel exchange functional (BR), which was fitted to model the shape of the Hartree-Fock exchange hole in a two-term Taylor expansion without any reference to the electron gas model. This functional (which depends on the density, its gradient and Laplacian as well as on the kinetic energy density) was tested in combination with the P86 correlation functional, employing a polarized triple-zeta TZ2P basis set, and atomization energies were obtained for a set of 27 diatomic first and second-row molecules with a mean error of 5.5 kcal/mol (compared to 6.0 kcal/mol for BP86 applied to the same set). When used in a refitted three parameter hybrid framework with some 20 % exact exchange admixture (BR3P86), the resulting atomization energies were improved with respect to the pure GGA, but slightly larger overall and maximum deviations occurred as compared with B3P86 results. As briefly mentioned in Chapter 6, Becke, 1997, proposed another functional, which contains exact exchange admixture and was derived from a systematic fitting to thermochemical data of the G2 set by adjusting 10 parameters, i. e., B97. Atomization energies were obtained with a mean absolute deviation of 1.8 kcal/mol and an absolute maximum error of 5.5 kcal/mol. This accuracy closely approaches that of the G2 extrapolation scheme, for which 1.2 and 5.1 kcal/mol result for mean and maximum absolute deviations, respectively. Hence, this fitting scheme created a functional which definitely surpasses the quality of hybrid functionals like B3LYP or B3PW91. However, this new method is awaiting the extensive testing which the latter two hybrids have seen in the recent past, and it remains to be verified whether its performance endures as favorably as found for the test set for which it has been parameterized. B97-1, the self-consistent reparameterization of the B97 functional leads to a slightly larger mean absolute error (2.2 kcal/mol) for atomization energies of a G2 subset of species if combined with a TZ2P basis set (Hamprecht et al., 1998). Rather impressive thermochemical results as documented in Table 9-5 have been reported by van Voorhis and Scuseria, 1998, for their VSXC functional. This functional was the outcome of a fitting procedure adjusting no less than 21 different parameters. In addition, it goes beyond the standard GGA functionals by depending also on the non-interacting kinetic energy density. Further developments along similar lines have been reported in the recent literature and are discussed in Chapter 6.

Neumann and Handy, 1996, implemented the recent B95 correlation functional and tested it on a G2 subset. This functional was originally proposed by Becke and obeys some physically motivated minimal requirements, thus representing our first example of path (b) among the lines of modern functional development. In Becke's original work (Becke, 1996a) the new method has been applied in a post-LDA manner (i. e., the functional was applied on KS orbitals and the corresponding density obtained from a converged SVWN calculation – as usually done by this author) whereas Neumann and Handy tested a fully selfconsistent implementation. Benchmarked for atomization energies of the G2 set (or a subset thereof by the latter authors), the new functional led to large overbinding for non-hydrogen species, exaggerating the above mentioned observations for the inclusion of gradient corrections to correlation even more. Species containing hydrogen atoms on the other hand, were described with a better accuracy. The error of the pure GGA (i. e., non-hybrid) form (BB95, 8.8 kcal/mol) was found inferior even to BP86 (6.0 kcal/mol) by Neumann and Handy, confirming the disappointing results of Becke's original investigation. Better results, however, were obtained for a fitted single parameter hybrid implementation, blending B with exact exchange (dubbed B1B95). Becke found this functional superior to his three parameter fit (mean unsigned error 2.0 vs. 2.4 kcal/mol). Correspondingly, a smaller error was also reported for B1B95 (2.6 kcal/mol) than for B3P86 (3.2 kcal/mol) in the study by Neumann and Handy. Another functional, which does not rely on empirical adjustments, is the PBE functional introduced by Perdew, Burke, and Ernzerhof, 1996. When applied to atomization energies for the G2 set of species (in combination with the rather flexible 6-311+G(3df,2p) basis set on top of MP2 optimized structures), this functional performs much better than SVWN, but does not reach the accuracy of BLYP (mean absolute errors are SVWN: 36.4, PBE: 8.6, BLYP: 4.7 kcal/mol). Admixing of 25 % exact exchange does ameliorate the performance but, as reported by Ernzerhof and Scuseria, 1999a, the resulting PBE1PBE hybrid functional still falls short of the B3LYP hybrid. For the extended G2 set (148 molecules) the errors amount to 4.8 and 3.1 kcal/mol for the PBE1PBE and B3LYP functionals, respectively. For the original G2 set consisting of 55 molecules the errors are reduced to 3.5 kcal/mol (PBE1PBE) and 2.4 kcal/mol (B3LYP). In a related study, Adamo and Barone, 1999, report an absolute mean error for the PBE1PBE functional combined with the 6-311++G(3df,3pd) basis set on the 32 molecule JGP subset of the G2 database of 2.6 kcal/mol. Finally we mention the recent contribution by Rabuck and Scuseria, 1999, who applied the B3LYP, VSXC, PBE1PBE and PBE functionals to the determination of enthalpies of formation for molecules which are not included in the typical density functional training sets and which are known to be problematic. As expected, the average errors are significantly larger than for the G2 or related references. The best performance is achieved with the VSXC functional (8.8 kcal/mol absolute average deviation, 24.3 kcal/mol maximum deviation) followed by B3LYP (10.6 and –37.2 kcal/mol, respectively) and PBE1PBE (11.5 and 39.5 kcal/mol, respectively). The pure GGA functional PBE works significantly worse and shows an average error of 38.2 kcal/mol and a maximum error exceeding 100 kcal/mol, rendering it fairly useless in this context. Remember that VSXC achieves its good performance without containing any Hartree-Fock exchange. Rather, it differs from regular GGA functionals by the fact that it depends not only on the density gradient but also on the kinetic energy density.

In conclusion, according to the results of a variety of systematic studies, the introduction of hybrid functionals can be considered a successful step towards the ultimate goal of chemical accuracy for the evaluation of atomization energies of main group species, provided that basis sets of polarized triple-zeta quality or better are used. Although functionals like B3LYP and B3PW91 do not quite reach the target accuracy of below 2 kcal/mol, they provide a pragmatic means to predict atomization energies with a pleasing accuracy. As such, they constitute highly efficient alternatives to far more demanding post-HF methods, which show comparable mean and maximum deviations in a variety of cases. These two hybrid methods in particular are available in several major computer codes and provide a significant improvement over results for pure GGA functionals with only few exceptions. From the data given above, the rough hierarchy of functionals given in Section 6-9, i. e., LDA < GGA < hybrid functionals, is confirmed. If we go one step further and also ask the question, which of the widely available functionals is to be recommended with respect to the quality of the computational prediction of atomization energies, we arrive at the following conclusion (with the expected accuracy increasing from left to right):

$$\text{SVWN} \ll \text{BP86} < \text{BLYP} \approx \text{BPW91} < \text{B3P86} < \text{B3LYP} \approx \text{B3PW91}.$$

In terms of basis sets, there is compelling evidence that sets smaller than polarized triple-zeta quality significantly reduce the accuracy that can be obtained with modern hybrid functionals and cannot be recommended if quantitative energetic results are the prime target.

In Table 9-5 we summarize the performance of various functionals discussed above as collected from many sources, which highlights the conclusions of the above discussion.

Table 9-5. Compilation of mean absolute and maximum absolute deviations (in parentheses) for atomization energies [kcal/mol] of small main group molecules from different sources.

32 1st row species, 6-31G(d) basis set, Johnson, Gill, and Pople, 1993					
HF	85.9		SVWN	35.7	
MP2	22.4		BVWN	4.4	
QCISD	28.8		BLYP	5.6	
33 1st and 2nd row diatomic molecules, TZ2P basis, Laming, Termath, and Handy, 1993					
LDA	43.6	(18.3)	CAM(A)-LYP	21.9	(14.5)
BLYP	9.5	(9.3)	CAM(B)-LYP	6.5	(12.0)
G2 set, B1 = 6-31G(d), B2 = 6-311+G(3df,2p), Bauschlicher, 1995					
HF/B1	80.5	(184.3)	HF/B2	74.5	(170.0)
MP2/B1	16.0	(40.3)	MP2/B2	7.3	(25.4)
BLYP/B1	5.3	(18.8)	BLYP/B2	5.0	(15.8)
BP86/B1	7.2	(24.0)	BP86/B2	10.3	(25.4)
B3LYP/B1	5.2	(31.5)	BP86/B2	2.2	(8.4)
B3P86/B1	5.9	(22.6)	BP86/B2	7.8	(22.7)

Table 9-5, continued.

G2 set, Bauschlicher and Partridge, 1995						
	6-31G(d)		aug-cc-pVTZ		6-311+G(3df,2p)	
B3LYP	5.2	(31.5)	2.6	(18.2)	2.2	(8.1)
44 1st and 2nd row species, Martell, Goddard, and Eriksson, 1997						
	BLYP	BP86	BPW91	B3LYP	B3P86	B3PW91
6-31G(d,p)	7.6	12.7	7.0	5.6	6.7	5.6
6-311G(d,p)	6.8	10.4	6.1	6.8	6.2	6.9
cc-pVDZ	7.3	9.9	7.2	8.5	5.4	8.4
cc-pVTZ	7.2	14.2	7.8	3.1	6.9	3.8
G2 set, Stewart and Gill, 1995						
BW/6-31G(d)	5.5	(25.5)	BLYP/6-31G(d)		4.5	(16.3)
BW/6-311+G(3df,2p)	4.9	(15.3)	BLYP/6-311+G(3df,2p)		4.7	(15.3)
27 1st and 2nd row diatomic molecules, TZ2P basis, Neumann and Handy, 1995, 1996						
BP86	6.0	(18.3)	B3P86		3.2	(9.3)
BRP86	5.5	(14.5)	BR3P86		3.1	(12.0)
B1B95	8.8	(24.1)	B1B95		2.6	(9.4)
19 species incl. 3rd row atoms, 6-311+G(3df,2p) basis, Redfern, Blaudeau, and Curtiss, 1997						
G2	1.2	(5.2)				
BLYP	5.3	(24.2)	B3LYP		3.3	(6.2)
BPW91	4.5	(24.7)	B3PW91		2.1	(5.7)
108 1st and 2nd row species, Scheiner, Baker, and Andzelm, 1997						
	$DZVP_{LDA}$	$TZVP_{LDA}$	DZP	6-31G(d)	TZ2P	UCC
SVWN	47.6	52.1	47.0	52.2	50.1	56.4
BLYP	7.4	6.9	10.2	7.0	7.4	7.1
BPW91	6.4	6.2	9.7	7.4	7.3	7.0
B3LYP	8.8	7.8	10.1	6.8	6.5	4.1
G2 set, 6-31+G(d) basis set, Adamson, Gill, and Pople, 1998						
EDF1	3.2	(15.3)	BLYP		4.4	(16.3)
B3LYP	5.9	(35.9)				
G2 (first 2 cols.) and ext. G2 set, 6-311+G(3df,2p) basis, Ernzerhof and Scuseria, 1999a						
SVWN	36.4	(84)			83.7	(216)
PBE	8.6	(26)			17.1	(52)
BLYP	4.7	(15)			7.1	(28)
B3LYP	2.4	(10)			3.1	(20)
PBE1PBE	3.5	(10)			4.8	(24)
VSXC	2.5	(10)			2.7	(8)

9.2 Atomic Energies

Now that we have considered the performance of DFT for the prediction of atomization energies for main group species in some detail, we focus a little closer on the right hand side of such reactions: the atoms. Atoms are not only the smallest subunits in chemistry, they are – seemingly paradoxically – also among the most difficult systems to describe for approximate density functional theory. The only exceptions which are completely unproblematic include closed-shell atoms, such as the ground states of the rare gases, but these are not the subject of this section. Although the errors seen in the preceding section stem at least to some extent from problems describing main group atoms in general and atomic states in particular, the main thrust of the following discussion will be geared towards transition-metal atoms and ions. The multifaceted chemistry of transition-metals is largely determined by their variable occupation of nd, (n+1)s, and (n+1)p valence orbitals which poses severe challenges for a theoretical treatment. From a physical point of view, subtle differential correlation and exchange effects of the various nd^p $(n+1)s^q$ occupations are realized in the different atomic states. A method which aims at an accurate description of atomic states must be capable of providing a balanced and unbiased representation of the many possible electronic situations. This is anything but an easy task for any current quantum chemical strategy, including sophisticated approaches such as configuration interaction or coupled cluster methods. One should therefore not be surprised that problems arise with Kohn-Sham methods based on approximate density functionals.

A second major reason why atoms are so difficult, in particular for methods rooted in approximate density functional theory, has been touched upon already in Chapter 5. In Kohn-Sham theory, by definition, we do not have access to the correct many-electron wave function and its symmetry requirements. It is therefore not clear how to deal with atomic terms whose wave functions are eigenfunctions of the $\hat{L}^2$, $\hat{S}^2$ and related operators (see Section 5.3.7). The usual way out is to select a single-determinantal non-interacting Kohn-Sham reference system for defining the values of the conserved quantum numbers. This leads to ambiguities and possible inconsistencies in the description of these states. Consider the high spherical symmetry of atomic species and recall from Chapter 5 the inability of current approximate functionals to properly account for the related degeneracy effects which occur in open-shell situations. A comprehensive computational study to investigate these problems has been reported by Baerends, Branchadell, and Sodupe 1997. They demonstrate that such difficulties already show up for main group atoms with partially occupied p-orbitals. Let us consider the example of a ^2P ground state for a boron atom with its [He] $(2s)^2$ $(2p)^1$ electron configuration. The energy differences between the spherical density with 1/3 of an electron in each of the three real p-orbitals and a non-spherical density derived from occupying the real p_z orbital[34] amounts to some non-negligible 0.2 eV if the BP86 protocol is used. If instead one of the complex p-orbitals is occupied (e. g., $p_x + ip_y$ which corresponds to $M_L = 1$), the resulting energy is roughly in-between the previous two

[34] This corresponds to the component of ^2P with $M_L = 0$. Occupying the real p_x or p_y orbital results in the same energy – but note that the real p_x and p_y orbitals are no longer eigenfunctions of the $\hat{L}^2$ operator.

results. The exact energy density functional would be invariant over the set of charge densities belonging to a degenerate ground state and would produce precisely the same energy for all these possible representations. However, none of the currently known approximate density functionals is able to meet this requirement. This type of problem is particularly prominent when it comes to describing transition-metal atoms or ions with partially filled d-shells. Here, the energy even depends on which of the real d-orbitals are selected for specifying the configuration, because the shape of the d_{z^2} orbital differs from the shape of the other four orbitals of this set. We already illustrated this problem in Chapter 5 for the d^1 configuration of the ten-fold degenerate ^2D ground state of the scandium dication. The important take home message here is that due to the deficiencies of the currently used density functionals, *there is no unambiguous reference energy for atoms in approximate density functional theory.*

What is obviously needed is a generally accepted recipe for how atomic states should be dealt with in approximate density functional theory and, indeed, a few empirical rules have been established in the past. Most importantly, due to the many ways atomic energies can be obtained, one should always explicitly specify how the calculations were performed to ensure reproducibility. From a technical point of view (after considerable discussions in the past among physicists) there is now a general consensus that open-shell atomic calculations should employ spin polarized densities, i. e. densities where not necessarily $\rho(\vec{r}) = \frac{1}{2}\rho_\alpha(\vec{r}) + \frac{1}{2}\rho_\beta(\vec{r})$. Note that this does not mean that the unrestricted Kohn-Sham formalism has to be used, restricted open-shell variants are in principle equally eligible (but recall the discussion in Section 5.3.5). All this condition states is that the α-density does not have to be equal to the β-density. Actually, this rule must seem trivial and enforces itself almost automatically, because spin unpolarized open-shell calculations are – if possible at all – usually difficult to perform with most current program packages. By the same token, densities that are allowed to be non-spherical should be used. The corresponding atomic orbitals should be occupied either by one or two electrons rather than distributing the N electrons equally over the n degenerate orbitals.[35] This rule is also – in principle – automatically obeyed in most calculations done with standard codes, since it represents the default way of performing such calculations. However, even in cases where a calculation is started with an integer occupation of d-orbitals, unphysical mixings between d-orbitals and the (n+1)s-orbital may occur, depending on the symmetry imposed. The solutions resulting from such scrambling of the original occupation pattern cannot usually be related to any physical state anymore, as outlined further below. When spin-polarized, non-spherical densities are allowed, the additional variational freedom leads to solutions which are usually significantly lower in energy than if these restrictions are enforced.

[35] The physical reasoning for why these densities were frequently employed in the earlier days of density functional theory was that in this way the degeneracy of the partially filled d-orbitals could be retained. A technical reason why these densities still have to be employed in some recent investigations is that calculations with integral orbital occupations simply do not converge in the self consistent field procedure (see, e. g., Blanchet, Duarte, and Salahub, 1997). Such densities correspond to a representation of a particular state ^{2S+1}L with $M_s = S$ and a spherical averaging over M_L.

Following the discussion in Section 5.3.7, among the possible occupations of the real atomic orbitals connected to a formal configuration one should select only those which correspond to a single-determinantal representation of the desired atomic term. As stated very early on by Ziegler, Rauk, and Baerends, 1977, only such single-determinantal states are valid for a description using the current Kohn-Sham technology and the corresponding approximate density functionals. This also means that there are states of atoms or molecules which cannot be computed directly owing to their inherent multi-determinantal character. In these cases alternative routes such as the sum method introduced in Section 5.3.7 must be used. For the transition-metal atoms and their positive ions, only the lowest multiplet components of a particular configuration are needed for the ground and first excited states and no such complications occur. Rather, all these terms can be represented by single determinants. However, with the exception of $m_l = 0$ all the $2l+1$ components of atomic orbitals for a given l are complex and therefore not directly accessible for a representation using real orbitals. Instead, linear combinations of the complex determinants sharing the same $\pm m_l$ need to be formed such that they lead to real representations. The adequate occupations of real d-orbitals, which correspond to single Slater determinants with the correct angular momentum and spin symmetry, have been summarized in an appendix of a frequently quoted paper by Hay, 1977, and are reproduced in Table 9-6.

It has been noted in Hay's paper that the occupations for the d^1, d^4, d^6, and d^9 states are in principle arbitrary. This does not strictly hold true for density functional applications because of the above-mentioned dependence of the energy on the shape of the occupied orbitals. The density generated from occupying the d_{z^2} differs from the one obtained from placing the electron in, e. g., the d_{xy} orbital. Feeding an approximate density functional with these two unequal densities may lead to non-identical energies (cf. Figure 5-2). In most practical applications, however, the errors introduced in this way should be much smaller than those caused by other limitations of the functional or basis set employed.

Table 9-6. Open-shell d-configurations after Hay, 1977.

Configuration	Ground State Term	Occupation
d^1	^2D	$(d_{z^2})^1$
d^2	^3F	$(d_{z^2})^1(d_{x^2-y^2})^1$
d^3	^4F	$(d_{xy})^1(d_{xz})^1(d_{yz})^1$
d^4	^5D	$(d_{x^2-y^2})^1(d_{xy})^1(d_{xz})^1(d_{yz})^1$
d^5	^6S	$(d_{x^2-y^2})^1(d_{z^2})^1(d_{xy})^1(d_{xz})^1(d_{yz})^1$
d^6	^5D	$(d_{z^2})^2(d_{x^2-y^2})^1(d_{xy})^1(d_{xz})^1(d_{yz})^1$
d^7	^4F	$(d_{z^2})^2(d_{x^2-y^2})^2(d_{xy})^1(d_{xz})^1(d_{yz})^1$
d^8	^3F	$(d_{z^2})^1(d_{x^2-y^2})^1(d_{xy})^2(d_{xz})^2(d_{yz})^2$
d^9	^2D	$(d_{z^2})^1(d_{x^2-y^2})^2(d_{xy})^2(d_{xz})^2(d_{yz})^2$

The generation of a clean occupation pattern for atomic d-orbitals is greatly facilitated if point-group symmetries can be exploited which prevent unphysical mixing. This has been discussed in some detail in a paper on density functional atomic calculations (using a restricted open-shell rather than the usual unrestricted strategy) of 3d transition-metal atoms by Russo, Martin and Hay, 1994. What one needs are symmetry constraints which allow the occupation of the nd- and (n+1)s-orbitals in such a way that any unwanted mixing leading to unclear, non-integer occupations is prevented by symmetry. Let us elucidate this by using the ^{5}F term of a d^3s^1 occupation, as realized in the first excited state of the V^+ ion as an example. If we are fortunate enough to have a program which supports non-Abelian point-group symmetries at our disposal, octahedral symmetry with inversion (point-group O_h) should be employed. The three singly occupied d-orbitals are – following Table 9-6 – chosen as the d_{xy}, d_{xz}, and d_{yz} orbitals which span the three-dimensional irreducible representation t_{2g}. Hence, these three orbitals are equivalent and each is occupied by one electron of the same spin. Since the Pauli principle excludes the occupation of each spin orbital by more than one electron and because the other d-orbitals belong to a different irreducible representation of O_h (namely e_g) the electrons deposited in the t_{2g} orbitals cannot move to any another d-orbital. The s-orbital belongs to the a_{1g} irreducible representation. Also since none of the d-orbitals transforms as a_{1g} and in particular the occupied d_{xy}, d_{xz}, and d_{yz} orbitals are separated from the s-orbital by symmetry, also s/d mixing is impossible. As a consequence the originally defined orbital assignment of electrons is frozen by symmetry and will not change during the course of the calculation. If instead we have only Abelian point-groups at our disposal and select C_{2v} symmetry (the typical choice) it is still possible to unambiguously assign the three d-electrons to the d_{xy}, d_{xz}, and d_{yz} orbitals. In this case, these three d-orbitals are all in an irreducible representation of their own, namely a_2, b_1 and b_2, respectively. Thus, no mixing between these or other d-orbitals can occur. However, the formally singly occupied 4s orbital belongs to the totally symmetric a_1 representation which happens to be the same irreducible representation in which the d_{z^2} and $d_{x^2-y^2}$ orbitals can be found. Hence, these three orbitals can now mix and there is no guarantee that the electron which was initially assigned to the 4s orbital stays there and does not partially move into the d_{z^2} or the $d_{x^2-y^2}$ orbital. Unfortunately, the use of high point-group symmetries and other symmetry arguments is not a panacea. First, many programs, such as for example the current versions of *Gaussian* simply do not support non-Abelian point-groups and O_h is thus out of reach. Second, there is also a number of cases where symmetry alone, even if symmetries such as O_h were accessible, does not help. A case in point is provided by the $3d^14s^1$ occupation of the ^{3}D ground term of the scandium cation. No point-group is available which could both exclude mixing of the s- and d-orbitals and still prevent unpaired d-electrons from moving between degenerate symmetry-equivalent d-orbitals. Neither the high O_h symmetry nor the Abelian C_{2v} point-group (nor any other point-group) assures that this electron distribution will persist throughout the calculation. Let us be specific: in O_h we can prevent the 4s-orbital from mixing with any of the 3d-orbitals because they are separated by symmetry. But because the 3d-orbitals transform as two- (e_g) or three-dimensional (t_{2g}) irreducible representations of O_h, the smearing of the one d-electron between symmetry equivalent orbitals cannot be ruled out. For example, if we initially place the electron in the d_{z^2} orbital, there is no symmetry related reason why this electron should not partially occupy the other d-orbital transforming as e_g, i. e., $d_{x^2-y^2}$. Fractional occupations of these orbitals would be the result. On the other hand, while a unique assignment of the 3d electron is possible in C_{2v} because there are three equivalent one-dimensional irreducible representations of C_{2v} in which this single electron could be uniquely accommodated, the 4s/3d mixing cannot be prevented in that point-group for the same reason as explained above. In such cases, one usually prefers C_{2v}, but the user has to observe carefully the progress of the calculation in order to ensure that the initially adjusted occupation pattern persists until convergence of the self consistent field, at least as much as possible. These guidelines can be summarized in the following rule of thumb for the calculation of atomic energies: accept only solutions with the correct occupation pattern and with integer d-orbital occupations (see Ricca and Bauschlicher, 1995a). Finally, we should note that by lifting all restrictions with respect to symmetry and occupation pattern, a further lowering of the atomic energy can sometimes be achieved, see Baerends, Branchadell, and Sodupe 1997. The physical meaning of such solutions, which are often characterized by fractional occupations of d- and s-orbitals is, however, questionable.

In addition to these more technical problems, there are other inconsistencies which restrict the quality of atomic energies. The most prominent issue in this context is the bias of current approximate density functionals towards preferentially occupying d- rather than s-orbitals (for detailed discussions see Gunnarsson and Jones, 1985, Ziegler and Li, 1994, Holthausen et al., 1995). This is just the opposite of what is generally seen for traditional wave function based approaches, which favor states with fewer d-electrons. We can rationalize the contrasting shortcomings of these two different schools of theory by the following considerations exemplified for the atomic excitation energies of 3d transition-metal cations. The separations between the d^n versus $d^{n-1}s^1$ states are determined by Coulomb repulsion, exchange energy, and electron correlation. As a consequence of the different sizes of d- and s-orbitals[36] the average interelectronic distances are larger for $d^{n-1}s^1$ configurations, resulting in a reduced Coulomb repulsion. From the traditional wave function based viewpoint, this means that the correlation problem is less severe for this occupation pattern. Hartree-Fock theory, which treats Coulomb and exchange interactions exactly but neglects correlation of electrons with antiparallel spin, indeed shows a pronounced bias towards $d^{n-1}s^1$ configurations. This is particularly so for the late transition-metals for which this change of the electronic configuration is accompanied by a spin flip and HF favors the high-spin states. The very same situation causes a different problem if density functional theory is invoked. Due to the more compact electron arrangements in the valence d-shells, the average exchange stabilization per d-electron pair, K_{dd}, is stronger than that between an s- and a d-electron, K_{sd}. As to the current situation, it has been shown that the LDA overestimates the absolute exchange terms K_{dd}, K_{sd}, as well as those between s- and d-electrons and core-orbitals. As discussed in a key paper by Gunnarsson and Jones, 1985, this exaggeration of exchange stabilizations is more pronounced for d^n than for $d^{n-1}s^1$ situations

---

[36] d-orbitals are generally more compact than s-orbitals of the following main quantum number by a factor of about 1.5 to 3.4. The $\langle r_s \rangle / \langle r_d \rangle$ ratio is particularly large for the 3d-series of elements, see Bauschlicher, 1998.

which leads to the bias mentioned above. Fortunately, these effects are compensated for in part by account of correlation energies, which operate in the opposite direction, so that even the LDA usually performs better than the HF approach. Absolute exchange energies, however, are much larger in transition-metal atoms than absolute correlation energies. For example, a SVWN treatment of the Cu^+ ion yields E_x in the order of $-60\ E_h$, whereas E_c is about $-3\ E_h$, which illustrates the predominant influence of shortcomings in the exchange part of functionals on observed errors.

We can already anticipate that gradient corrections to exchange or an admixture of HF exchange in the hybrid functional scheme can help to improve agreement with experiment. This has indeed been observed, but a tendency to artificially stabilize d^n over $d^{n-1}s^1$ configurations remains. The inclusion of atomic state splittings of transition-metals into the databases used to construct new functionals appears to be a logical consequence, but as far as we are aware, this has not been done as of yet. As a consequence, even using the most advanced functionals, there are a number of cases where either wrong atomic ground states are predicted by density functional theory, or where s/d mixing results in intermediate occupations, which cannot be connected to physically reasonable configurations. For example, the atomic ground state term of cobalt is ^4F, dominated by a $3d^74s^2$ occupation. The second ^4F term with a $3d^84s^1$ occupation is – after correcting for the differential relativistic effects – 0.17 eV higher in energy. Density functional calculations of various flavor give the reverse result with the $3d^84s^1$ occupation being lower in energy than the $3d^74s^2$ one.

Finally we need to mention that heavier elements exhibit strong relativistic effects, which also have a significant influence on the physical properties of the d-block elements. While common wisdom has it that relativistic contributions have even qualitative consequences for bonding or electronic state splittings of 5d transition-metals,[37] their influence is not that dramatic, but still non-negligible, for the 4d elements. The relativistic contributions to 3d elements are often ignored, but yet, for the later elements of this row, they are larger than inexperienced newcomers to the field might anticipate. In particular for Cu there is quite a deal of influence on the stability of atomic states. The experimental value for the $3d^{10}(^1S) \rightarrow 3d^94s^1(^3D)$ state splitting in the copper cation is 2.81 eV. However, a sophisticated post-HF treatment of these two states gives a value of 3.11 eV.[38] Such a deviation is outside the expected error range for a high quality level of theory, and indeed, the inclusion of scalar relativistic effects by means of perturbation theory (mass-velocity and Darwin corrections) gives a value of 2.85 eV in good agreement with experiment. Hence, the $3d^94s^1$ configuration is stabilized by 0.26 eV with respect to $3d^{10}(^1S)$ due to relativistic contributions. This is certainly a non-negligible source of error for non-relativistic calculations if triplet copper cations are involved. Hence, while non-relativistic calculations on 3d elements are very well acceptable for many purposes, one has to take into account relativity if a higher accuracy is aimed at, at least for the later elements in the row. The most convenient way to

[37] In fact, for 5d transition-metals relativistic contributions, and in particular spin-orbit coupling, can be of the same order of magnitude as chemical bonding.

[38] At the highly correlated CASSCF-AQCC level of conventional ab initio theory using a very large [7s6p4d3f2g] ANO basis set.

include scalar relativity to some extent into the calculations is the use of relativistic effective core potentials or explicit one-component schemes, but the neglect of spin-orbit coupling effects can be problematic for the heavier elements.

A typical set of results for excitation energies of 3d transition-metal atoms and their cations based on the results of Koch and Hertwig, 1998, is summarized in Tables 9-7 and 9-8. The SVWN, BLYP and B3LYP functionals were combined with a sufficiently flexible contracted GTO basis set of 8s7p4d2f quality to expand the Kohn-Sham orbitals. In these calculations the rules outlined above were followed. For clarity the occupations used in the respective point-group symmetries are also included in the Tables. Since the computationally predicted excitation energies have been obtained in a completely non-relativistic scheme, they are compared to experimental energies that have been empirically corrected for differential scalar relativistic effects taken from Raghavachari and Trucks, 1989a and 1989b. These corrections are based on approximate calculations and neglect the influence of relativity on the correlation energy (and vice versa), but have been shown to provide a good approximation. The preference of current approximate density functionals for d-rich occupations – in particular with the LDA – can easily be inferred from these results. Use of the gradient-corrected BLYP protocol or the hybrid B3LYP approach does lead to a significant reduction of the deviations but also to a less systematic behavior. Nevertheless, as compared to other strategies the overall accuracy of these results, in particular for the B3LYP functional, is satisfying. For example, the mean absolute deviations for the neutral excitation energies (i. e., d^ns^2 versus $d^{n+1}s^1$ configurations) as determined with the Hartree-Fock model, second order Møller-Plesset perturbation theory or the QCISD(T) model amount to 0.86, 0.55 eV, and 0.14 eV, respectively (Raghavachari and Trucks, 1989a). The corresponding density functional results are 0.74 eV for the SVWN, 0.55 eV for the BLYP functional, and 0.33 eV if the B3LYP scheme is used. For the cations (i. e., d^ns^1 versus d^{n+1}) the density functional approaches perform significantly better with mean unsigned errors of 0.32, 0.18 and only 0.16 eV for LDA, BLYP, and B3LYP, respectively. For comparison, Raghavachari and Trucks, 1989b report mean deviations of the HF, MP2 and QCISD(T) schemes of 1.32, 0.35, and 0.23 eV, respectively. It should be noted that due to the above mentioned limitations in the symmetry treatment for some of the difficult cases, the converged wave functions were not absolutely clean. A case in point is provided by the Fe ^5D ground term. The corresponding d^6s^2 occupation cannot be treated in a unique, point-group symmetry determined way. As indicated in Table 9-7, one has to resort to C_{2v} symmetry and face the problem of s/d mixing. In fact, in all final Kohn-Sham wave functions some of the s-electron population had moved to the symmetry related d-orbitals, creating a slightly fuzzy picture and adding to the inherent uncertainty of the results. Overall, this uncertainty in atomic Kohn-Sham calculations is at least in the order of some tenths of an eV, but larger deviations may also occur using even state-of-the-art functionals.

Table 9-7. Excitation Energies [eV] of 3d-Transition Metals.

Atom	n	$d^n s^2$	Point group	Occupation	$d^{n+1}s^1$	Point group	Occupation	LSD	BLYP	B3LYP	Exp.[a]
Sc	1	^2D	C_{2v}	$a_1{}^1a_1{}^2$	^4F	O_h	$e_g{}^2a_{1g}{}^1$	0.56	0.67	0.79	1.33
Ti	2	^3F	O_h	$e_g{}^2a_{1g}{}^2$	^5F	O_h	$t_{2g}{}^3a_{1g}{}^1$	–0.26	0.06	0.24	0.69
V	3	^4F	O_h	$t_{2g}{}^3a_{1g}{}^2$	^6D	C_{2v}	$a_1{}^1b_1{}^1b_2{}^1a_2{}^1a_1{}^1$	–0.94	–0.39	–0.19	0.11
Cr	4	^5D	C_{2v}	$a_1{}^1b_1{}^1b_2{}^1a_2{}^1a_1{}^2$	^7S	O_h	$e_g{}^2t_{2g}{}^3a_{1g}{}^1$	–1.15	–0.81	–0.79	–1.17
Mn	5	^6S	O_h	$e_g{}^2t_{2g}{}^3a_{1g}{}^2$	^6D	C_{2v}	$a_1{}^2a_1{}^1b_1{}^1b_2{}^1a_2{}^1a_1{}^1$	1.00	1.00	1.42	1.97
Fe	6	^5D	C_{2v}	$a_1{}^2a_1{}^1b_1{}^1b_2{}^1a_2{}^1a_1{}^2$	^5F	O_h	$e_g{}^4t_{2g}{}^3a_{1g}{}^1$	0.22	0.21	0.28	0.65
Co	7	^4F	O_h	$e_g{}^4t_{2g}{}^3a_{1g}{}^2$	^4F	O_h	$e_g{}^2t_{2g}{}^6a_{1g}{}^1$	–0.72	–0.47	–0.10	0.17
Ni	8	^3F	O_h	$e_g{}^2t_{2g}{}^6a_{1g}{}^2$	^3D	C_{2v}	$a_1{}^2a_1{}^1b_1{}^2b_2{}^2a_2{}^2a_1{}^1$	–1.26	–0.80	–0.41	–0.33
Cu	9	^2D	C_{2v}	$a_1{}^2a_1{}^1b_1{}^2b_2{}^2a_2{}^2a_1{}^2$	^2S	O_h	$e_g{}^4t_{2g}{}^6a_{1g}{}^1$	–2.57	–2.14	–1.82	–1.85

[a] Corrected for relativistic effects, taken from Raghavachari and Trucks, 1989a.

Table 9-8. Excitation Energies [eV] of 3d-Transition Metal Cations.

Cation	n	$d^{n+1}s^1$	Point group	Occupation	d^{n+2}	Point group	Occupation	LSD	BLYP	B3LYP	Exp.[a]
Sc^+	0	^3D	C_{2v}	$a_1{}^1a_1{}^1$	^3F	O_h	$e_g{}^2$	0.36	0.33	0.30	0.44
Ti^+	1	^4F	O_h	$e_g{}^2a_{1g}{}^1$	^4F	O_h	$t_{2g}{}^3$	–0.39	–0.26	–0.21	–0.07
V^+	2	^5F	O_h	$t_{2g}{}^3a_{1g}{}^1$	^5D	C_{2v}	$a_1{}^1b_1{}^1b_2{}^1a_2{}^1$	–0.94	–0.65	–0.55	–0.55
Cr^+	3	^6D	C_{2v}	$a_1{}^1b_1{}^1b_2{}^1a_2{}^1a_1{}^1$	^6S	O_h	$e_g{}^2t_{2g}{}^3$	–1.99	–1.70	–1.67	–1.78
Mn^+	4	^7S	O_h	$e_g{}^2t_{2g}{}^3a_{1g}{}^1$	^5D	C_{2v}	$a_1{}^2a_1{}^1b_1{}^1b_2{}^1a_2{}^1$	1.21	0.95	1.22	1.54
Fe^+	5	^6D	C_{2v}	$a_1{}^2a_1{}^1b_1{}^1b_2{}^1a_2{}^1a_1{}^1$	^4F	O_h	$e_g{}^4t_{2g}{}^3$	–0.28	–0.41	–0.18	–0.07
Co^+	6	^5F	O_h	$e_g{}^4t_{2g}{}^3a_{1g}{}^1$	^3F	O_h	$e_g{}^2t_{2g}{}^6$	–1.15	–0.96	–0.73	–0.80
Ni^+	7	^4F	O_h	$e_g{}^2t_{2g}{}^6a_{1g}{}^1$	^2D	C_{2v}	$a_1{}^2a_1{}^1b_1{}^2b_2{}^2a_2{}^2$	–0.80	–1.48	–1.22	–1.48
Cu^+	8	^3D	C_{2v}	$a_1{}^2a_1{}^1b_1{}^2b_2{}^2a_2{}^2a_1{}^1$	^1S	O_h	$e_g{}^4t_{2g}{}^6$	–3.63	–3.17	–2.94	–3.26

[a] Corrected for relativistic effects, taken from Raghavachari and Trucks, 1989b.

9.3 Bond Strengths in Transition-Metal Complexes

The knowledge of accurate thermochemical data for individual metal-ligand bond strengths is of major importance for the rational design of catalytic processes. Such data is unfortunately rather limited for larger species in solution, but the last decade has provided an ever growing list of accurate results for small, unsaturated complexes in the gas-phase by elaborate experimental techniques (see, e. g., Freiser, 1996). Today, however, computational thermochemistry is an important tool to fill those gaps, which cannot be covered by experimental means alone. Of course, a theoretical method, which shall be used to predict unknown binding energies of metal-ligand bonds, first has to master a suite of known benchmarks typical for the problem under investigation, in order to prove that it can handle the delicate and subtle electronic problems in yet unexplored species. Experimental gas-phase data are very well suited for this purpose due to the exclusion of severe complications, like solvent effects. On the traditional side, theories like Hartree-Fock, the MP perturbation methods and also the complete-active-space SCF (CASSCF) approach have been shown not to recover enough electron correlation and do not yield accurate geometries or binding energies for first-row transition-metal complexes. The better players in the post-HF field are the modified coupled-pair functional (MCPF) and especially the more rigorous CCSD(T) approach. Also second-order perturbation theory based on CASSCF references (CASPT2) has been shown to yield fairly accurate results even in complicated cases (Roos et al., 1996). Quantitative accuracy, however, is hard to obtain even at the highest levels of multireference treatment. Often, even such extraordinarily expensive methods need empirical corrections to afford 'best estimates', in order to approach an accuracy better than 5 kcal/mol. For an overview of this area of research the reader is referred to reviews by Bauschlicher, 1995b, and Siegbahn, 1996a, and references cited therein. Lately, as a more efficient alternative, parameterized extrapolation schemes have been developed, which – much like G2 and related theories – empirically scale the correlation energy (obtained from a moderate level of theory within a limited basis set) to estimate the results of a more complete, but prohibitively expensive treatment. A prominent example is the PCI-80 approach developed by Blomberg and Siegbahn (for a review see Blomberg, 1998).

In Section 9.1 we have outlined the difficulties of modern density functionals to reach chemical accuracy for atomization energies of main group species, and we concluded that this goal is eventually not completely out of reach for the hybrid functional approach. In this respect, we have looked at atomization processes as the worst case for a density functional treatment, because the changes from molecular to atomic densities are most severe. We can therefore expect a better description of the thermochemistry connected to the rupture of individual bonds leading to larger (non-atomic) fragments due to a more complete cancellation of errors. Density functional theory appears to be in a reasonably good shape to further conquer this terrain. In this section, we concentrate on the strength of individual bonds in small, coordinatively unsaturated, and mostly open-shell transition-metal complexes, a worst case scenario in this field. Let us consider a specific example. If we want to compute a metal-ligand bond strength according to M–X $\rightarrow$ M + X, then of course the isolated metal atom M becomes an integral part of the computational problem, including the large errors discussed in the previous section. From what we know about the way density functional theory deals with transition-metal atoms, we should prepare to be less ambitious with our expectations as to the accuracy of computations if these elements are involved. Furthermore, a remnant of inorganic chemistry text book knowledge suffices to confirm that this feeling is adequate not only for the right hand side of the above-mentioned computational problem: the binding situation of transition-metals within complexes in one way or another involves a redistribution of electrons between the valence nd, (n+1)s, and (n+1)p orbitals on the metal, too. We can view this as a mixture of contributions from ground, excited, and ionic states of the isolated atom.[39] It can be energetically favorable, for example, to (formally) promote the central metal atom M in a complex M-X to an excited state in order to improve electrostatic interactions with the ligand. As an example, we mention a particularly prominent binding mechanism which involves s/d hybridization. Imagine a doubly occupied $3d_{z^2}$ orbital and an empty 4s orbital as the (hypothetical) ground state of a metal atom in a complex M-X. As illustrated in Figure 9-1, the formation of a $(4s + 3d_{z^2})$ hybrid-orbital maximizes the orbital overlap between metal and ligand X. For an efficient charge transfer from the ligand to the metal this orbital has to be empty in order to accept the corresponding electron density. The electron pair of the metal occupies the $(4s - 3d_{z^2})$ hybrid-orbital instead. Thereby, the charge density is efficiently polarized out of the metal ligand bond axis into an equatorial orbital lobe, which reduces the electron-electron repulsion between the formerly doubly occupied $3d_{z^2}$ and ligand orbitals. If both hybrid orbitals are being occupied in such a way, the energetic costs for this formal promotion of electron density into the 4s orbital pays off very well in many cases, and an overall energy gain can be obtained by such a bonding mechanism.

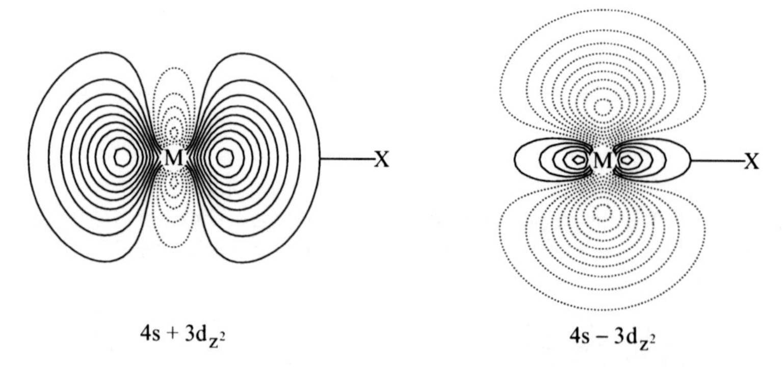

Figure 9-1. Schematic plot of the $(4s + 3d_{z^2})$ and $(4s - 3d_{z^2})$ orbitals of a third-row transition-metal involved in a M-X bond.

[39] For a guide to a detailed understanding of the subject, the reader is encouraged to study the somewhat dated but highly enlightening contribution of Carter and Goddard, 1988.

Table 9-9. Experimental binding energies D_0 [kcal/mol] for the cationic hydrides of the 3d elements and deviations from these data obtained at various levels of theory.

Method	ScH^+	TiH^+	VH^+	CrH^+	MnH^+	FeH^+	CoH^+	NiH^+	CuH^+	MAD[e]
Exp.[a]	56	53	47	32	48	49	46	39	21	±2
SVWN[b]	+9	+7	+9	+16	+11	+10	+14	+14	+21	12
BP86[b]	+5	+4	+6	+8	+6	+10	+10	+11	+14	8
B3LYP[c]	+3	+4	−2	+4	+1	+9	+5	+3	+4	4
B3LYP[d]	+4	+3	+7	+5	+1	+9	+3	+7	—	5
MCPF[d]	−6	−5	−5	−4	−10	−4	−9	−6	—	6
PCI-80[d]	0	+2	+2	−1	−3	+3	+1	+3	—	2

[a] Armentrout and Kickel, 1996; [b] Ziegler and Li, 1994; [c] TZVP+f basis, corrected for zero point vibrational energies (Barone, Adamo, and Mele, 1996); [d] Single point energy calculations using the 6-311+G(2d,2p) basis set. Geometries have been obtained using the smaller Hay and Wadt ECP/basis combinations (Blomberg, Siegbahn, and Svensson, 1996); [e] Mean absolute deviation from experiment; for experiment, the average of the cited experimental error bars is given.

During the last years, more and more researchers have applied density functional theory to small transition-metal complexes and benchmarked the results against either high level wave function based methods or experimental data. A particular set of systems for which reasonably accurate benchmark data are available are the cationic M^+-X complexes, where X is H, CH_3 or CH_2. Let us start our discussion with the cationic hydrides of the 3d transition-metals.

From the data in Table 9-9 we see clearly the recurring picture of LDA-typical overbinding, which is ameliorated to some extent by inclusion of gradient corrections. Still, the BP86 results show a large overbinding in particular for the later elements of the series. Ziegler and Li, 1994, attributed this latter finding to a general overestimation of exchange interactions between the electron on the hydrogen atom and the same-spin electrons present on the metal atom, which leads to a particularly large error for copper. The B3LYP hybrid functional performs much better and halves the errors present in the GGA results. We note a reasonable agreement between the corresponding results of Barone, Adamo, and Mele, 1996, on one hand and Blomberg, Siegbahn, and Svensson, 1996, on the other except for the vanadium hydride, where the results of both groups differ. The reason for this discrepancy might be attributed to the differences in basis sets, geometries and zero point vibrational energies, but these should be reflected in the other data as well. Thus, one cannot fully exclude the possibility that discrepancies with respect to molecular or atomic states are present in this data. However, the PCI-80 approach, which performs best with a mean deviation of 2 kcal/mol for this series, gives a value for VH^+ right in the middle of the two B3LYP results – so no conclusive statement can be made as to whether or not this is an intrinsic shortcoming of the method. In any case, the large deviation for the copper hydride cation found at the BP86 level is not seen in the B3LYP data of Barone, Adamo and Mele (this species has not been considered by Blomberg, Siegbahn and Svensson[40]). A conspicuously large deviation of +9 kcal/mol occurs for the FeH^+ species in the B3LYP results of both groups. Also the PCI-80 approach shows the largest deviation for this species, although the error is much smaller (3 kcal/mol).

Now we turn to the MCH_3^+ cations, which have been studied by Holthausen et al., 1995, and Blomberg, Siegbahn, and Svensson, 1996. The former authors used the BHLYP and B3LYP hybrid functionals in combination with a rather limited Hay and Wadt ECP/basis set. With the B3LYP hybrid functional, quite large overestimations of the binding energies were obtained for some species (cf. Table 9-10). The mean absolute deviation was as high as 9 kcal/mol, similar to the MCPF results of Blomberg, Siegbahn, and Svensson. The unsatisfactory performance of the MCPF approach, however, has been attributed by Blomberg Siegbahn and Svensson to the near-degeneracy effects, which are stronger in the methyl complexes than in the hydrides. The much improved results obtained with the BHLYP functional are remarkable, although it seems fortuitous in the present context: the much larger Wachters/6-311+G(2d,2p) basis set combination used by Blomberg, Siegbahn, and Svensson improves the B3LYP values substantially. However, none of the DFT methods tested reaches the good performance of the scaled PCI-80 approach.

If we finally look at the corresponding methylene complexes (see Table 9-11) the poor MCPF performance suggests that these systems pose a severe near-degeneracy problem, and an appropriate rigorous wave function based approach would have to invoke a multireference treatment in combination with large basis sets. Therefore the more interesting it is to see that the B3LYP hybrid performs quite well. If used in combination with the small Hay and Wadt ECP/basis, larger overestimations are seen for the Mn^+, Fe^+ and Co^+ methylene complexes, but the overall quality of the results is better than that achieved for the methyl species. The mean unsigned errors are significantly smaller if the larger Wachters basis set is used, quite independent of the contraction scheme or basis employed on the CH_2 unit – however, some binding energies deviate quite substantially. The BHLYP approach underestimates the binding energies systematically, irrespective of the basis set used. In fact, the pronounced tendency for overbinding seen in the methyl results above leads for the methylene complexes to a somewhat better performance for the smaller Hay and Wadt ECP/basis set combination. This clearly demonstrates error compensation effects. From these findings we conclude that neither the use of the BHLYP functional, nor that of the small Hay and Wadt effective core potential can be recommended for the evaluation of binding energies of coordinatively unsaturated, open-shell transition-metal complexes. As noted already in Section 9.1, the B3LYP hybrid should be used with basis sets of at least

---

[40] At this point we like to add a note of caution: Blomberg et al. used the 6-311+G(2d,2p) basis set for single point energy calculations. This notation was probably chosen by the authors of *Gaussian* to maintain the standard nomenclature of the Pople-type basis sets but retrieves a modified Wachters, 1970, basis for the 3d elements out of the basis set library. Somewhat irritatingly, the request for d-polarization functions in the above notation actually augments the metal with f-functions. While this makes physical sense, the notation is at best inconsistent. We additionally note that – at least up to *Gaussian98*, Rev. A7 – the internally stored basis set for Cu mixes the d-function exponents and contraction coefficients optimized for the ^2D atomic term with the s- and p-functions determined for the ^2S term. Use the 'gfinput' keyword in order to check the basis set used in *Gaussian* calculations.

Table 9-10. Experimental binding energies D_0 [kcal/mol] for the cationic methyl complexes of the 3d elements and deviations from these data obtained at various levels of theory. Calculated values include a uniform zero point vibrational energy correction of 2.6 kcal/mol.

Method	$ScCH_3^+$	$TiCH_3^+$	VCH_3^+	$CrCH_3^+$	$MnCH_3^+$	$FeCH_3^+$	$CoCH_3^+$	$NiCH_3^+$	$CuCH_3^+$	MAD[d]
Exp.[a]	56	51	46	26	49	55	49	45	27	±3
B3LYP[b]	+18	+21	0	+5	+10	+16	+5	−2	+1	9
BHLYP[b]	+7	+9	−6	−5	−1	0	−2	−11	−4	5
QCISD(T)[b]	−2	2	−3	−4	−8	−5	−3	−13	−16	6
B3LYP[c]	+5	+8	+4	+9	+1	+8	+5	+5	—	6
MCPF[c]	−8	−8	−9	−6	−12	−7	−9	−13	—	9
PCI-80[c]	0	+2	−1	3	−3	2	0	−3	—	2

[a] Armentrout and Kickel, 1996; [b] Hay and Wadt ECP/D95(d,p) ECP/basis set combination (Holthausen et al., 1995); [c] Single point energy calculations using the 6-311+G(2d,2p) basis set. Geometries have been obtained using the smaller Hay and Wadt ECP/basis combinations (Blomberg, Siegbahn, and Svensson, 1996); [d] Mean absolute deviation from experiment; for experiment, the average of the cited experimental error bars is given.

Table 9-11. Experimental binding energies D_0 [kcal/mol] for the cationic methylene complexes of the 3d elements and deviations from these data obtained at various levels of theory.

Method	$ScCH_2^+$	$TiCH_2^+$	VCH_2^+	$CrCH_2^+$	$MnCH_2^+$	$FeCH_2^+$	$CoCH_2^+$	$NiCH_2^+$	$CuCH_2^+$	MAD[e]
Exp.[a]	89	91	78	52	68	82	76	73	61	±2
B3LYP/HW[b]	+8	+5	−3	+4	+14	+14	+10	+2	+3	7
B3LYP/W[b]	−4	−4	−3	+7	+3	+4	+2	+2	+2	4
BHLYP/HW[b]	−13	−7	−14	−9	−1	−7	−5	−16	−10	9
BHLYP/W[b]	−24	−21	−15	−7	−13	−18	−15	−17	−12	16
B3LYP[c]	+1	−2	−3	+8	+5	+6	+3	+4	+5	4
B3LYP[d]	−5	−6	−2	+5	0	−6	+2	−2	—	4
MCPF[d]	−18	−20	−17	−14	−25	−26	−22	−21	—	20
PCI-80[d]	−1	+1	+7	+14	+2	−2	0	+1	—	4

[a] Taken from Armentrout, 1989; [b] HW: Hay and Wadt ECP/D95(d,p) ECP/basis set combination; W: [8s4p3d] contraction of the Wachters basis, D95(d,p) on CH_2; a uniform zero point vibrational energy correction of 2 kcal/mol has been added (Holthausen, Mohr, and Koch, 1995); [c] Sc, Ti: [8s7p4d2f], V-Cu: [8s6p4d2f] contraction of the Wachters basis; 6-31G(d,p) on CH_2 (Ricca and Bauschlicher, 1995a); [d] Single point energy calculations using the 6-311+G(2d,2p) basis set. Geometries have been obtained using the smaller Hay and Wadt ECP/basis combinations (Blomberg, Siegbahn, and Svensson, 1996); [e] Mean absolute deviation from experiment; for experiment, the average of the cited experimental error bars is given.

polarized triple-zeta quality. A comparison of results from the systematic studies presented so far shows that the B3LYP functional in combination with extended basis sets gives a rather stable mean absolute deviation on the order of 4–6 kcal/mol, whereas maximum deviations can reach 9 kcal/mol. Thereby, the B3LYP functional outperforms the MCPF approach. The former produces smaller mean absolute errors and is more robust even for problematic cases. However, as noted above, the density functional treatment does not reach the good performance of the PCI-80 extrapolation scheme, at least not for the cases tested so far.

We note in passing that empirical correction schemes have been proposed to account for the bias towards d-rich situations of current density functionals. Ricca and Bauschlicher, 1995a, compute the dissociation energies of cationic transition-metal carbenes $M{=}CH_2^+ \rightarrow M^+ + CH_2$ as a weighted average of the energies of the $3d^n4s^1$ and the $3d^{n+1}$ asymptotes with the weights determined by the metal 3d occupation in the carbene. For example, the $FeCH_2^+$ ion has at its equilibrium geometry a 3d occupation of 6.52 electrons according to a Mulliken population analysis. Ricca and Bauschlicher compute the Fe^+/CH_2 binding energy as an average of 48 % of the ^6D $(3d^64s^1)$ and 52 % of the ^4F $(3d^7)$ asymptote. However, this correction has only a marginal influence and in fact yields inferior results in some cases. The universal applicability of such correction schemes seems therefore to be questionable. It is probably fair to say that as of today there seems to be no patent remedy in sight which would reduce the inherent uncertainty in the atomic energies.

Related studies include the binding energies of 3d-transition-metal monocarbonyls (Fournier, 1993, Ricca and Bauschlicher, 1994, Barone, 1994, Ricca and Bauschlicher, 1995b), MCH_2^+ for 4d-transition row elements (Eriksson et al., 1994), $Fe(H_2O)_n^+$ (Ricca and Bauschlicher, 1995c), $Co(H_2)_n^+$ (Bauschlicher and Maitre, 1995), ethylene complexes of Cu^+, Ag^+, and Au^+ (Hertwig et al., 1996), neutral metal hydrides (Barone and Adamo, 1997a), small titanium/oxygen compounds (Bergström, Lunell, and Eriksson, 1996), M^+-CO_2 and OM^+CO complexes for the 3d-elements (Sodupe et al., 1997a), 3d-element dihalides (Wang and Schwarz, 1998), binding of nitric oxide to 3d-metal atoms (Blanchet, Duarte, and Salahub, 1996) and 3d-metal cations (Thomas, Bauschlicher, and Hall, 1997). All these studies (and many more, which we have not mentioned) essentially validate the applicability of density functional theory approaches for transition-metal problems, and most of them carry the same essential message: the minimum errors for the density functional treatment of binding energies are of the same order of magnitude as the deviations found for computed atomic state splittings. In light of the preceding discussion this comes to no surprise. We have also seen above that the particular amount of exact exchange admixture can have dramatic consequences for computed binding energies. Hence, as long as there are no better physically motivated approaches the only obvious way to improve the results of modern functionals in the transition-metal area seems to be the inclusion of atomic state splittings into the training sets used for the empirical construction of these functionals.

Finally, we should mention that experimental data serving as a benchmark for the appraisal of computational methods must be highly accurate. Setting the goal of 2 kcal/mol for useful accuracy of calculated thermochemical data means of course, that a still better level of accuracy must be reached by experimental measurements. The high accuracy of experimental data is in fact a most attractive property of the G2 set of molecular systems, and is certainly not standard. The bulk of the metal-ligand binding energies determined experimentally, usually obtained by elaborate mass spectrometric techniques, are often far less accurate. As outlined by Blomberg and Siegbahn, 1998, a more realistic, but still useful goal of accuracy in transition-metal chemistry can be set to 4 to 6 kcal/mol. We have shown above that this goal still has to be regarded as quite ambitious, but it seems well within reach of modern density functional techniques. However – and this is certainly one of our main conclusions of this chapter – for every system under investigation, the importance of calibrating the performance of the chosen strategy against appropriate reference results cannot be overemphasized.

9.4 Ionization Energies

Let us now turn to the determination of ionization energies in the density functional framework. This property, i. e., the energy required to remove an electron from a bound state to infinite separation, is one of the most important characteristics for atoms. But also for molecules ionization processes have attracted much attention from the early nineteenseventies onwards when photoelectron spectroscopy emerged as a new and exciting experimental technique. In photoelectron spectroscopy experiments monoenergetic radiation is used to specifically eject electrons from any of the occupied levels in a molecule (of course, only if sufficiently energetic radiation is used). Since each of these levels has a different energy, the analysis of the ionization energies (as well as intensity and angular distribution of the emitted electrons) serves as a microscopic probe for the detailed electronic structure of molecules. It is therefore not surprising that this technique has served as a testing ground for theoretical methods and nowadays, a strong synergy between experiment and theory renders this field of research highly efficient. In the present context we note that the assignment of photoelectron spectra was among the first successful applications of density functional schemes.

Correlation effects are of particular importance for a proper description of ionization processes since the number of electrons changes during the ionization. As a consequence, the Hartree-Fock approximation usually underestimates ionization energies, since correlation effects – which are neglected at the HF level – affect the neutral system usually more strongly than the ionized cation with its one electron less. Obviously, the same observations apply to exchange-only density functionals, such as the old X_α method. For example, the experimental ionization energy of the oxygen atom is 13.61 eV, but both, Hartree-Fock and X_α severely undershoot this target and yield only 12.02 and 12.44 eV, respectively (using a cc-pVQZ basis set). Of course, in all modern applications of density functional theory combined exchange-correlation functionals are employed and the accuracy of the computed predictions depends on the balanced description of the exchange and correlation contributions in the neutral system and the ion. A number of systematic studies on the performance of Kohn-Sham functionals regarding ionization energies, mostly based on the G2 data or extensions thereof, have been published recently. Let us begin by quoting results from Becke, 1992b. This author applied various local and GGA functionals in his numerical, basis set free scheme (i. e., the results mirror the capability of the functional and are not blurred by basis set deficiencies or biases) to the 42 accurately determined ionization energies of small species contained in the original G2 set. The LDA (where the Slater exchange part was combined with the parameterization of the uniform electron gas data due to Perdew and Wang, 1992) afforded mean absolute and maximum deviations of 0.23 and 0.62 eV, respectively. The inclusion of Becke's gradient corrections to the local exchange part actually yields results significantly *inferior* to those of the plain local functional, increasing the errors by about a factor of two: the average error amounts to 0.41 eV and the maximum error increased to 1.3 eV. Inclusion of the PW91 gradient-corrected correlation functional (i. e., ending up with BPW91) significantly reduces the average error to 0.15 eV with a maximum deviation of 0.44 eV. This situation is in striking contrast to the observations made for atomization energies. There, the LDA was completely useless due to dramatic overbinding and gradient corrections to exchange were found essential, while gradient-corrected correlation functionals actually spoiled the error statistics (see above). In his paper introducing the three-parameter hybrid functionals, Becke, 1993b, reports an average error of 0.14 eV for his original B3PW91 functional with a maximum error of 0.41 eV. Thus, the beneficial effect of admixing a certain amount of exact exchange is much less pronounced for ionization than for atomization energies. Very similar conclusions can be extracted from recent papers by, e. g., De Proft and Geerlings, 1997, or Curtiss et al. 1998. The latter authors established the accuracy of predicted ionization energies for seven popular exchange-correlation functionals combined with the 6-311+G(3df,2p) basis set with respect to an extended set of more than 80 atoms and molecules. Unlike Becke in his above-mentioned study, these authors employed the standard SVWN implementation of the LDA. Interestingly, this functional performs rather disappointingly with an average unsigned error of 0.59 eV, confirming similar results by De Proft and Geerlings. The maximum deviation occurs for CN whose ionization energy is overestimated by 1.74 eV. For the GGA functionals BLYP, BP86, and BPW91, mean absolute deviations of 0.26, 0.20, and 0.22 eV, respectively, were reported. The errors are slightly larger than those reported earlier by De Proft and Geerlings but show the same trend in that BLYP seems to be the least accurate GGA functional. Among the three hybrid functionals tested, B3LYP and B3PW91 perform rather well with mean errors of 0.18 and 0.19 eV, respectively. Very surprisingly, if the P86 correlation functional is employed in the hybrid scheme (leading to the B3P86 functional), a very large mean error of 0.57 eV results. The origin of this dramatic effect is unclear. Hence, the winner in this contest turns out to be the B3LYP functional. To put this result into perspective we need to point out that the G2 procedure performs much better with a mean absolute deviation of only 0.06 eV and a maximum error of 0.32 eV (for B_2F_4). Finally, we would like to mention the work of Ernzerhof and Scuseria, 1999a, who mostly focus on the new PBE1PBE functional whose performance is comparable but slightly better than B3LYP. However, these authors also show that the local VWN5 correlation functional yields very satisfactory ionization energies (mean and maximum absolute errors of 0.22 and 0.6 eV) very close to the LDA results reported by Becke, while using VWN they reproduced the large errors mentioned above. Thus it is obvious that even though the VWN parameterization for the local correlation functional is very close to other implementations such as VWN5 or the local PW91 parameterization and unproblematic for almost all applications, it is not suited for computing ionization energies.

The important conclusion of this section is that ionization energies can be determined with an average error of around 0.2 eV. The usual hierarchy of functionals, i. e., hybrid functionals better than GGA better than LDA, however, does not strictly apply. Already the local approximation provides good results as long as the VWN parameterization is avoided. The GGA and hybrid functionals show only a small improvement over the LDA. On the other hand, the B3P86 hybrid functional is an exception in that it performs very poorly and should therefore not be applied to the determination of ionization energies. We have not yet said anything about basis set requirements. It seems that ionization energies are less sensitive in this respect than atomization energies. For example, Curtiss et al., 1998, show that the mean absolute error for the B3LYP functional of 0.18 eV does not necessarily require the large 6-311+G(3df,2pd) basis set. This accuracy is already reachable with the smaller sets 6-311+G(2df,p) and even 6-31+G(d). A summary of the various benchmark results discussed in this section is given in Table 9-12.

Table 9-12. Compilation of mean absolute (maximum) deviations for ionization energies [eV] of small main group molecules from different sources.

42 atoms and molecules, numerical, basis set free, Becke, 1992b and 1993b			
LDA	0.23 (0.62)	BPW91	0.15 (0.44)
BVWN	0.41 (1.26)	B3PW91	0.14 (0.41)
38 atoms and molecules, De Proft and Geerlings, 1997			
SVWN, aug-cc-pVTZ	0.69	BLYP, aug-cc-pVTZ	0.19
BP86, aug-cc-pVTZ	0.17	B3LYP, cc-pVDZ	0.18
B3LYP, aug-cc-pVTZ	0.15	B3PW91, cc-pVDZ	0.20
B3PW91, aug-cc-pVTZ	0.15		
83 atoms and molecules, Curtiss et al., 1998			
SVWN, 6-311+G(3df,2p)	0.59 (1.74)	B3LYP, 6-31+G(d)	0.18
BLYP, 6-311+G(3df,2p)	0.26 (1.02)	B3LYP, 6-311+G(2df,p)	0.18
BPW91, 6-311+G(3df,2p)	0.22 (1.17)	B3LYP, 6-311+G(3df,2p)	0.18 (1.65)
BP86, 6-311+G(3df,2p)	0.20 (1.20)	B3PW91, 6-311+G(3df,2p)	0.19 (1.67)
B3P86, 6-311+G(3df,2p)	0.57 (2.22)	G2	0.06 (0.32)
38 atoms and molecules, 6-311+G(3df,2p) basis, Ernzerhof and Scuseria, 1999			
SVWN	0.69 (1.2)	PBE	0.16 (0.5)
SVWN5	0.22 (0.6)	PBE1PBE	0.16 (0.7)
BLYP	0.20 (0.6)	VSXC	0.13 (0.4)
B3LYP	0.17 (0.8)		

9.5 Electron Affinities

The electron affinity of a neutral system is the energy gained upon attaching an additional electron, thereby generating the corresponding anion. The addition of an electron does not in all cases lead to energetically more favorable anions. Rather, there are many atoms or molecules, where the energy of the anion is higher (i. e., less favorable) than that of the parent neutral, i. e., where the excess electron is not bound but will auto-detach immediately. These species do not have a positive, but a negative electron affinity. While there are sophisticated experimental techniques to probe the transient anionic species resulting from neutrals with negative electron affinities we will in the following only consider stable anions characterized by positive electron affinities. The computational prediction of electron affinities has always been a particularly difficult task for wave function based methods. The correlation energy of the anion with the additional, albeit weakly bound electron is larger than that of the neutral molecules and the anions are in addition usually characterized by a very diffuse charge density. As a consequence, sophisticated treatments of the correlation energy combined with one-particle basis sets augmented by diffuse and high angular momentum functions are required for an adequate description of the extra electron rendering such calculations intrinsically prohibitive for larger molecules. Density functional methods would therefore be a highly welcomed alternative. Before we enter a quantitative discussion about the capability of approximate density functionals to determine electron affinities we need to address a more general point. One frequently reads that for principal reasons local functionals such as the LDA simply do not bind the extra electron and are therefore not suitable for studying anions and related properties like electron affinities. Of course, the Hohenberg-Kohn theorem and hence density functional theory as such applies to all bound systems, irrespective of whether the atom or molecule is neutral or charged, be it positively or negatively. If the exact exchange-correlation was known, the exact solutions for cations, neutrals and anions would be available. Having said this, it is also true that local approaches like the LDA or the GGA functionals have indeed some intrinsic problems when it comes to treating systems with an excess negative charge. The physical reasoning behind this is related to the self-interaction problem and the incorrect asymptotic behavior of current approximate exchange-correlation potentials briefly introduced in Sections 6.7 and 6.8. Recall that in none of the currently used density functional implementations is the spurious repulsion of the probe electron with itself included in the Coulomb term $J[\rho]$, precisely cancelled by the exchange-correlation energy, $E_{XC}[\rho]$. In terms of the related potentials this applies in particular to distances far from the system because the corresponding exchange-correlation potentials all die out too fast with increasing distance. Hence, for large r, the repulsive Coulomb potential prevails and the approximate potential is less attractive than it should be. As a consequence the excess electron of an anion is too weakly bound, if it is bound at all, as clearly pointed out by Rösch and Trickey, 1997. These problems due to the incomplete cancellation of self-repulsion should be the more significant the more localized the extra electron is. Hence, the most problematic cases are expected to be atoms, followed by diatomics etc., while large molecules should be much more well-behaved. If we, however, adopt the DFT-typical pragmatic point of view, things are not quite so bad in real applications. In almost all Kohn-Sham calculations finite basis sets are used to describe the electron density and the additional electron is forced to remain within the spatial area defined by the basis functions. It simply cannot escape the atom or molecule leading to an artificial stabilization which counterbalances the self-interaction based error. As we will see in this section, there have been several systematic studies of electron affinities whose conclusions may be best summarized by quoting from the title of a paper by Tschumper and Schaefer, 1997: 'Some positive results for negative ions'. In their extensive study Curtiss et al, 1998, not only examined the performance of approximate density functionals with respect to ionization energies but also looked at electron affinities. Their extended G2 set included a total of 58 electron affinities for atoms and small molecules (with up to three non-hydrogen main group atoms). Combined with the large 6-311+G(3df,2p) one-electron basis, the G2 method sets the standard with a mean absolute error of only 0.061 eV. The best density functional is BLYP with an average deviation of 0.11 eV, closely followed by BPW91 (0.12 eV), B3LYP (0.13 eV), and B3PW91 (0.15 eV). 59 % of the BLYP electron affinities are within 0.1 eV and almost 85 % are within 0.2 eV of the experimental ones, not showing a systematic over- or underestimation. The maximum error occurs for the C_2 molecule whose electron affinity is overestimated by 0.69 eV. Just as with ionization energies, the B3P86 hybrid functional is lagging behind with an error of 0.60 eV, which is only 0.1 eV less than with SVWN (0.70 eV). In all cases these two functionals overestimate the stability of the anion and yield electron affinities that are too large. Errors obtained with smaller basis sets such as 6-31+G(d) were only marginally larger. De Proft and Geerlings, 1997, come to similar conclusions with respect to 27 electron affinities from the original G2 test set. While they find B3PW91 to be the most accurate functional with an average error of 0.11 eV, B3LYP and BLYP are of comparable accuracy with average deviations of 0.12 and 0.14 eV, respectively. The BP86 functional has a slightly larger mean error of 0.23 eV and the SVWN functional is again far off, its error amounts to 0.77 eV. There have been other studies as well, which essentially confirm the results reported from these two representative investigations. We conclude that for the calculation of electron affinities there seems to be no noteworthy beneficial effect of mixing in exact exchange. This is in distinct contrast to atomization energies but parallels the conclusions for ionization energies drawn in the preceding section. Thus we emphasize that exact exchange is of large importance for bond-breaking processes but hardly of relevance in processes where only the number of electrons is being changed. These conclusions remain valid also outside the domain of the small molecules in the G2 set. In a number of investigations, Schaefer and coworkers tested the applicability of DFT for the determination of electron affinities of a variety of molecules containing first- and second-row main group elements. They reported overall mean errors for 49 electron affinities of 0.21 eV for BYLP and B3LYP and of 0.22 eV for BP86. Also in their test set B3P86 and SVWN turn out to be not well suited for computing electron affinities, with mean errors amounting to 0.62 and 0.70 eV, respectively (Brown, Rienstra-Kiracofe, and Schaefer, 1999, and references cited therein). An important point is that these encouraging results were obtained with fairly small basis sets of polarized double-zeta quality augmented by diffuse functions, in agreement with the above-mentioned data presented by Curtiss et al., 1998. This

Table 9-13. Compilation of mean absolute deviations (maximum deviation in parentheses) for electron affinities [eV] of small main group molecules from different sources.

27 atoms and molecules, De Proft and Geerlings, 1997			
SVWN, aug-cc-pVTZ	0.77	BLYP, aug-cc-pVTZ	0.14
BP86, aug-cc-pVTZ	0.23	B3LYP, aug-cc-pVDZ	0.15
B3LYP, aug-cc-pVTZ	0.12	B3PW91, aug-cc-pVDZ	0.13
B3PW91, aug-cc-pVTZ	0.11		
58 atoms and molecules, Curtiss et al., 1998			
SVWN, 6-311+G(3df,2p)	0.70 (1.31)	B3LYP, 6-31+G(d)	0.16
BLYP, 6-311+G(3df,2p)	0.11 (0.69	B3LYP, 6-311+G(2df,p)	0.14
BPW91, 6-311+G(3df,2p)	0.12 (0.77)	B3LYP, 6-311+G(3df,2p)	0.13 (1.08)
BP86, 6-311+G(3df,2p)	0.19 (0.88)	B3PW91, 6-311+G(3df,2p)	0.15 (1.06)
B3P86, 6-311+G(3df,2p)	0.60 (1.61)		
25 atoms and molecules, 6-311+G(3df,2p) basis, Ernzerhof and Scuseria, 1999			
SVWN	0.74 (1.2)	PBE	0.11 (0.3)
SVWN5	0.30 (0.7)	PBE1PBE	0.13 (0.3)
BLYP	0.11 (0.4)	B3LYP	0.11 (0.5)

makes the DFT based approach to electron affinities much more versatile and applicable to larger molecules than wave function based strategies, for which significantly larger basis sets are required. Table 9-13 summarizes some of the numerical data discussed in this section.

Finally, we mention the recent study by de Oliveira et al., 1999, who reported atomic electron affinities from a variety of methods including a number of density functionals. While the general conclusions of this section are substantiated, one of their results will be explicitly mentioned here: these authors show that for the first-row atoms the choice of the exchange functional is decisive while the quality of the correlation functional is less important. For the second-row analogs it is the other way round and the correlation functional seems to be of greater importance. Even more oddly, the quality of various correlation functionals varies and depends on the atoms. For example, the LYP functional works best for first-row atoms, while PW91 turns out to be preferable for second-row ones. If nothing else, these results point to the highly empirical character that all these functionals still have, and that one has to be careful when judging the predictive capabilities of present functionals.

9.6 Electronic Excitation Energies and the Singlet/Triplet Splitting in Carbenes

We conclude this chapter with an overview of how modern density functional theory deals with electronic excitation energies. From the very beginning, electronically excited states have been identified as a very difficult area for Kohn-Sham density functional theory, which is in principle limited to ground states only. Methods which offer accurate results paired with theoretical soundness have therefore long been high up on the DFT whish list. In Section 5.3.7 we presented some of the more recent developments in that area. The greatest impetus in that regard certainly came from the time-dependent formulation of density functional theory which opened the avenue of a formally rigorous extension of regular Kohn-Sham DFT to a time-dependent scheme. By this token, the frequency dependent response of the charge density becomes available which in turn can be directly related to excitation energies. Probably the most important and elegant characteristics of this approach is that it is based on ground state properties of the system and no extension of the Kohn-Sham formalism into the grey area of excited states is necessary. We already mentioned in Section 5.3.7 that in general TDDFT provides excitation energies with good accuracy and in the following we will substantiate these claims, focussing on the choice of functional and basis set but also pointing out some problematic areas.

All the systematic investigations published so far essentially agree that TDDFT provides accurate excitation energies that rival more sophisticated and much more costly wave function based approaches, as long as we are dealing with low-energy transitions involving valence states. Differences between functionals are not very pronounced. For example, Bauernschmitt and Ahlrichs, 1996b, report excitation energies for N_2, formaldehyde, ethylene and pyridine derived from the SVWN, BP86 and B3LYP functionals combined with a Gaussian basis set specifically designed for static polarizabilities, the POL basis set due to Sadlej (which contains diffuse functions). If only low energy transitions with excitation energies below half the ionization threshold are considered, mean absolute deviations of 0.36, 0.29 and 0.30 eV for these three functionals were obtained. These results are superior to those from Hartree-Fock based approaches such as the random-phase-approximation (RPA, which essentially is time-dependent HF) or the configuration interaction with single excitations (CIS) schemes, which yield errors of 0.88 and 1.65 eV, respectively. Similarly, Casida et al., 1998, report average errors of a few tenths of an eV for the LDA as long as only low-lying states are included in the statistics. Stratmann, Scuseria, and Frisch, 1998, also include larger molecules such as benzene, porphin and up to C_{70} in their investigation. Their conclusions are completely in line with what we have presented so far. Excited states which lie well below the ionization threshold are described very well and hybrid functionals like B3LYP seem to yield slightly more accurate results than simple GGA ones. Their results also turned out to be fairly insensitive to the basis set, in particular for the low-lying transitions. Extending the basis set from 6-31+G(d) to aug-cc-pVTZ changed the energies of two $\pi \rightarrow \pi^*$ transitions in benzene by less than 0.1 eV. This promising performance of TDDFT also applies to excitations from radical species, which pose severe problems to conventional, Hartree-Fock based methods such as RPA or CIS. As shown by Hirata and Head-Gordon, 1999, valence states with both, single and double excitation character, are described uniformly reasonable with errors within a few tenths of an eV for small radicals such as BH, CH_3, CN, or CO^+, but also for bigger systems including benzyl, anilino, and phenoxyl radicals. While for the former group the performance increased from SVWN to BLYP and B3LYP, no significant differences between these three functionals were observed for the latter set of radicals. This sharply contrasts with the performance of HF based techniques. While the RPA and CIS schemes are only slightly inferior to TDDFT for excitations with dominant single excitation character, they show very large errors for transitions with an appreciable double excitation character, rendering these methods rather useless in such applications.

We always made the restriction in the above discussion that the excitations are to low lying states. The reason is simply that as soon as higher lying states, which are frequently of Rydberg character, are included, the picture changes dramatically and the TDDFT results deteriorate significantly. Quoting again from the landmark paper by Bauernschmitt and Ahlrichs, 1996b, the mean errors rise to 0.53 eV (SVWN), 0.55 eV (BP86) and 0.49 eV (B3LYP) if excitations up to the molecular ionization energy are included, and similar observations have been reported by many other authors. The origin of this surprising phenomenon was first uncovered by Casida et al., 1998, and later expounded by Tozer and Handy, 1998. These authors convincingly demonstrate that this shortcoming is intimately connected to the incorrect asymptotic behavior of approximate exchange-correlation potentials. Let us try to understand this. We saw in Section 6.8 that a common feature of current approximate functionals is that the corresponding potentials are not attractive enough. Among the reasons for this, we noted the wrong asymptotic decay of the potentials (which is faster than the correct $-1/r$ behavior) and the fact that they vanish at infinity whereas the correct potential should converge to a positive, constant value. As a consequence, the orbital energy of the highest occupied orbital, which for the exact exchange-correlation functional equals the ionization energy and therefore defines the ionization threshold, comes out too high with approximate functionals, making the ionization threshold too low. Likewise, the description of the virtual orbitals suffers, leading to orbital energies which are too high. As a consequence, only a few virtual orbitals have negative energies. Since the TDDFT formalism makes use of orbital energy differences, the quality of the virtual orbitals and their energies is mirrored in the quality of the corresponding excitation energies. Now it is easy to explain why excitations to energetically low lying states are described reasonably well. They involve low lying virtual orbitals whose energies are negative, i. e., which are bound with current approximate functionals. On the other hand, high energy Rydberg excitations involve higher virtual orbitals whose positive energies are not even qualitatively correct anymore with these functionals. Casida et al., 1998, present two criteria which an excitation must fulfill in order that a standard functional can be used without being negatively affected by the problems due to the incorrect asymptotic behavior. First, the excitation energy must be significantly smaller than minus the orbital energy of the highest occupied orbital. Second, the transition should not involve major contributions from promotions to virtual orbitals which are only weakly bound or even unbound in the selected functional.

An obvious remedy to this situation is to use potentials that by construction exhibit the correct asymptotic behavior. Indeed, using the LB94 or the HCTH(AC) potentials yields significantly improved Rydberg excitation energies. As an instructive example, we quote the detailed study by Handy and Tozer, 1999, on the benzene molecule. These authors computed a number of singlet and triplet $\pi \rightarrow \pi^*$ valence and $\pi \rightarrow n = 3$ Rydberg excitations employing their asymptotically corrected HCTH(AC) procedure. Over all energies a mean absolute deviation from experimental or accurate theoretical data of only 0.12 eV resulted, only slightly higher than the error obtained with the much more demanding CASPT2 wave function approach. Most importantly the Rydberg excitations, so difficult for the conventional functionals, came out just as accurately as the valence transitions. Actually, the highest error of 0.62 eV in their computed transition energies was not due to a Rydberg excitation but occurred for the $\pi \rightarrow \pi^*$ valence excitation to the $^3B_{2u}$ state. Further examples, which corroborate these conclusions but also add some small grains of salt can be found in Tozer et al., 1999. While the HCTH(AC) procedure reproduced Rydberg and valence excitations in most cases to within a few tenths of an eV, these authors noted that excitations which involve a considerable charge transfer have significantly larger errors.

In a recent contribution, Adamo, Scuseria, and Barone, 1999, have studied the performance of the PBE GGA functional and the corresponding PBE1PBE hybrid functional in the framework of time-dependent DFT and compared it to results obtained with the standard B3LYP technique and with Handy and Tozer's HCTH(AC) data. Interestingly, if the PBE functional is blended with 25 % exact exchange the resulting PBE1PBE scheme provides good excitation energies across the board, not only for valence but also for Rydberg excitations close to the ionization threshold. On the one hand this is due to the admixture of Hartree-Fock exchange in the functional, which improves the asymptotic behavior of the exchange-correlation potential by introducing a discontinuity in the potential as it increases through an integer number of electrons and by generating an asymptotic decay of $-a/r$ which is closer to the correct one than the GGA asymptotic decay (see Section 6.8). On the other hand, part of the success of PBE1PBE must also be intrinsic to the PBE functional itself. Other hybrid functionals like B3LYP are not as successful by far and already the pure PBE functional (i. e., without exact exchange) is much better than BLYP or other GGA implementations, even though it was not designed for any asymptotic features. Why this is so remains, however, to be uncovered. The conclusions of this discussion are amply demonstrated in Table 9-14 and Figure 9-2 which summarize the corresponding excitation energies obtained from various functionals using the ethylene molecule as an example.

Note in particular the sharp deterioration of the computationally predicted excitation energies as one moves closer to the ionization threshold (exp.: 10.51 eV) for the standard functionals, which is the most eye-catching feature in Figure 9-2. The other statements discussed above also find their confirmation in the electronic excitations of ethylene. Both, the local SVWN and the gradient-corrected BLYP functionals show non-negligible deficiencies, with BLYP being even significantly inferior to SVWN. Switching to B3LYP improves matters somewhat. These disappointingly average absolute errors are, however, only due to the higher energy excitations, as indicated in Figure 9-2. The low energy region of the spectrum is unproblematic for all these functionals. A substantial progress in terms of performance is, however, seen with the HCTH(AC) functional. Most notably, because this functional is augmented with an improved long-range potential, it performs satisfactorily for all excitations included, not only for the low energy transitions, rivaling sophisticated and computationally much more expensive CASPT2 calculations. Finally, we point to the similarity of the PBE1PBE results obtained with two different basis sets. This indicates a

Table 9-14. Electronic excitation energies for single electron excitations from the π (b_{3u}) orbital of C_2H_4 [eV] using an augmented POL basis set.

Transition	SVWN[a]	BLYP[a]	B3LYP[a]	HCTH(AC)[a]	PBE[b]	PBE1PBE[b]	PBE1PBE[b,c]	CASPT2[d]	Exp[a]
$\pi \rightarrow \pi^*$ ($^3B_{1u}$)	4.70	4.32	4.07	4.33	4.34	3.92	3.94	4.39	4.36
$\pi \rightarrow 3s$ ($^3B_{3u}$)	6.51	6.13	6.50	7.10	6.48	6.79	6.85	7.05	6.98
$\pi \rightarrow 3s$ ($^1B_{3u}$)	6.55	6.18	6.57	7.16	6.52	6.93	6.99	7.17	7.11
$\pi \rightarrow 3p\sigma$ ($^3B_{1g}$)	7.03	6.63	7.05	7.76	7.06	7.36	7.42	7.80	7.79
$\pi \rightarrow 3p\sigma$ ($^1B_{1g}$)	7.05	6.65	7.08	7.78	7.08	7.51	7.59	7.85	7.80
$\pi \rightarrow 3p\sigma$ ($^1B_{2g}$)	7.04	6.63	7.09	7.77	7.03	7.52	7.62	7.95	7.90
$\pi \rightarrow \pi^*$ ($^1B_{1u}$)	7.39	7.12	7.36	7.61	7.46	7.58	7.60	8.40	8.00
$\pi \rightarrow 3p\pi$ (^3A_g)	7.27	6.90	7.33	8.15	8.04	8.08	8.34	8.26	8.15
$\pi \rightarrow 3p\pi$ (^1A_g)	7.31	6.96	7.41	8.33	8.11	8.04	8.08	8.40	8.28
$\pi \rightarrow 3d\sigma$ ($^3B_{3u}$)	7.10	6.66	7.34	8.64	8.65	8.55	8.66	8.57	8.57
$\pi \rightarrow 3d\sigma$ ($^1B_{3u}$)	7.11	6.68	7.36	8.70	8.25	8.65	8.69	8.66	8.62
$\pi \rightarrow 3d\delta$ ($^1B_{3u}$)	7.63	7.21	7.75	8.95	8.26	8.76	8.82	9.03	8.90
$\pi \rightarrow 3d\delta$ ($^1B_{2u}$)	7.72	7.35	7.87	9.04	8.78	9.02	9.09	9.18	9.05
$\pi \rightarrow 3d\pi$ ($^1B_{1u}$)	7.80	7.75	8.14	9.32	8.85	9.33	9.30	9.31	9.33
mean abs. error	0.94	1.26	0.85	0.07	0.43	0.21	0.19	0.09	

[a] Taken from Tozer and Handy, 1998; [b] taken from Adamo, Scuseria and Barone, 1999; [c] with the 6-311++G(d,p) basis set; [d] taken from Serrano-Andrés, et al., 1993.

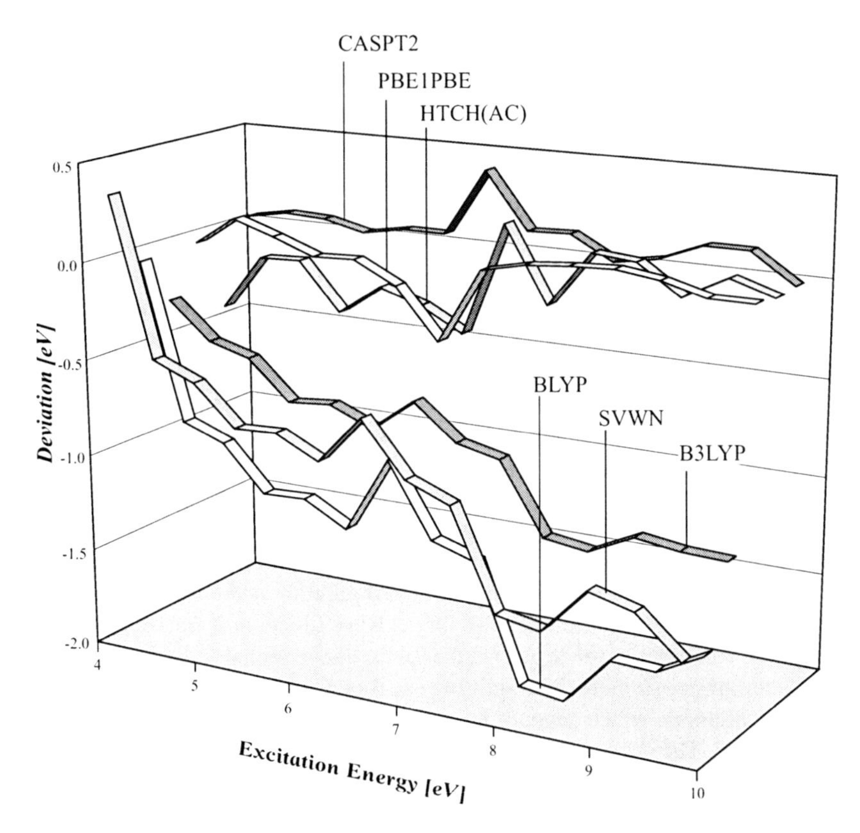

Figure 9-2. Performance of various functionals in the framework of time-dependent DFT for excitation energies of ethylene.

certain robustness of the computed excitation energies with regard to the choice of the basis set, provided basis sets of at least polarized triple-zeta quality augmented by diffuse functions are used.

A particular class of excitation energies is provided by the relative stability of the lowest lying singlet and triplet states of carbenes and related species. Even though the energy difference between these two states can easily be computed by the ΔSCF approach, because they both represent the lowest states in their respective multiplicity, the computation of reliable excitation energies for such species is a long standing problem in quantum chemistry (for general overviews see Bettinger et al., 1997 and 1998). Let us take methylene as the simplest example to illustrate the peculiarities and concomitant problems for the theoretical treatment of this group of molecules. A carbene is characterized by two electrons not engaged in bonding, and two non-bonding orbitals to accommodate them, i. e., the π-type (b_1 in case of the C_{2v} symmetric CH_2) and the lower lying σ-type orbital (a_1 for CH_2). In the singlet $^1A_1$ state, the two electrons are spin paired while in the $^3B_1$ triplet the two electrons of like spin occupy both orbitals with one electron each. Depending on the energetic spacing of the two orbitals, the ground state of a carbene will be either singlet or triplet. The closer together the a_1 and b_1 orbitals are, the more favored the triplet is (Hund's rule), while for well separated orbitals a singlet ground state emerges. From a computational point of view an adequate and unbiased description of the two states is paved with obstacles. Due to the different multiplicity the correlation contributions in the singlet and the triplet states differ. In addition, singlet carbene is one of the prototypes were a single-determinantal approach is inadequate and non-dynamical correlation plays a decisive role. While the double occupation of the a_1 orbital is more favored than putting both electrons into the b_1 MO, this second alternative will nevertheless play an important role in the exact wave function. Hence, in order to adequately account for this situation both configurations need to be included in the approximate wave function. The triplet state, on the other hand, is well described by one determinant. This situation is schematically depicted in Figure 9-3.

The simple Hartree-Fock ansatz is doomed to fail for two reasons. First, the Fermi correlation due to parallel spins is included while the Coulomb correlation is not. Therefore the triplet with its two open-shell parallel spin electrons will be described better than the singlet. Second, the HF scheme uses only one configuration and completely neglects the second determinant needed to describe the singlet wave function. A further destabilization of the singlet as compared to the triplet results. In the conventional wave function arena the problem of the carbene singlet-triplet splitting can therefore only be solved if sophisticated and expensive methods which account for dynamical and non-dynamical correlation effects are employed. The singlet-triplet gap in methylene is therefore for good reasons known as a 'testing ground for electronic structure methods' (Bettinger et al., 1998). How does approximate density functional theory fare in this complicated situation? Table 9-15 shows $\Delta E_{S\text{-}T}$ as computed with various functionals using a standard 6-311+G(d,p) and a series of

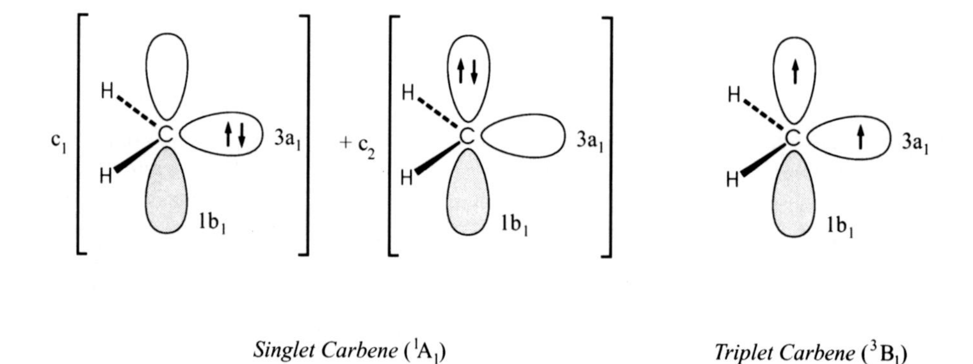

Figure 9-3. Schematic representation of the occupations of the highest occupied orbitals dominating the wave functions of singlet and triplet carbene. For the singlet, $c_1 > c_2$.

Table 9-15. Singlet ($^1A_1$) – triplet ($^3B_1$) energy gaps [kcal/mol] for methylene, CH_2 using different basis sets.

Functional	6-311+G(d,p)	aug-cc-pVTZ	aug-cc-pVQZ	aug-cc-pV5Z
HF	28.7	28.0	28.1	28.1
SVWN	14.0	12.9	12.9	12.9
SLYP	15.5	14.2	14.2	14.2
BVWN	9.3	8.5	8.6	8.6
BLYP	10.9	10.0	10.0	10.0
B3LYP	12.2	11.3	11.3	11.3
BP86	14.5	13.7	13.8	13.8
B3P86	15.7	14.8	14.9	14.9
BPW91	16.7	15.9	15.9	15.9
B3PW91	17.1	16.3	16.4	16.4
exp.		9.4		

augmented correlation-consistent basis sets of increasing quality.[41] Experimentally, the triplet is more stable than the singlet by 9.4 kcal/mol.[42]

As expected from the above discussion, the HF method significantly overshoots this value. It describes the triplet state as much too stable, being almost 20 kcal/mol off the target. All DFT entries in Table 9-15 are of much better quality. Before we analyze the performance of the density functionals we note that already the triple-zeta basis set, i. e., aug-cc-pVTZ, seems to provide converged results with respect to the basis set. Increasing the basis set further does hardly change the computed singlet-triplet gaps, the results from the quadruple and quintuple sets are completely identical (even though the number of basis functions is much larger in the latter). On the other hand, going from 6-311+G(d,p) to aug-cc-pVTZ systematically lowers $\Delta E_{S\text{-}T}$ by roughly 1 kcal/mol. With the correlation consistent basis sets, already the local SVWN approach gives a very reasonable splitting (12.9 kcal/mol). Applying the Becke gradient correction for exchange together with the local VWN correlation functional, the triplet is found to be 8.6 kcal/mol more stable than the singlet, i. e., the reference value is underestimated by 0.8 kcal/mol. If only the correlation functional is gradient-corrected in the SLYP functional instead, the singlet-triplet gap deteriorates to 14.2 kcal/mol. The GGA BLYP functional yields the most satisfactory singlet-triplet splitting, missing the experimental value by only 0.6 kcal/mol. Using the B3LYP hybrid functional, the much treasured champion among the currently popular recipes in many areas, the agreement with experiment worsens and we end up with a 11.3 kcal/mol energy difference between the $^3B_1$ and $^1A_1$ states of CH_2, i. e., an erroneous stabilization of triplet carbene. If instead of LYP the P86 or the PW91 correlation functionals are used,

[41] These basis sets consist of 34 (6-311+G(d,p)), 92 (aug-cc-pVTZ), 172 (aug-cc-pVQZ), and finally 287 (aug-cc-pV5Z) contracted Gaussian functions and include polarization functions up to d- (6-311+G(d,p)), f- (aug-cc-pVTZ), g- (aug-cc-pVQZ), and even h-character (aug-cc-pV5Z).

[42] This is the adiabatic T_e reference value, i. e., without zero-point vibrational energy corrections, see Bettinger et al., 1997 and 1998.

similar but more pronounced trends emerge. Let us try to understand this behavior. As noted above, a balanced description of the two states is only possible if the near-degeneracy effects of the singlet are properly taken into account. The large beneficial effect of the gradient-corrected exchange functional is in line with one important argument put forward earlier, namely that it is the exchange and not the correlation GGA functional which represents the non-dynamical correlation! The somewhat disappointing performance of the hybrid schemes is of course a simple reflection of the bias of the HF approximation towards the triplet. We have discussed a similar situation in Section 6.6 were the ground state of the ozone molecule was introduced as an example for a situation dominated by non-dynamical correlation. Also in that example we found pure GGA functionals to perform much better than hybrid ones.

The conclusions obtained for the parent methylene can be generalized for other carbenes and related species. Worthington and Cramer, 1997, for example, studied a variety of substituted carbenes and vinylidenes (i. e., where the carbene carbon is part of a double bond, RR'C=C:) at the BVWN5 and BLYP levels in combination with a cc-pVTZ basis set. The singlet-triplet splittings were usually accurate to within a few kcal/mol of the experimental reference data. For the carbenes the BLYP functional showed slightly, but consistently larger errors than the BVWN5 protocol with a bias towards stabilizing the triplet. In the case of the vinylidenes the errors are less systematic and no clear-cut conclusions can be drawn. Using a polarized double-zeta basis set, Vargas, Galván, and Vela, 1998, studied a number of halocarbenes CXY with X, Y = H, F, Cl, Br, and I. In harmony with the conclusions drawn above, it turned out to be the general trend that the SVWN functional achieved the best agreement with the experimental reference data. The BPW91 GGA functional was second and the worst performance was obtained systematically with the B3PW91 hybrid scheme. In all cases the triplet was computed as too stable. Very satisfying agreement between $\Delta E_{S\text{-}T}$ obtained from sophisticated CCSD(T) calculations and the B3LYP hybrid functional were reported by Gonzalez et al., 1998, for several carbenes and isoelectronic nitrenium ions, $R'NR^+$ and by Holthausen, Koch, and Apeloig, 1999, for silylenes, SiRR' using the BLYP and B3LYP functionals. The common denominator of these and several related studies is that today's approximate density functionals are in fact capable of treating both, the singlet and triplet states in a balanced way. The delicate and subtle effects due to the near-degeneracy in the singlet and the differential dynamical correlation effects of the two multiplicities, which challenge conventional, wave function based ab initio theory, do not seem to pose crucial problems to DFT. Note however that due to the special electronic situation in carbenes and related species pure density functionals are usually to be preferred over hybrid ones.

10 Electric Properties

In this chapter we will focus on the distribution of the electrons in a molecule and on the properties related to the response of the charge distribution to an applied external field. We will commence with a brief survey of how the standard methods for population analysis assign partial charges to the individual atoms if used together with present day density functional methods. Next, we will investigate the performance of current functionals for the determination of typical molecular electric properties such as the static dipole moment of a molecule, which reflects the molecular charge distribution and is therefore related to the quality of the ground state electron density. If a molecule is exposed to an electric field $\vec{F}$ the charge density will respond to this perturbation and the energy of the system will be modified. The energy can then be described in terms of a Taylor expansion relative to its field-free energy (i, j, k and l run over the Cartesian components, i. e., x, y and z):

$$E(\vec{F}) = E(\vec{0}) + \sum_i \left(\frac{\partial E}{\partial F_i}\right)_0 F_i + \frac{1}{2}\sum_i\sum_j \left(\frac{\partial^2 E}{\partial F_i \partial F_j}\right)_0 F_i F_j \\ + \frac{1}{6}\sum_i\sum_j\sum_k \left(\frac{\partial^3 E}{\partial F_i \partial F_j \partial F_k}\right)_0 F_i F_j F_k \\ + \frac{1}{24}\sum_i\sum_j\sum_k\sum_l \left(\frac{\partial^4 E}{\partial F_i \partial F_j \partial F_k \partial F_l}\right)_0 F_i F_j F_k F_l + \cdots \tag{10-1}$$

The derivatives of the energy taken at zero field define the static[43] response properties of the molecule and are a measure of the interaction between the applied electric field and the system. In particular we are dealing with the following properties, which are sometimes also classified as first, second and higher order properties, depending on the degree of differentiation of the energy:

$$\mu_i = -\left(\frac{\partial E}{\partial F_i}\right)_0 \tag{10-2}$$

the i'th component of the *dipole moment* vector $\vec{\mu}$,

$$\alpha_{ij} = -\left(\frac{\partial^2 E}{\partial F_i \partial F_j}\right)_0 \tag{10-3}$$

---

[43] We only consider *static* response properties in this chapter, which arise from a *fixed* external field. Their *dynamic* counterparts describe the response to an oscillating electric field of electromagnetic radiation and are of great importance in the context of non-linear optics. As an entry point to the treatment of frequency-dependent electric response properties in the domain of time-dependent DFT we recommend the studies by van Gisbergen, Snijders, and Baerends, 1998a and 1998b.

the i,j'th component of the *polarizability* tensor $\ddot{\alpha}$,

$$\beta_{ijk} = -\left(\frac{\partial^3 E}{\partial F_i \partial F_j \partial F_k}\right)_0 \tag{10-4}$$

the i,j,k'th component of the *first hyperpolarizability* tensor $\ddot{\beta}$, and

$$\gamma_{ijkl} = -\left(\frac{\partial^4 E}{\partial F_i \partial F_j \partial F_k \partial F_l}\right)_0 \tag{10-5}$$

the i,j,k,l'th component of the *second hyperpolarizability* tensor $\ddot{\gamma}$.

In the following we will concentrate on the quality of results obtained for these quantities from density functional theory. A more general discussion of polarizabilities, hyperpolarizabilities etc., is beyond the scope of the present book, but can be found in many textbooks on physical or theoretical chemistry, such as Atkins and Friedman, 1997.

Besides these response properties of a molecule we will also devote one section in this chapter to the experimentally important infrared intensities, which are needed to complement the theoretically predicted frequencies for the complete computational simulation of an IR spectrum. This discussion belongs in the present chapter because the infrared intensities are related to the derivative of the permanent electric dipole moment μ with respect to geometrical parameters.

Many of these properties are known to be very sensitive to the effects of electron correlation. Therefore simple wave function based approaches such as the Hartree-Fock method are often inadequate and need to be augmented by some strategy to include these contributions. In addition, it is well known from conventional approaches that flexible basis sets augmented with polarization and diffuse functions are needed for a successful description of response properties involving perturbations from an electric field. Among the reasons for this demand is that the basis sets are usually independent of the perturbation. In other words, the basis set itself does not change under the influence of the electric field. Hence, we must augment the basis set by additional functions such that it acquires enough flexibility to allow the density to properly respond to the perturbation. This reasoning already applies to the computation of dipole moments but even more so for higher derivatives of the energy with respect to the electric field or mixed derivatives, such as the infrared intensities discussed below. Because of the need for including electron correlation and large basis sets, conventional methods soon become too demanding in terms of computer resources and cannot be applied to larger, chemically relevant molecules. Thus, an assessment of the applicability of density functional methods as a potential alternative is of great interest.

10.1 Population Analysis

Atomic partial charges are a difficult concept in quantum chemistry. On the one hand, assigning charges to individual atoms in a molecule is very close to the classical interpretations so successfully used in organic or inorganic chemistry and allows these inherently qualitative models to be quantified. However, on the other hand it is well known that this very concept is totally artificial since atomic charges in a molecule do not represent a physical observable. Stated in other words, questions such as 'what charge does the oxygen atom in water carry?' are simply not allowed and lie outside the physics of the problem. Such fundamental reasoning notwithstanding, the availability of charges of individual atoms in a molecule would offer a lot of interpretational power and would be of enormous help in analyzing quantum chemical calculations. Hence, a lot of different schemes to partition the total electron density (which of course is a physical observable) onto the individual atoms have been developed in the past. Because of the lack of a fundamental physical basis, all these recipes necessarily contain a certain amount of ambiguity and there cannot be a 'best' population analysis. Also, because no unequivocal reference values exist to which a population analysis must finally converge, it is important to keep in mind that one must not compare results obtained from different schemes or even values obtained within one method but with different basis sets or numerical integration grids. We will not pursue these principle aspects of population analysis any further, since extensive discussions of this subject can be found in most quantum chemical textbooks. As a good and still up-to-date source of further information we refer the reader to the review of Bachrach, 1995.

Geerlings, De Proft, and Martin, 1996, have compared the atomic populations of 15 small molecules as obtained from Bader's Atoms-in-Molecules (AIM, see Bader, 1994) approach for population analysis using a variety of traditional, wave function based methods and approximate density functionals. As a reference they selected QCISD, as a fairly sophisticated correlated wave function based technique. The AIM scheme is probably the theoretically soundest way of partitioning the electron density between the atoms in a molecule. It is solely based on the properties of the charge density and makes no recourse to basis functions and the like. The integral of the density assigned to a particular atom (where the area belonging to an atom, the so-called *atomic basin* is defined through the gradient paths of the density) gives the corresponding number of electrons on that center and hence its partial charge. The deviation between the atomic populations of the various methods from the QCISD reference is of course an indicator of the differences between the corresponding charge densities. A word of warning not to over-interpret these numbers is, however, in order. The authors chose a rather modest cc-pVDZ basis set and the resulting QCISD/cc-pVDZ level is certainly not sophisticated enough to produce highly accurate electron densities. The deviations in the atomic populations therefore reflect only the internal consistency of these different quantum chemical methods and do not include information about the quality of the charge density on an absolute scale. Electron correlation has important consequences for the atomic populations as evidenced by the large mean absolute deviation of 0.133 |e| for the uncorrelated Hartree-Fock ansatz. The absolute values of the atomic charges are usually too large at this level, as expected from the known trend of Hartree-Fock to produce charge distributions that are often too polar. In the DFT domain, the uniform electron gas based SVWN functional also exhibits fairly large deviations with an average error of 0.105 |e|. The two GGA functionals studied, BP86 and BLYP, are closer to the QCISD values with mean deviations of 0.071 and 0.067 |e|, respectively, while the hybrid functionals B3LYP and B3PW91 are closer still, the deviations are reduced to 0.035 and 0.038 |e|, respectively. The DFT charges are in most cases smaller than the QCISD ones, indicating less polar charge distributions. The error reduction upon going from LDA to GGA functionals is easily rationalized if one considers the main difference between the corresponding densities. As shown, e. g., by Fan and Ziegler, 1995, the changes in $\rho(\vec{r})$ when going from the LDA to the GGA are small. The main effect is that $\rho(\vec{r})$ increases in the core region and the valence tails, while it is depleted in the intermediate region. Hence, the polarization in the charge distribution is usually somewhat increased. Overall, the compatibility of correlated wave function based and density functional results enhances the level of credibility that can be attributed to the many studies found in the literature which make use of population analysis based on density functional methods, in particular if hybrid functionals are being used.

10.2 Dipole Moments

The permanent moments of a molecule are important descriptors of the ground state charge distribution. The ability of a certain theoretical model to reproduce experimental permanent dipole moments is therefore very helpful in assessing the quality of the corresponding electron probability distribution. The dipole moment is in addition an important physical property. For example, the electrostatic interaction of two molecules with non-vanishing permanent dipole moments is dominated by the dipole-dipole term. There have been numerous investigations of dipole moments at the Hartree-Fock and post-HF levels which underline the importance of electron correlation for reaching accurate results. The most famous example in this respect is probably the carbon monoxide molecule, whose experimental dipole moment amounts to 0.11 D with the negative end at the carbon atom, i. e., being of $^{-}C{=}O^{+}$ polarity. The Hartree-Fock approximation is not even able to reproduce this qualitative orientation. Near the HF limit (i. e., using a very large, almost complete one-electron basis set) a dipole moment of –0.28 D is calculated, i. e., the HF dipole moment has the reverse orientation, $^{+}C{=}O^{-}$. Modern approximate density functional theory performs much better. Table 10-1 summarizes some typical results for dipole moments[44] $\mu = \sqrt{\mu_x^2 + \mu_y^2 + \mu_z^2}$ along the molecular axis from several recent investigations involving typical approximate exchange-correlation functionals (the negative sign in case of CO indicates a $^{+}C{=}O^{-}$ polarity).

Dickson and Becke, 1996, use a basis set free numerical approach for obtaining their LDA dipole moments, which defines the complete basis set limit. In all other investigations basis sets of at least polarized triple-zeta quality were employed. Some of these basis sets have been designed explicitly for electric field response properties, albeit in the wave function domain. In this category belong the POL basis sets designed by Sadlej and used by many authors as well as basis sets augmented by field-induced polarization (FIP) func-

[44] The dipole moment is a vector property, but usually only its magnitude is given.

Table 10-1. Dipole moments for selected molecules [D, 1 D = 0.3934 a.u.].

Molecule	HF POL[a]	MP2 POL[a]	SVWN numerical[b]	SVWN TZVP-FIP[c]	BLYP TZVP-FIP[c]	BLYP POL[a]	BLYP 6-31G(d)[d]	B3LYP cc-pVTZ	B3LYP POL[a]	Exp.
CO	–0.25	0.31	0.23	0.24	0.19	0.19	0.15	0.13	0.10	0.11
H_2O	1.98	1.85	1.86	1.88	1.83	1.80	2.04	1.92	1.86	1.85
H_2S	1.11	1.03		1.15	1.07	0.97		1.19	1.01	0.97
HF	1.92	1.80	1.80	1.81	1.76	1.75	1.81	1.83	1.80	1.83
HCl	1.21	1.14				1.08		1.21	1.12	1.11
NH_3	1.62	1.52	1.53	1.57	1.52	1.48	1.90	1.59	1.52	1.47
PH_3	0.71	0.62				0.59		0.53	0.62	0.57
SO_2	1.99	1.54				1.57		2.01	1.67	1.63

[a] Cohen and Tantirungrotechai, 1999; [b] Dickson and Becke, 1996; [c] Calaminici, Jug and Köster, 1998, [d] Johnson, Gill and Pople, 1993.

tions, which are rather diffuse s, p, d and possibly also f-type functions. It is clear from the representative data given in Table 10-1 that smaller basis sets cannot be recommended because their performance can be significantly worse. For example, with the popular and often used 6-31G(d) basis set all density functionals, including SVWN and BLYP, employed in the early study of Johnson, Gill, and Pople, 1993, significantly overestimated the NH_3 dipole moment by 0.4 to 0.5 D, while all the other data in Table 10-1 are within some 0.1 D for this molecule. The POL and FIP basis sets customized for electric response properties do significantly better than the regular polarized triple-zeta sets. For example, the numerical results by Dickson and Becke are in very good agreement with the local density approximation data obtained by Calaminici, Jug, and Köster, 1998, with the TZVP-FIP basis set and the dipole moments obtained with the POL basis are closer to the experimental μ than those computed with the standard cc-pVTZ expansion of the KS orbitals. The data in Table 10-1 demonstrate that, overall, the density functionals yield rather good dipole moments, provided they are combined with flexible enough basis sets. These techniques are indisputably much better than the Hartree-Fock approach but at comparable costs. Notably, even the problematic CO molecule poses no difficulty to DFT. All functionals employed, including the simple local density approximation provide dipole moments for carbon monoxide which not only show the correct $^{-}C{=}O^{+}$ orientation, but also give very satisfactory quantitative results. For the totally uncorrelated HF and the partially correlated MP2 wave function methods, Cohen and Tantirungrotechai, 1999, report mean absolute deviations of 0.17 and 0.05 D for the dipole moments of the ten small molecules of their test set. The two GGA functionals included in their study, BLYP and HCTH, clearly outperform the HF approximation but fall short of MP2 with an average error of 0.09 D for both DFT procedures. The winners among the standard functionals are once more the hybrid functionals. The commonly used B3LYP protocol is characterized by a mean absolute deviation of only 0.04 D. Cohen and Tantirungrotechai also tested very recent functionals, such as two different flavors of Becke's 10-parameter functional alluded to in Chapter 6, namely B97 and B97-1. These functionals also perform very satisfactorily, but not better than B3LYP, with errors around 0.04 D. With the larger JGP set of 32 molecules selected from the G2 data base and with the use of the rather large 6-311++G(3df,3pd) basis set, which should be close to being saturated, Adamo, di Matteo, and Barone, 1999, report in their recent review mean absolute errors which underline these conclusions and confirm the expected hierarchy of functionals. Quoting from their results, the SVWN dipole moments are on average off by 0.25 D, all the GGA functionals perform similarly to each other but considerably better than the local approximation with mean unsigned deviations scattering between 0.10 to 0.12 D. Finally, the three-parameter hybrid functionals work best with errors of only 0.08 D. This latter result is particularly noteworthy since it indicates that for this particular set of molecules the hybrid functionals surpass even highly correlated and computationally expensive wave function methods in accuracy: with the same basis set the Hartree-Fock, MP2, and CCSD(T) techniques yield mean unsigned errors of 0.29, 0.28, and 0.10 D, respectively.

10.3 Polarizabilities

The static electric polarizability $\ddot{\alpha}$ is a measure of the ease with which the electronic distribution, i. e., the charge density of a system will get distorted by an external electric field. Atomic and molecular polarizabilities are important properties in many areas. For example a large fraction of the electrostatic intermolecular interaction energy is related to this quantity, in particular for systems without a permanent dipole moment. Similarly, polarizabilities are crucial for an understanding of many properties in molecular optics and spectroscopy. From a practical point of view, polarizabilities and hyperpolarizabilities as defined in equations (10-3) to (10-5) can be computed by numerical differentiation of the field-dependent energy or dipole moment. This so-called *finite field* approach is easy to implement but prone to errors due to problems with numerical stability in the differentiation – errors which are very difficult to control. The alternative is to compute the derivatives analytically, which is more elaborate in terms of programming since it involves solving the so-called coupled-perturbed Kohn-Sham equations (see Colwell et al., 1993, and Lee and Colwell, 1994) but is numerically considerably more stable. It is the latter method which is implemented in most major codes, such as *Gaussian 98*, but the finite field approach is still being used, see, e. g., Calaminici, Jug, and Köster, 1998. If we take the additional electric field as a perturbation to the original Hamiltonian of the system, the polarizabilities can also be expressed through perturbation theory. One can then show that the polarizability is inversely proportional to the excitation energies of the system, i. e. $\alpha \propto \sum_{n=1} \frac{1}{E_0 - E_n}$ (0 is the ground state, n denotes the n'th excited state and the sum must of course exclude n = 0). It is not our intention to make an excursion into perturbation theory, we only mention this result because it allows a pictorial insight into some of the factors controlling the polarizability of a system and helps to understand why density functional methods perform the way they do. The energy differences between the ground and excited states can to a first approximation be expressed in terms of the corresponding Kohn-Sham orbital energy differences. The largest contribution will obviously be from the excitations involving high-lying occupied and low-lying unoccupied Kohn-Sham orbitals which give the smallest denominator, but the other energy differences also contribute. We have seen in the preceding discussions that due to the incorrect asymptotic behavior of the exchange-correlation potentials, all regular density functionals have problems in describing those orbital energies. In particular, because the long-range potentials are not attractive enough, the energies of the highest occupied orbitals usually come out significantly too high and the orbital differences are hence too low. One should therefore expect that these functionals should predict polarizabilities that are exaggerated. And indeed, that is exactly what is observed. Before we enter an in-depth discussion of the performance of the various approximate density functionals we should point out that the computed static polarizability data are not strictly comparable with experimental ones, because it is difficult to deduce the pure static, frequency independent values from the experimental data and to remove the unwanted contributions from vibrational, rotational or other effects. Generally it is expected that accounting for such effects would increase the computed value. In other words, a slight underestimation of the experimental data by the computational methods – which are strictly frequency independent – is actually a desirable feature. Already in the early study of Guan et al., 1993, who analyzed the performance of the local density approximation with regard to molecular polarizabilities the tendency of the LDA to overestimate the mean polarizabilities[45] <α> was noted. These authors were probably also the first to underline the need for flexible basis sets, which must include polarization and diffuse functions to properly describe polarizabilities. Subsequently, McDowell, Amos, and Handy, 1995, reported mean polarizabilities <α> of a series of small molecules obtained with the SVWN and BLYP functionals in comparison with Hartree-Fock, correlated wave function and experimental data employing Sadlej's POL basis set especially optimized for this purpose. While the HF polarizabilities were systematically too low by about 5–10 %, exactly the opposite trend was found for the density functionals. Both functionals studied overestimated <α> fairly significantly. In terms of absolute errors, the density functional methods were not superior to the HF approximation. Furthermore and in contrast to many other properties, in this case the simple SVWN functional generally provided results closer to the experimental or highly correlated reference data than the gradient-corrected BLYP technique. In order to assess the basis set dependence of their results, the authors repeated the calculations with a larger set, where in particular the diffuse part was improved. Irrespective of the method this led to an *increase* of the calculated polarizabilities (including the HF and related, more sophisticated models). This is in perfect agreement with the numerical, basis-set free calculations of Dickson and Becke, 1996, employing the SVWN functional. Their average polarizabilities are in most cases larger than any results using the same functional but with a finite basis set. In other words, increasing the basis in general also increases <α>. Since the LDA and GGA functionals already systematically overestimate this quantity, this means that by increasing the basis set the error for the density functional methods also gets larger. Thus, we have another typical example of getting the right answer for the wrong reason. Using only a small basis set might give overoptimistic results for polarizabilities, which do not mirror the intrinsic accuracy of density functional theory but are only due to a fortuitous error cancellation. Other studies essentially confirmed these conclusions, see, e. g., Calaminici, Jug, and Köster, 1998. Fuentalba and Simón-Manso, 1997, included the B3PW91 and B3LYP hybrid functionals in their study and showed that they lead to a significant improvement. Since these functionals contain some Hartree-Fock exchange and HF is known to give too small polarizabilities, the resulting <α> should indeed benefit from error compensation. Another interpretation for the better performance is that including exact HF exchange improves the asymptotic decay of the exchange potential, which is so important for polarizabilities. These authors also included the LB94 potential, which is characterized by the correct –1/r asymptotic decay. Supporting the above reasoning, this choice also yielded improved results, in harmony with the qualitative discussion above that functionals which give better asymptotic potentials should partially correct the usually observed overestimation of polarizabilities. Similar observations regarding

[45] Usually, the individual components of the polarizability tensor $\ddot{\alpha}$ are not given, but only the average value of its diagonal elements which is defined as $\langle\alpha\rangle = 1/3\ (\alpha_{xx} + \alpha_{yy} + \alpha_{zz})$.

the LB94 potential and the importance of an accurate potential can be extracted from the work of van Gisbergen, Snijders, and Baerends 1998b (see also the related earlier contribution by van Gisbergen et al., 1996 and the more recent study by Banerjee and Harbola, 1999), who used accurate exchange-correlation potentials derived from high quality charge densities from configuration interaction calculations. For the six small molecules included in their study, they found the LDA to overestimate the experimental $<\alpha>$ by 8.8 %, which improved to an average absolute error of 3.6 % for LB94 (no systematic error, the computed data scatter below and above the reference values and the average signed error amounts to only 0.6 %). The role of asymptotically correct exchange-correlation functionals was also the focus of the polarizability calculations by Tozer and Handy, 1998. These authors showed that the improved asymptotic behavior of the HCTH functional was mirrored in the quality of the computed mean polarizabilities. Inclusion of the 'asymptotic correction' (cf. Section 6.8) resulted in even better data. For the 14 molecules included in their test set, mean absolute deviations of 0.83 a. u. or 5.1 % (SVWN), 0.90 a. u. or 5.9 % (BLYP), 0.33 a. u. or 2.0 % (B3LYP), 0.36 a. u. or 2.5 % (HCTH) and 0.26 a. u. or 1.3 % (HCTH(AC)) are obtained with the POL basis set. SVWN and BLYP systematically overestimate the mean polarizabilities. B3LYP and HCTH have the same trend, but to a lesser extent and with two exceptions, CO_2 and CH_4. Use of the adiabatically corrected HCTH functional/potential always lowers $<\alpha>$, usually improving the agreement with the experiment. Only for carbon dioxide and methane – which were already too low at HCTH – and ethylene is no improvement as compared with the HCTH data observed. For comparison, the MP2 treatment is very close to the HCTH(AC) results with a mean absolute error of 0.24 a. u. or 1.4 %. Van Caillie and Amos, 1998, and Cohen and Tantirungrotechai, 1999, also used the POL basis sets and studied the mean polarizabilities of a number of species including slightly larger molecules such as ethane, ethene, ethyne and butadiene. Their qualitative results overall confirm the above conclusions. In particular the B3LYP functional led to average polarizabilities with an accuracy approaching (but not surpassing) that of sophisticated wave function techniques. Finally, we mention a recent study by Adamo et al., 1999. These authors applied the two hybrid functional B3LYP and PBE1PBE for a larger variety of molecules, including some aliphatic and aromatic hydrocarbons, using different basis sets. In line with the surprisingly good performance of the PBE1PBE functional for excitation energies noted in Chapter 9, this functional also works very well for polarizabilities. While for B3LYP and HCTH(AC) mean absolute errors in $<\alpha>$ of 0.39 and 0.29 a. u. are obtained, for the PBE1PBE functional an error of only 0.20 a. u. with respect to a small set of twelve reference molecules has been noted. Also, this functional does not show a systematic error, rather the results scatter irregularly around the experimental target numbers. Actually, Adamo et al. also included fairly sophisticated wave function approaches such as second and fourth order perturbation theory (MP2 and MP4) or the Brueckner doubles with perturbational estimate of triple excitations (BD(T)) strategy. These methods were characterized by mean absolute deviations of 0.25, 0.28, and 0.23 a. u., respectively. Interestingly, none of these did surpass the accuracy obtained from the PBE1PBE functional, in spite of being significantly more expensive computationally. For the eight hydrocarbons, including fairly large molecules such as naphthalene, the PBE1PBE functional continues to provide

Table 10-2. Average Polarizablities <α> for selected molecules [a.u., 1 a.u. = 0.1482 Å^3].

Molecule	HF POL[a]	MP2 POL[a]	SVWN numerical[b]	SVWN POL[a]	SVWN TZVP-FIP[c]	BLYP POL[a]	BLYP TZVP-FIP[c]	B3LYP POL[a]	B3LYP d-augTZ[d]	HCTH(AC) POL[e]	PBE1PBE POL[f]	Exp.[a,d]
HF	4.88	5.67	6.23	6.17	5.94	6.26	6.00	5.83	5.79	5.60	5.67	5.60
HCl	16.67	17.37		18.43		18.54		17.90	17.99	17.77	17.58	17.39
F_2	8.58	8.22		8.82		8.96		8.69	8.46	8.43	8.37	8.38
Cl_2	29.89	30.56		31.70		31.97		31.16	31.27	30.84	31.05	30.35
CH_4	15.91	16.54	17.70	18.01		17.59		17.03	17.25	16.51	16.86	17.27
SiH_4	29.97	31.04		34.28		33.14		32.25	32.13			31.90
NH_3	12.94	14.42	15.54	15.57	14.96	15.62	14.94	14.73	14.73	14.25	14.34	14.56
PH_3	29.93	30.69		32.52		32.13		31.35	31.32	31.07	31.03	30.93
H_2O	8.51	9.80	10.60	10.54	10.20	10.64	10.26	9.96	10.01	9.67	9.69	9.64
H_2S	23.77	24.70		26.13	24.95	26.07	24.72	25.24	25.19	24.99	24.60	24.71
CO	12.23	13.09	13.70	13.62	13.39	13.66	13.43	13.18		13.03		13.08

[a] Cohen and Tantirungrotechai, 1999 and McDowell, Amos and Handy, 1995; [b] Dickson and Becke, 1996; [c] Calaminici, Jug and Köster, 1998; [d] van Caillie and Amos, 1998; [e] Tozer and Handy, 1998; [f] Adamo et al., 1999.

Table 10-3. Compilation of mean absolute deviations for static average polarizabilities [a.u.] of small main group molecules from different sources.

13 molecules, POL basis set, McDowell, Amos and Handy, 1995			
HF	1.18	BD(T)	0.36
MP2	0.36	LDA	0.99
MP4	0.40	BLYP	0.95
19 molecules, augmented TZP STO basis set, van Gisbergen et al., 1996			
LDA	0.92	LB94	0.63[a]
BP86	0.43		
8 molecules, numerical, basis set free, Dickson and Becke, 1996			
LDA	0.60		
16 molecules, POL basis set, Van Caillie and Amos, 1998			
HF	1.06	B3LYP	0.38
LDA	0.99		
16 molecules, d-aug-cc-pVTZ basis set, Van Caillie and Amos, 1998			
HF	1.07	B3LYP	0.39
LDA	0.98		
14 molecules, POL basis set,Tozer and Handy, 1998			
LDA	0.83	HCTH	0.36
BLYP	0.90	HCTH(AC)	0.26
B3LYP	0.33	MP2	0.24
5 molecules,TZVP+FIP basis set, Calaminici, Jug and Köster, 1998			
HF	1.29	BLYP	0.41
LDA	0.33	CCSD(T)	0.31
12 molecules, POL basis set, Adamo et al., 1999			
MP2	0.25	B97	0.42
MP4	0.28	B3LYP	0.39
BD(T)	0.23	HCTH	0.29
PBE1PBE	0.20		
20 molecules, POL basis set, Cohen and Tozer, 1999			
HF	1.76	HCTH	1.38
MP2	0.95	B3LYP	1.79
BD	1.29	B97	1.50
BLYP	2.25	B97-1	1.53

[a] Note that the *signed* average error is considerably smaller. The LB94 potential shows no systematic errors.

promising results and significantly outperforms MP2 values, as far as the latter are available. It should be noted that in this case the standard B3LYP functional performs significantly worse on average than the correlated methods, including MP2. Unfortunately, none of the currently popular GGA functionals were included in this comparison. A representative collection of the data that we discussed in this section can be found in Table 10-2, while Table 10-3 summarizes some of the recent literature studies on the accuracy of DFT methods for $<\alpha>$.

The bottom line of this section is that density functional methods certainly provide a promising tool for predicting static average polarizabilities. However, simple functionals such as SVWN or GGA approaches like BLYP systematically and significantly overestimate these quantities. There are probably several reasons for this, among which the erroneous asymptotic decay of the corresponding exchange-correlation potentials holds a prominent place. Hybrid functionals and approaches specifically designed for dealing with the asymptotic decay problem lead to much better results. The workhorse of today's DFT practitioners, B3LYP, performs satisfactorily, even though the results are frequently somewhat less accurate than from standard MP2 theory. In terms of basis sets one should make sure that they include polarization as well as diffuse functions. The POL basis sets, even though optimized for wave function approaches, seems to offer a particularly well suited compromise between size and computational economy for density functional approaches too.

10.4 Hyperpolarizabilites

The tensor of the static first hyperpolarizabilities $\ddot{\beta}$ is defined as the third derivative of the energy with respect to the electric field components and hence involves one additional field differentiation compared to polarizabilities. Implementations employing analytic derivatives in the Kohn-Sham framework have been described by Colwell et al., 1993, and Lee and Colwell, 1994, for LDA and GGA functionals, respectively. If no analytic derivatives are available, some finite field approximation is used. In these cases the $\ddot{\beta}$ tensor is preferably computed by numerically differentiating the analytically obtained polarizabilities. In this way only one non-analytical step, susceptible to numerical noise, is involved. Just as for polarizabilities, the individual tensor components are not regularly reported, but rather the average hyperpolarizability defined as $\langle\beta\rangle = \frac{3}{5}(\beta_{xxz} + \beta_{yyz} + \beta_{zzz})$. It is well known from many wave function and density functional based studies that hyperpolarizabilities are even more sensitive towards basis sets than lower-order electric properties. Hence, large basis sets including polarization and diffuse functions are a must if reasonable calculations on $<\beta>$ are to be performed. The hyperpolarizabilities are largely determined by the electron density at long range and thus by the corresponding exchange-correlation potentials far from the nuclei. Therefore the dependence on the particular functional employed should be similar to or even more pronounced than what we discussed above for the polarizability $\ddot{\alpha}$. Indeed, normal LDA and GGA functionals provide results for $<\beta>$ which are similar to each other but in general much too large if compared to experimental or accurate, correlated wave function data. We note in passing that for the hyperpolarizabilities the comparison to experimental data is even more problematic than for $\overleftrightarrow{\alpha}$, because they usually do not refer to the static limit and are severely contaminated by vibrational, rotational and other contributions. As shown in the recent study by Cohen, Handy, and Tozer 1999, however, the extent to which $\langle\beta\rangle$ is overestimated varies significantly from molecule to molecule. For example, the BLYP average hyperpolarizability of formaldehyde, $H_2C{=}O$, is too large by some 125 % whereas for the hydrogen fluoride and carbon monoxide molecules the error is less than 25 % if measured against accurate numbers from correlated wave function based calculations. Similar results are obtained with SVWN as demonstrated by many studies including the basis-free approach of Dickson and Becke, 1996. Better results are furnished by hybrid functionals like B3LYP, probably due to the improved asymptotic behavior brought about by the admixture of some exact, non-local Hartree-Fock exchange. Special strategies, which either include a large flexibility like the highly parameterized B97-1 hybrid functional, or which are designed with the asymptotically correct potential in mind such as the LB94 potential or the HCTH(AC) protocol, are better still. But these are also certainly not a panacea and some problematic cases remain. For example, a reliable computational result of $\langle\beta\rangle$ for acetonitrile, CH_3CN, is –40.4 a. u. The SVWN and BLYP functionals show the typical overestimation and give values of –62 and –64 a. u., respectively. As expected, the admixture of exact exchange in the B3LYP protocol reduces the error and yields –53 a. u. Rather than reducing the value for $\langle\beta\rangle$ further and thus bringing it closer to the reference, the asymptotically corrected HCTH(AC) procedure predicts $\langle\beta\rangle = -145.5$ a. u., miles away from the target. The reason is probably that the parent HCTH functional produces a lowest unoccupied orbital of CH_3CN with Rydberg character which apparently the asymptotic correction is unable to cope with and the whole thing blows up. A compilation of representative results for computed hyperpolarizabilities from the recent literature can be found in Table 10-4.

Little is known about the predictive power of density functionals for second hyperpolarizabilities, $\ddot{\gamma}$. But the trends discussed so far for $\ddot{\beta}$ can be safely extrapolated to $\ddot{\gamma}$. However, the basis set requirements are larger still and more subtle. More functions in the polarization space are required as well as a very flexible description of the outer regions with diffuse functions in order to achieve converged results. For example, Calaminici, Jug, and Köster, 1998, show that upon adding a set of field-induced f-functions to their d-polarized triple-zeta FIP basis set, the value of $\langle\gamma\rangle$ computed with the SVWN functional increased dramatically for most of the small molecules included in their study (depending on the molecule by 50 to more than 100 %). The overall accuracy seems to be less satisfactory than for the lower order properties. LDA and GGA functionals severely overestimate the individual tensor components and the average of γ due to the wrong asymptotic behavior of the corresponding potentials. The LB94 potential improves the results but usually errs in the opposite direction if compared to accurate coupled-cluster (CCSD(T)) data as shown by van Gisbergen, Snijders, and Baerends, 1998b and Banerjee and Harbola, 1999.

Table 10-4. Average first hyperpolarizablities <β> for selected molecules [a.u.].

Molecule	Accurate ab initio[a]	SVWN numerical[a]	SVWN aug-POL[c]	SVWN TZVP-FIP[d]	SVWN aug-STO-VTZP[e]	BLYP aug-POL[c]	BLYP TZVP-FIP[d]	BLYP aug-STO-VTZP[e]	B3LYP aug-POL[c]	B77-1 aug-POL[c]	HCTH(AC) aug-POL[c]	LB94 aug-STO-VTZP[e]
HF	–7.3	–9.2	–8.9	–9.6	–9.2	–8.7	–9.9	–9.5	–7.4	–7.2	–7.4	–6.9
H_2O	–18.0	–24.8	–24.8	–22.4	–25.7	–24.4	–22.3	–27.1	–19.2	–18.7	–18.3	–16.7
H_2S	–7.7		–14.7	–9.1		–10.5	–9.8		–7.2	–7.4	–10.2	
NH_3	–34.3	–55.6	–55.7	–34.1	–55.4	–58.6	–34.9	–64.6	–41.0	–38.9	–33.3	–30.5
H_2CO	40.3		90.7			90.9			67.1	64.1	43.1	
CH_3F	40.3		63.7			62.3			50.6	49.7	34.3	
CH_3CN	–40.4		–61.6			–64.2			–52.9	–46.3	–145.5	
CO	23.5, 26.6	30.5	29.5	22.9	30.5	29.3	22.0	31.1	26.9	26.3	22.6	20.6

[a] From different sources, quoted in Cohen, Handy and Tozer, 1999 and van Gisbergen, Snijders and Baerends, 1998b; [b] taken from Dickson and Becke, 1996; [c] taken from Cohen, Handy and Tozer, 1999; [d] taken from Calaminici, Jug, and Köster, 1998; [e] taken from Gisbergen, Snijders and Baerends, 1998b.

10.5 Infrared and Raman Intensities

The computational prediction of vibrational spectra is among the important areas of application for modern quantum chemical methods because it allows the interpretation of experimental spectra and can be very instrumental for the identification of unknown species. A vibrational spectrum consists of two characteristics, the frequency of the incident light at which the absorption occurs and how much of the radiation is absorbed. The first quantity can be obtained computationally by calculating the harmonic vibrational frequencies of a molecule. As outlined in Chapter 8 density functional methods do a rather good job in that area. To complete the picture, one must also consider the second quantity, i. e., accurate computational predictions of the corresponding intensities have to be provided.

In the case of an infrared spectrum, the intensity is related to the square of the infinitesimal change of the electric dipole moment μ with respect to the normal coordinates,[46] q,

$$I_a = C\left|\frac{\partial \mu}{\partial q_a}\right|^2 . \tag{10-6}$$

C is a numerical constant and includes the degeneracy of the vibration a. Recall from equation (10-2) that $\mu = -\left(\frac{\partial E}{\partial F}\right)$, i. e. with respect to the energy the I_a are the mixed second derivatives with respect to the electric field and to the nuclear coordinates and hence also a second-order property. In the realm of conventional wave function techniques it is well established that the reliable prediction of intensities is more demanding than locating the frequencies of the vibrational transitions. Very akin to the response properties discussed above, electron correlation turns out to be a particularly important factor for obtaining useful predictions. Consequently, the Hartree-Fock approximation is not a useful alternative. In addition, basis sets with polarization and diffuse functions are needed to sample the tail regions of the wave functions and allow the basis set to respond to the electric field perturbation and are therefore a second prerequisite. As we know well by now, density functional theory implicitly includes electron correlation through the exchange-correlation functional. Indeed, successful computations of DFT frequencies and corresponding infrared intensities are reported in a large number of papers (for representative examples see Stepanian et al., 1999, and Bauschlicher, Hudgins, and Allamandola, 1999). The most comprehensive systematic investigation of the performance of modern density functionals, including hybrid ones, is probably the recent study of Halls and Schlegel, 1998. These authors compared the local SVWN functional with the most popular GGA implementations (BLYP, BP86, and BPW91) and their 3-parameter hybrid counterparts (B3LYP, B3P86, and B3PW91) on a test set of twelve small, mostly organic molecules. The quantitative

[46] To be precise, this expression employs the so-called *double-harmonic approximation*, where cubic and higher force constants as well as second and higher dipole derivatives are ignored. This approximation is common to all current implementations of calculating IR and Raman intensities. For details see Amos, 1987.

determination of absolute experimental IR intensities is difficult and usually only accurate to within ±10%. In addition, the double harmonic approximation introduces another uncertainty of about ±10% and hence no strict comparability between experimental and computationally predicted data exists. To circumvent these ambiguities the authors chose intensities computed at the conventional QCISD level as reference data. With the rather modest 6-31G(d) basis set, the local SVWN functional had a mean absolute error of 9.5 km/mol, less than half of the mean deviation of the Hartree-Fock values (24.2 km/mol) from the QCISD reference with the same basis set. Interestingly, the GGA functionals did not come out any better. On the contrary, although in the same ballpark, the mean absolute errors of BLYP (12.1 km/mol), BP86 (10.6 km/mol), and BPW91 (10.2 km/mol) are all larger than the SVWN result. A big difference is connected with the admixture of some exact exchange. The hybrid functionals cut the error down to half and win the competition with rather low deviations of 5.8 km/mol (B3LYP), 4.9 km/mol (B3P86), and 4.8 km/mol (B3PW91). For comparison, the post-HF method MP2 has an error of 6.3 km/mol. If the basis set is increased to the rather large 6-311+G(3df,3pd) set, these conclusions remain by and large valid. With this basis set the HF, SVWN, BLYP, and B3LYP mean absolute errors with respect to the QCISD level amount to 27.1, 10.9, 10.6, and 5.0 km/mol, respectively. If we take the QCISD/6-311+G(3df,3pd) data as close to the converged final and hence idealized experimental values, we finally ask how big a basis must be used to get close enough to the above errors. Not unexpectedly, the 6-31G(d) set is too small and not well suited. Its SVWN and BLYP errors with respect to QCISD/6-311+G(3df,3pd) amount to 20.4 and 24.3 km/mol. Only after augmenting this set with several polarization and diffuse functions an acceptable deviation close to 10-11 km/mol results. For example with the 6-31+G(3d,2p) set, the SVWN and BLYP errors are 12.1 and 11.5 km/mol, respectively, which are quite satisfactory. In spite of the very elaborate comparisons presented in that study one must keep in mind that the quantitative analysis should not be overvalued. IR intensities are very sensitive quantities and using sophisticated coupled cluster CCSD(T) calculations rather than QCISD together with large basis sets actually leads to deviations of up to 10% and more, in spite of the formal similarity of the two approaches. In their study, De Proft, Martin, and Geerlings, 1996, also included IR intensities computed with the B3LYP functional and compare these with coupled cluster and QCISD results. Their results corroborate the above conclusions that this functional is well suited for predicting infrared intensities if combined with d,f-polarized triple zeta basis sets. Smaller basis sets are on average (but not in all cases) worse but still allow a semiquantitative analysis of spectra. We should also mention the early study by Fan and Ziegler, 1992, who applied the LDA and BP86 functionals together with various STO basis sets. Their conclusion about the quality of DFT intensities and the need for sufficiently large basis sets is similar. However, unlike Halls and Schlegel, they report a small but consistent improvement in the computed intensities when going from local to GGA functionals. Some representative numerical data from the studies cited are collected in Table 10-5.

The intensities of Raman scattering depend on the square of the infinitesimal change of the polarizability α with respect to the normal coordinates, q. Since the polarizability itself is already the second derivative of the energy with respect to the electric field – see equa-

Table 10-5. Infrared intensities for selected molecules [km/mol].

Molecule (symm.)	Vibration (irrep)	Exp.[a]	CCSD(T) TZ2Pf[b]	QCISD Large[c]	SVWN 6-31G(d)[c]	SVWN TZP[d]	SVWN Large[c]	BP86 TZP[d]	BLYP 6-31G(d)[c]	BLYP Large[c]	B3LYP 6-31(d)[b]	B3LYP cc-pVTZ[b]	B3LYP Large[c]
HCN	ω_1 (σ)	54	64	65.5	61.1		76.6		46.2	58.6	52.5	60.7	67.4
($C_{\infty v}$)	ω_2 (π)	46	71	69.4	71.2		73.8		65.8	75.6	71.0	72.6	72.8
	ω_3 (σ)	0.1	0.1	0.66	0.8		0.7		0.3	0.2	2.0	1.3	1.4
CO_2	ω_2 (π_u)	48	59	68.0	47.4		51.8		43.2	48.1	61.4	63.7	64.1
($D_{\infty h}$)	ω_3 (σ_u)	548	634	708.4	442.5		563		428.1	562.3	545.8	629.2	677.1
H_2O	ω_1 (a_1)	2.2	4.7	4.8	2.1	3.8	5.3	0.7	0.0	1.6	1.7	3.2	4.5
(C_{2v})	ω_2 (a_1)	63.9	69.5	68.0	79.1	79.5	72.2	68.6	62.7	63.6	75.8	69.5	72.1
	ω_3 (b_2)	48.2	48.4	56.1	28.0	65.9	72.3	42.0	8.0	45.4	19.4	40.8	60.2
H_2CO	ω_1 (a_1)	75.5±7	59.4	62.6	57.5	70.7	72.4	79.5	63.1	81.1	55.9	69.2	72.3
(C_{2v})	ω_2 (a_1)	74±5	74.5	88.7	91.7	110.9	106.6	107.0	87.1	102.6	98.7	107.1	112.5
	ω_3 (a_1)	11.2±1	10.6	12.4	4.7	9.6	9.3	10.4	6.6	11.1	6.4	9.9	12.0
	ω_4 (b_1)	6.5±0.6	4.4	5.8	2.9	6.1	7.1	4.9	1.5	5.8	1.4	3.2	5.7
	ω_5 (b_2)	87.6±8	108.4	91.5	170.7	117.5	115.7	145.1	195.4	137.3	164.7	145.1	114.2
	ω_6 (b_2)	9.9±1	12.0	11.8	7.9	8.3	8.4	9.4	9.9	9.4	12.6	12.8	11.7
NH_3	ω_1 (a_1)	7.6±0.9	2.3	3.8	0.3	4.1	3.2	7.4	3.6	7.3	0.8	2.4	4.0
(C_{3v})	ω_2 (a_1)	138±6	147	141.4	218.1	139.5	156.8	120.6	120.8	131.3	156.1	146.7	150.0
	ω_3 (e)	3.8±0.8	3.8	6.6	1.2	17.6	23.1	5.1	7.0	4.2	1.0	1.1	9.2
	ω_4 (e)	28.2±0.5	31	27.6	39.8	36.7	38.8	33.4	23.2	29.4	30.6	33.6	32.8

[a] From different sources, quoted in De Proft, Martin, and Geerlings, 1996; [b] taken from De Proft, Martin, and Geerlings, 1996; [c] taken from the supplementary material of Halls and Schlegel, 1998, Large = 6-311++G(3df,3pd); [d] taken from Fan and Ziegler, 1992 using STO type basis sets.

tion (10-3) – Raman scattering intensities are a third order property and hence are expected to be even more sensitive to the functionals and basis sets used. In their systematic evaluation of density functional methods for the calculation of Raman intensities, Halls and Schlegel, 1999, report results for 12 small molecules obtained from SVWN, BLYP, B3LYP, *m*PW1PW with various basis sets and compare them to experimental as well as Hartree-Fock and MP2 data. The hybrid functionals work better than the BLYP functional which, in turn, offers no improvement as compared to the SVWN approximation. This somewhat disappointing performance of the GGA functionals for the determination of Raman intensities was already noted in an earlier study by Stirling, 1996. As indicated by the mean absolute deviations from experimental data collected in Table 10-6, the POL basis set provided an accuracy very similar to the much larger aug-cc-pVTZ set while the errors from using 6-31G(d) were more significant by about a factor of two. Based on these results Halls and Schlegel recommend the B3LYP/POL combination for computing Raman intensities. In a subsequent study Van Caillie and Amos, 2000, used time-dependent DFT to compute the Raman intensities for a similar set of molecules with the POL basis set. Among the functionals investigated were also recent exchange-correlation functionals such as PBE, PBE1PBE, and B97. Their conclusions substantiate the results of Halls and Schlegel. GGA functionals such as the PBE functional do not improve the SVWN results while the hybrid functionals B3LYP, B97 and in particular PBE1PBE perform much better, see Table 10-6. Van Caillie and Amos report also calculations where not the derivative of the static polarizability is employed but in which they use dynamic (i. e., frequency dependent) polarizabilities for

Table 10-6. Compilation of mean absolute deviations for Raman intensities [$Å^4$ amu^{-1}]

12 molecules, Halls and Schlegel, 1999			
HF/6-31G(d)	30.3	B3LYP/6-31G(d)	30.2
SVWN/6-31G(d)	31.0	*m*PW1PW/6-31G(d)	29.7
BLYP/6-31G(d)	31.2	MP2/6-31G(d)	30.2
HF/POL	15.1	B3LYP/POL	13.8
SVWN/POL	16.9	*m*PW1PW/POL	13.6
BLYP/POL	17.7	MP2/POL	11.5
HF/large[a]	14.6	B3LYP/large[a]	12.0
SVWN/large[a]	14.2	*m*PW1PW/large[a]	11.8
BLYP/large[a]	13.6	MP2/large[a]	10.0
10 molecules, POL basis set, Van Caillie and Amos, 2000			
HF	16.9	B97	13.2
SVWN	18.1	PBE1PBE	12.6
PBE	18.7	MP2	11.0
B3LYP	13.9		
HF[b]	17.3	B97[b]	14.5
SVWN[b]	19.1	PBE1PBE[b]	13.7
PBE[b]	19.6	B3LYP[b]	15.2

[a] large = aug-cc-pVTZ; [b] using dynamic polarizabilities.

determining the Raman intensities. This frequency dependence, which is usually neglected, increases the computed intensities on average by some 15% leading to an overall deterioration of the average errors. Hence, as the authors state, whether including the frequency dependence represents an improvement cannot be concluded at this time.

The take home message of this section is that local and GGA functionals perform more or less similarly for IR and Raman intensities, whereas the hybrid ones offer a significant improvement, yielding results comparable or even better than MP2 for significantly less computational cost. In terms of basis sets, at least double-zeta sets augmented by flexible polarization and diffuse functions are needed. The POL basis set seems to offer a particularly good price/performance ratio.

11 Magnetic Properties

When a molecule is under the influence of an external magnetic field, this perturbation gives rise to some very important effects, which all involve the interaction of a nuclear or electron spin with the local electronic currents induced by the externally applied magnetic field. Specifically, the interaction of a 'magnetic' nucleus, i. e., a nucleus whose spin $I \neq 0$, with an external magnetic field results in the well known *chemical shift* which is the prime observable in nuclear magnetic resonance (NMR) experiments. Similarly, the second important source of information from an NMR spectrum, i. e., nuclear *spin-spin coupling* effects is related to the interaction of a spin at one nucleus with the electronic currents brought about by a second magnetic nucleus, i. e., the interaction of two nuclear magnetic moments mediated by the electronic spin density. On the other hand, if it is not a nuclear spin, but the spin of an *unpaired* electron that interacts with the magnetic field induced currents we enter the domain of electron spin resonance (ESR, also known as electron paramagnetic resonance, EPR). The central quantities in these techniques are the so-called *g-tensor*, which in a way resembles the NMR chemical shifts as it describes differences in the interaction due to the chemical environment and the *hyperfine coupling constants* which probe the amount of unpaired spin density at the nuclear position.

Both spectroscopic methods enjoy an ever-increasing popularity in many chemical applications since they offer extremely valuable information about the geometrical and electronic structures of the system. While NMR is a well-established standard tool for probing virtually any closed-shell molecule, ESR techniques represent a central source of information for open-shell systems such as radicals. An important area where ESR data are of great value is, for example, the investigation of enzymes containing non closed-shell transition-metals, even though the connection between the experimental spectrum and the electronic and geometrical details of the particular system are frequently not fully understood. This great importance of NMR and ESR spectroscopy indicates that the capability of reliable computational predictions of the corresponding properties is in high demand. However, as will we outline succinctly further below, the theory behind the determination of magnetic properties is fairly involved. Hence, it is only in the last two decades that meaningful chemical shift calculations at the simple Hartree-Fock level have become available on a routine basis. Unfortunately, it soon turned out that in many instances, namely whenever electron correlation is an issue, HF results are not of sufficient quality to provide useful information for guiding and interpreting experimental activities. The obvious solution to this problem is to employ wave function based strategies that explicitly include electron correlation effects. The high computational demands of such methods, however, limit their application to small, chemically less relevant molecules. The only conventional correlated strategy which has gained some popularity because it is still affordable if applied to medium sized molecules is based on second order Møller-Plesset perturbation theory (MP2). However, as we will see from the examples discussed below, this method fails miserably in cases where near-degeneracy problems are present and has to be used with great care. As in other areas discussed in this book it seems therefore to be a natural suggestion to combine the strategies for computing magnetic properties with density functional methods, where the electron correlation effects are implicitly accounted for via the exchange-correlation functional. Indeed, the past few years have witnessed a tremendous development of implementations for the density functional theory based determination of NMR and ESR parameters. At the time of writing these techniques seem to be the only avenue leading to meaningful predictions, for example, for large molecules[47] or transition-metal compounds with a potential which is growing at an impressive pace.

In this chapter we are going to introduce the most common strategies aimed at computing magnetic properties in a density functional theory context. We commence by briefly reviewing the qualitative aspects of the theoretical foundations and present the implementations available. The reader should be aware that in this case we again face the dilemma of deciding between a formal and theoretically thorough presentation of the material, digging deep into the complex physics, and a more qualitative introduction which aims at providing the reader with a feeling of how the methods work and how well they perform in practical applications. Following the general philosophy of this book we intentionally decided on the latter option. For readers who wish to learn more about this fascinating field, in particular the physics governing this whole area, we have included key references to the relevant literature. We then go on by discussing the degree of accuracy that can be expected from such calculations and how the choice of the functional and the one-electron basis set affects the performance. Our main focus will be on the most active field in this area, i. e., the computational prediction of NMR chemical shifts. The methodological aspects and representative calculations of nuclear spin-spin couplings as well as ESR g-tensors and hyperfine structures of radicals will be mentioned more briefly. In addition to presenting results for lighter main group elements, the discussion will be extended to transition-metal compounds which represent a considerable challenge for any method and where density functional theory seems to be particularly successful and sometimes the only choice.

11.1 Theoretical Background

The general theory of the quantum mechanical treatment of magnetic properties is far beyond the scope of this book. For details of the fundamental theory as well as on many technical aspects regarding the calculation of NMR parameters in the context of various quantum chemical techniques we refer the interested reader to the clear and competent discussion in the recent review by Helgaker, Jaszunski, and Ruud, 1999. These authors focus mainly on the Hartree-Fock and related correlated methods but briefly touch also on density functional theory. A more introductory exposition of the general aspects can be found in standard text books such as McWeeny, 1992, or Atkins and Friedman, 1997. As mentioned above we will in the following provide just a very general overview of this

[47] For example, despite the high symmetry of C_{60}, no calculation of its NMR properties at the MP2 level has been reported so far. DFT-based methods, on the other hand, can readily be applied to this and to larger fullerenes, see Bühl et al., 1999.

subject. Similar to the dipole polarizability $\vec{\alpha}$ introduced in the preceding chapter, NMR chemical shifts and spin-spin couplings as well as ESR *g*-tensors are second-order properties and can be expressed as a mixed second derivative of the total electronic energy with respect to two perturbations,

$$\sigma_{st} \propto \left. \frac{\partial^2 E}{\partial X_s \partial Y_t} \right|_{\vec{X}=\vec{Y}=0} . \tag{11-1}$$

When $\vec{X}$ corresponds to the magnetic field $\vec{B}$ the σ_{st} are elements of the nuclear shielding tensor or of the ESR *g*-tensor if $\vec{Y}$ is the nuclear magnetic moment or an electronic spin, respectively. If, on the other hand, the derivative is taken with respect to two different nuclear magnetic moments, we arrive at an expression for the (reduced) spin-spin coupling constants of NMR spectroscopy. The usual way to tackle such equations is by employing stationary perturbation theory. In the conventional Hartree-Fock approach this leads to the coupled perturbed Hartree-Fock equations, which describe the linear response of the molecular orbitals that define the Slater determinant of the corresponding HF ground state to the external perturbation. Due to the non-local character of the exchange operator in HF theory, the response of a particular molecular orbital depends on the linear response of all other occupied orbitals. On the other hand, if the same reasoning is applied to Kohn-Sham density functional theory based on the commonly used local or gradient-corrected exchange-correlation functionals, this linear response vanishes exactly because of the local character of the corresponding exchange operators. There is no coupling between the linear responses of different molecular orbitals and one ends up with the so-called *uncoupled density functional theory* (UDFT) equations. The loss of the coupling is a shortcoming which reflects the fact that the standard functionals depend only on the charge density and are designed for 'normal' situations, i. e., where no magnetic field is present. It is very important to realize that for a proper description of molecules in the presence of an external magnetic field, the usual Hohenberg-Kohn theorems as outlined in Chapter 4 do not hold any more and the corresponding exchange-correlation functional not only has to depend on the electron density $\rho(\vec{r})$ but also on the current density $j(\vec{r})$ induced by the magnetic field,

$$E_{XC}[\rho(\vec{r})] \xrightarrow{\text{presence of magnetic field}} E_{XC}[\rho(\vec{r}), j(\vec{r})] . \tag{11-2}$$

Approaches employing such *current density functionals* have been around for some time (see Vignale, Rasolt, and Geldart, 1990) and first implementations have been reported recently, for example by Lee, Handy, and Colwell, 1995, who used a local density type approximation, while Becke, 1996b, outlined the formalism for gradient-corrected current density functionals. These techniques are, however, still in the early development phase and have not reached the maturity to be of any practical relevance. Rather, all present-day implementations of computing magnetic properties are based on regular density functionals, i. e., they employ the approximation that the dependence on the current density can be neglected,

$$E_{XC}[\rho(\vec{r}), j(\vec{r})] \approx E_{XC}[\rho(\vec{r})]. \tag{11-3}$$

Fortunately, as shown by Lee, Handy, and Colwell, 1995, it seems that the consequences of this approximation with regard to the accuracy of the computed chemical shifts are rather modest and are of less significance than the general shortcomings inherent in the functionals used. Hence, from an application-oriented, pragmatic point of view one does not need to worry too much about using functionals which are formally inadequate because they neglect the required dependence on $j(\vec{r})$.

Another fundamental problem which significantly plagues all attempts to compute magnetic properties using a finite one-electron basis set approach, be it in the realm of density functional theory or otherwise, is the so-called *gauge-problem*. To explain in simple terms what that means we have to mention a few key aspects of how magnetic properties enter the Schrödinger equation. The central *observable* in this context is the magnetic field $\vec{B}$. However, in the *operators*, not $\vec{B}$, but the related vector potential $\vec{A}$ of the field enters the appropriate equations. The connection between these two quantities is that

$$\vec{B} = \nabla \times \vec{A}, \tag{11-4}$$

i. e., the magnetic field is defined as the *curl* of the vector potential. For the non-expert, we only mention that the curl is a particular kind of gradient that can be applied to a vector field (for details see McWeeny, 1992, or Atkins and Friedman, 1997). It is not necessary in the present context to enter an in-depth discussion of what this means, we only need to convey the decisive point: the addition of the gradient of an arbitrary function to this vector potential $\vec{A}$ leaves the magnetic field $\vec{B}$ unchanged. Or, in more pictorial terms, two different choices of origin would give two alternative values of $\vec{A}$ at any point in space, while the field $\vec{B}$ is of course independent of the arbitrarily chosen origin. Hence, many vector potentials give rise to *same* magnetic field and there is no unique definition for the choice of $\vec{A}$ corresponding to a particular magnetic field $\vec{B}$. Since expectation values such as NMR chemical shifts only depend on the observable, i. e. $\vec{B}$, the results must of course be independent of the actual choice of the vector potential $\vec{A}$ (as long as it yields $\vec{B}$). It is this requirement which is meant if one states that the magnetic field is *gauge invariant*.

In all our computational strategies we are limited to approximate schemes and the use of finite one-electron basis sets. One of the outgrowths of these approximations is that gauge invariance is not fulfilled. The unpleasant consequence is that the computationally predicted magnetic properties depend on the choice of the coordinate system. In the coupled Hartree-Fock scheme, for example, gauge invariance is only assured with a complete, viz., infinite basis set. In fact, the breakthrough which triggered the fast growth of high accuracy computational studies of NMR and ESR properties during the past two decades was the development of efficient strategies to cope with this gauge problem. The two most widely used techniques are the so-called *individual gauges for localized orbitals* (IGLO) developed by Kutzelnigg and coworkers (for a review see Kutzelnigg, Fleischer, and Schindler, 1990) and the *gauge-invariant atomic orbital* (GIAO, this acronym has been criticized since the orbitals are actually gauge dependent and an alternative name frequently used is *gauge-including atomic orbital*) approaches first put forward by London as early as 1937. In the IGLO ansatz, gauge-dependent factors are used on localized orbitals, while in the GIAO framework the explicit field-dependence is built into the atom-centered basis functions. For details the reader should consult Helgaker, Jaszunski, and Ruud, 1999, and the many references given therein. The first chemically significant calculations of NMR shielding constants in the context of density functional theory appeared a few years ago by Malkin, Malkina, and Salahub, 1993, and employed the IGLO approach, while most subsequent implementations use the GIAO technique, as described in detail by Rauhut et al., 1996, and Cheeseman et al., 1996. As we will document presently, using uncoupled density functional theory in either the IGLO or GIAO implementation leads in general to good agreement with experimental data in spite of the formal shortcomings of this approach. However, in particular for systems characterized by low-lying excited states, but also in other cases, shielding values which are systematically too low have been noted. As pointed out by Bühl et al., 1999, it is not completely understood, whether this is due to general deficiencies in the exchange-correlation functionals or whether it is due to the missing current dependence. In best tradition of pragmatic solutions in density functional theory, Malkin et al., 1994, suggested to introduce an ad hoc correction term to the corresponding expressions as a first attempt to introduce some current dependence. The resulting modified *sum-over-states density functional perturbation theory* (SOS-DFPT) approach indeed often leads to improved accuracy, even though it lacks the physical rigor to make it a promising and serious technique from a purists' point of view. Before we begin a detailed discussion of the actual performance of density functional methods in this arena, let us summarize the current state of affairs with regard to the computational prediction of magnetic properties in the Kohn-Sham framework. (i) The current dependence of the functionals required for a theoretically correct description is neglected altogether; instead the usual exchange-correlation functionals which depend solely on the charge density are employed. (ii) The gauge problem is tackled with the same IGLO, GIAO or similar techniques as in wave function based approaches. In most current implementations, for example in *Gaussian*, *Turbomole* or *ADF*, the latter is realized. Among the problems of the former is that there are examples, where the results depend significantly on the localization scheme chosen. (iii) A special implementation aimed at reducing the systematic errors in the chemical shieldings and thus attempting to implicitly introduce a current dependence is the SOS-DFPT technique. This method is, however, fairly proprietary since it is implemented in combination with the IGLO choice of origin only in the *deMon* program and is not available in any other generally accessible quantum chemical software.

11.2 NMR Chemical Shifts

The most important magnetic property by far is the chemical shift of NMR spectroscopy. While proton (^{1}H) and ^{13}C shieldings hold a prominent place in organic chemistry, other magnetic nuclei such as ^{15}N, ^{29}Si, or ^{31}P but also heavier nuclei such as transition-metals are increasingly important in many areas of chemistry. Obviously, all these nuclei are equally amenable to computational investigations. In the following we will give an overview of the level of confidence that can be expected for NMR shieldings computed for the various relevant magnetic nuclei using different exchange-correlation functionals, basis sets and implementations. Before we enter this discussion we need to point out that computed magnetic properties are in general extremely sensitive to the geometry chosen. This applies in particular to chemical shift calculations and already small changes in bond lengths or angles may lead to significant deviations in the computed shifts. Hence, reliable chemical shifts can only be expected if these calculations are based on good geometries. Actually, this strong response of the chemical shift to structural variation has already frequently been used as a means to identify the geometrical parameters of the target molecule, see Bühl, 1998. As the bottom line one should always keep in mind that the use of reliable structures is an important prerequisite for obtaining meaningful chemical shift, spin-spin coupling, *g*-tensor or hyperfine structure information.

Let us begin with ^{1}H chemical shifts. Computational studies in this area are less frequent than for other nuclei because proton shifts span only some ten ppm and rovibrational and solvent effects might be comparable to the range of chemical shifts itself. Still, there have been several reports on the successful application of density functional theory to proton chemical shifts, see Dejaegere and Case, 1998, or Alkorta and Elguero, 1998, for representative examples. In a recent detailed comparative study on ^{1}H chemical shifts, Rablen, Pearlman, and Finkbiner, 1999, reported the performance of the three popular hybrid functionals B3P86, B3PW91 and B3LYP for reproducing ^{1}H chemical shifts of 80 small to modest sized organic molecules. With the GIAO scheme and using a 6-311++G(2df,p) basis set relative shieldings of pleasing quality were obtained. While all three functionals performed similarly, B3LYP was rated best. Increasing the basis set further did not improve the overall performance significantly and already the smaller 6-311++G(d,p) basis set gave results only marginally inferior to the 6-311++G(2df,p) ones. These authors also pointed out that a linear scaling of the computed NMR chemical shifts improved the results, very akin to the procedures established for harmonic frequencies. For example, a scaling factor of 0.9422 was established for chemical shifts computed with the GIAO B3LYP/6-311++G(2df,p) level based on B3LYP/6-31+G(d) optimized geometries.

Using fairly large Gaussian type basis sets to expand the Kohn-Sham orbitals and the GIAO technique, Rauhut et al., 1996, and Cheeseman et al., 1996, tested the performance of some popular exchange-correlation functionals ranging from LDA through GGA functionals such as BPW91 and BLYP up to the related three-parameter hybrid functionals B3PW91 and B3LYP for ^{13}C, ^{15}N, and ^{17}O chemical shifts. In particular, the density functional results were compared to the conventional Hartree-Fock ansatz and the MP2 extension for incorporating (dynamical) electron correlation effects, which represent the standard non-density functional methods used for this purpose. Some representative results taken from Cheeseman et al., 1996, and Adamo, Cossi, and Barone, 1999, are shown in Table 11-1.

For ^{13}C shieldings the test set of simple molecules included in Table 11-1 reveals that the LDA cannot be recommended. This is because absolute deviations for both, absolute and relative shifts are significantly higher than those of the HF method and much higher than the rather small errors of the MP2 approach. As expected, the BLYP generalized gradient

Table 11-1. Absolute and relative (in square brackets, with respect to CH_4, NH_3, and H_2O) ^{13}C, ^{15}N, and ^{17}O NMR chemical shifts [ppm].

Molecule	Nucl.	HF[a]	MP2[a]	LDA[a]	BLYP[a]	B3LYP[a]	PBE1PBE[b]	Exp.[a]
CH_4	C	195.7	201.5	193.7	187.5	189.6	194.0	195.1
NH_3	N	262.6	276.2	266.1	259.2	260.3	263.1	264.5
H_2O	O	326.9	344.8	332.3	326.4	325.7	328.9	344.0
C_2H_6	C	84.0	188.0	176.7	169.7	173.6	179.7	180.9
		[11.7]	[13.5]	[17.0]	[17.8]	[16.0]	[14.3]	[14.2]
C_2H_4	C	59.9	71.2	42.3	47.1	48.7	58.4	64.5
		[135.8]	[130.3]	[151.4]	[140.4]	[140.9]	[135.6]	[130.6]
C_2H_2	C	113.9	123.3	100.0	105.7	106.3	114.0	117.2
		[81.8]	[78.2]	[93.7]	[81.8]	[83.3]	[80.0]	[77.9]
CH_2CCH_2	C_{term}	114.0	120.9	103.2	103.0	104.5	112.5	115.2
		[81.7]	[80.6]	[90.5]	[84.5]	[85.1]	[81.5]	[79.9]
	C_{centr}	–44.3	–26.0	–53.0	–51.7	–51.7	–36.6	–28.9
		[240.0]	[227.5]	[246.7]	[239.2]	[241.3]	[230.6]	[224.0]
C_6H_6	C	55.0	64.0	41.7	43.7	45.2	55.3	57.2
		[140.7]	[137.5]	[152.0]	[143.8]	[144.4]	[138.7]	[137.9]
CH_3F	C	124.5	121.8	103.2	101.2	106.6	116.5	116.8
		[71.2]	[79.7]	[90.5]	[86.3]	[83.0]	[77.5]	[78.3]
CF_4	C	79.2	64.4	39.2	38.3	46.5	59.2	64.5
		[116.5]	[137.1]	[154.5]	[149.2]	[143.1]	[134.8]	[130.6]
HCN	C	68.1	87.3	63.0	68.7	67.2	76.6	82.1
		[127.6]	[114.2]	[130.7]	[118.8]	[122.4]	[117.4]	[113.0]
	N	–56.0	1.0	–60.2	–49.2	–53.1	–34.9	–20.4
		[318.6]	[275.2]	[326.3]	[308.4]	[313.4]	[298.0]	[284.9]
CH_3CN	C_{term}	190.9	193.6	182.3	177.1	180.4	187.7	187.7
		[4.8]	[7.9]	[11.4]	[10.4]	[9.2]	[6.3]	[7.4]
	C_{centr}	60.6	76.1	54.7	57.8	57.4	68.2	73.8
		[135.1]	[125.4]	[139.0]	[129.7]	[132.2]	[125.8]	[121.3]
	N	–46.6	–13.2	–44.7	–36.5	–40.7	–24.4	–8.1
		[309.2]	[289.4]	[310.8]	[295.7]	[301.0]	[287.5]	[272.6]
CH_3NH_2	C	163.8	164.9	151.1	145.3	150.1	157.1	158.3
		[31.9]	[36.6]	[42.6]	[42.2]	[39.5]	[36.9]	[36.8]
	N	250.0	261.2	244.7	233.1	238.4	244.0	
		[12.6]	[15.0]	[21.4]	[26.1]	[21.9]	[19.1]	
N_2	N	–128.7	–44.9	–104.8	–97.1	–105.4	–76.8	–61.6
		[391.3]	[321.1]	[370.9]	[356.3]	[365.7]	[339.9]	[326.1]
CH_3OH	C	143.7	142.2	126.1	122.0	127.4	136.5	136.6
		[52.0]	[59.3]	[67.6]	[65.5]	[62.2]	[57.5]	[58.5]
	O	274.7	350.6	334.5	313.9	321.6	334.7	
		[52.2]	[–5.8]	[2.2]	[10.9]	[4.1]	[–5.8]	

Table 11-1, continued.

Molecule	Nucl.	HF[a]	MP2[a]	LDA[a]	BLYP[a]	B3LYP[a]	PBE1PBE[b]	Exp.[a]
CH_2O	C	–9.2	6.7	–41.0	–27.6	–25.4	–11.1	
		[204.9]	[194.8]	[234.7]	[215.1]	[215.0]	[205.1]	
	O	–461.2	–341.9	–509.2	–459.7	–469.8	–422.2	
		[788.1]	[686.7]	[841.5]	[784.5]	[795.5]	[–751.1]	
H_3CCOCH_3	C_{methyl}	163.5	164.5	148.8	146.9	150.4	157.0	158.0
		[32.2]	[37.0]	[44.9]	[40.6]	[39.2]	[37.0]	[37.1]
	C_{CO}	–23.2	–5.8	–44.4	–37.4	–35.7	–11.1	–13.1
		[218.9]	[207.3]	[234.5]	[224.9]	[225.3]	[205.1]	[208.2]
	O	–340.5	–279.8	–375.5	–351.5	–358.1	–330.2	
		[667.4]	[624.6]	[707.8]	[676.3]	[683.8]	[659.1]	
CO	C	–29.2	11.1	–23.9	–17.3	–21.7	–7.8	1.0
		[224.9]	[190.4]	[217.6]	[204.8]	[211.3]	[201.8]	[194.1]
	O	–95.0	–47.4	–93.7	–82.9	–87.8	–70.0	–42.3
		[421.9]	[392.2]	[426.0]	[407.7]	[413.5]	[398.9]	[386.3]
CO_2	C	47.8	63.5	47.2	47.9	46.9	56.8	58.5
		[147.9]	[138.0]	[146.5]	[139.6]	[142.7]	[137.2]	[136.6]
	O	214.8	241.0	203.3	206.5	206.9	220.0	243.4
		[112.1]	[103.8]	[129.0]	[118.3]	[118.8]	[108.9]	[100.6]
Mean abs. deviation	C	8.5	5.6	15.2	15.0	13.0	3.1	
		[8.1]	[1.6]	[14.4]	[7.8]	[7.9]	[2.5]	

[a] Cheeseman et al., 1996 using a doubly polarized quadruple zeta basis; [b] Adamo, Cossi, and Barone, 1999, using a standard 6-311+G(2p,d) basis set.

functional improves the quality of the results, in particular the error in the relative shifts is reduced to about one half of the LDA value. Interestingly, switching to the B3LYP hybrid functional does not lead to the usually observed further betterment but furnishes results very similar to the parent pure density functional. Rauhut et al. 1996, arrived at similar conclusions even though some of their quantitative data differ due to the use of slightly different basis sets and geometries. A promising recent alternative seems to be the PBE1PBE functional. Using this protocol but employing a slightly different basis set, the mean absolute errors in the absolute and relative shieldings drop to 3.1 and 2.5 ppm, respectively, as compared to the MP2 result of 5.6 and 1.6 ppm, respectively (Adamo, Cossi, and Barone, 1999). Similarly, the *m*PW1PW functional shows very promising results for ^{13}C shieldings as demonstrated by Wiberg, 1999, for 18 organic molecules. Also the recent B98 functional was shown to be competitive both with the MP2 approach and the best density functionals by Bienati, Adamo, and Barone, 1999. Qualitatively similar conclusions can be drawn for ^{15}N shieldings. However, for the few data for this nucleus included in Table 11-1 the LDA already outperforms the HF scheme, even though for all methods the errors themselves are considerably larger than for the ^{13}C case. For ^{17}O shieldings in main group element compounds density functional approaches – either of the GGA or the hybrid type – seem to be

Table 11-2. Absolute ^{17}O NMR chemical shifts [ppm] for ozone.

Nucleus	HF[a]	MP2[a]	CCSD(T)[b]	LDA[a]	BLYP[a]	B3LYP[a]	PBE1PBE[c]	Exp.
$O_{terminal}$	−2793	+1055	−1208	−1520	−1454	−1673	−1453	−1290[b], −1254[c]
$O_{central}$	−2717	+2675	−754	−914	−892	−1115	−1040	−724[b], −688[c]

[a] At the experimental geometry (R_{OO}=1.272 Å, α = 116.8°) using an aug-cc-pVQZ quadruple-zeta basis set; [b] taken from Gauss and Stanton, 1996, using R_{OO}=1.2693 Å, α = 117.0° and a large pentuple-zeta basis set; [c] taken from Adamo and Barone, 1999 using R_{OO}=1.2406 Å, α = 118.3° and a triple-zeta 6-311(d,p) basis set.

insufficient to yield reliable results. For this nucleus, the conventional MP2 methods is clearly superior, but also significantly more expensive computationally. A special, important benchmark case is again provided by the ozone molecule because of the significant effects arising from non-dynamical correlation. As shown in Table 11-2, only highly correlated and computationally extremely demanding methods, such as CCSD(T) in combination with very large basis sets, are able to supply chemical shifts reasonably close to experimental ones. Both, the HF and in particular the MP2 method, yield disastrously poor and completely useless results, missing the experimental target data by some 2000 and 3500 ppm, respectively! On the other hand, just as in the previously discussed example of the vibrational frequencies of O_3, pure density functionals perform best, while in particular the chemical shift of the central oxygen poses a significant problem to hybrid functionals. The density functional results in Table 11-2 have been obtained within the GIAO scheme, but as shown by Kaupp, Malkina, and Malkin, 1997, calculations using the IGLO method for dealing with the gauge problem, either in the simple uncoupled or the SOS-DFPT picture lead to very similar conclusions.

All the above results were obtained using comparably large basis sets and we must ask whether the choice of the basis set influences these general conclusions and which kinds of standard basis sets can be recommended for routine calculations of NMR chemical shifts. Even though no systematic studies exploring the basis set dependence of NMR chemical shift calculations in the Kohn-Sham framework seem to be available, the general consensus appears to be that the basis set requirements are similar to those of the Hartree-Fock scheme and less than for post-HF approaches. Typically, sets of polarized triple-zeta quality are employed, such as the 6-311+G(2d,p) standard set which also includes an additional set of diffuse functions on the non-hydrogen atoms. This basis set was recommended in particular by Cheeseman et al., 1996, in their evaluation of the applicability of density functional theory to NMR chemical shifts. These authors also advocated the use of smaller sets such as 6-31G(d) for larger molecules, albeit with some loss in accuracy. However, one should keep in mind that there are indications that the rather good performance of these basis sets may be due to fortuitous error cancellations and that bigger, more flexible sets are required for arriving at the correct answer for the correct reason. Another frequently employed and probably better suited class of basis sets is the IGLO-III set originally derived by Kutzelnigg and coworkers (Kutzelnigg, Fleischer, and Schindler, 1990) for the calculation of NMR chemical shieldings in the context of Hartree-Fock theory. These are loosely contracted sets augmented by diffuse and polarization functions. For example, for carbon the IGLO-III basis set consists of 11s, 7p and 2d primitive Gaussian functions contracted to a final basis set of 7s6p2d quality.

To summarize, for compounds containing elements of the first-row, GGA and hybrid functionals provide chemical shifts of roughly similar accuracy, usually better than corresponding Hartree-Fock results, while the LDA cannot be recommended. However, also the GGA and hybrid approaches in many cases fall short of the precision that can be expected from MP2 theory, let alone more elaborate techniques for accounting for electron correlation such as CCSD(T). On the other hand, there are examples such as the ^{13}C chemical shifts in *ortho*-benzyne, where conventional methods such as HF and MP2 fail while density functional theory provides high quality predictions, see Orendt et al., 1996. Hence, density functional based methods are a valuable addition to the quantum chemist's tool box but in most cases do not represent a major breakthrough if small molecules containing only light elements are being studied. Rather, the use of density functional techniques has its particular merits when it comes to large systems due to its computational efficiency or to species which contain heavier elements, such as transition-metal compounds. As noted before, for these molecules electron correlation is of great importance and often relativistic effects can no longer be ignored. If the latter are incorporated through the use of relativistic effective core potentials, the valence orbitals of the heavy elements show by construction the wrong nodal behavior near the core. Hence, the straightforward use of RECPs for the prediction of magnetic properties is limited to the ligand nuclei and cannot be used for the metal. Another possibility for including relativistic effects is the quasirelativistic scheme implemented in the *ADF* program. In this approach, the valence orbitals are orthogonalized against the frozen core orbitals which ensures correct tails of the valence orbitals close to the nucleus. As pointed out for example by Schreckenbach and Ziegler, 1998, chemical shifts are mostly determined by the core tails of the valence orbitals and not by the core orbitals themselves. Therefore, unlike RECPs, this technique can also be used to study chemical shifts of the metal atom. In addition, very recently, the *zeroth order regular approximation* (ZORA) technique for incorporating spin free relativistic effects was implemented in the context of NMR shielding constants by Wolff et al., 1999.

An ever growing number of chemical shift calculations for transition-element containing molecules has been carried out in the past few years as eloquently summarized in a couple of recent reviews, such as Schreckenbach and Ziegler, 1998, Kaupp, Malkin, and Malkina, 1998, and Bühl et al., 1999. An instructive example for ligand chemical shifts is given by the ^{17}O chemical shifts of neutral and charged tetrahedral d^0 transition-metal oxo complexes MO_4. These systems are particularly well suited for a comparative study, since the ^{17}O chemical shifts cover a large range and are very sensitive on the metal atom and the bonding situation. Table 11-3 contains the oxygen shielding constants of various MO_4 complexes obtained using a variety of density functional methods. For comparison, results from the conventional, wave function based HF and MP2 techniques are also included.

Each of the functionals included in Table 11-3 shows a comparably good performance for all oxo complexes. The differences between the functionals, be they of pure GGA or hybrid type are in general small. This is even more remarkable if one takes into account that

Table 11-3. Absolute ^{17}O NMR chemical shifts [ppm] in transition metal oxo complexes.

Molecule	HF GIAO BS I[b]	MP2 GIAO BS I[b]	BP86 GIAO STO[c]	BP86 GIAO BS I[b]	BP86 IGLO[a] BS I[b]	B3LYP GIAO BS I[b]	B3LYP GIAO BS II[d]	PBE1PBE GIAO BS II[d]	Exp.[e]
WO_4^{2-}	−194	−21	−140	−157	−138	−183	−108	−102	−129
MoO_4^{2-}	−335	−60	−216	−251	−231	−289	−201	−193	−239
CrO_4^{2-}	−1308	2173	−446	−508	−490	−640	−480	−479	−544
ReO_4^-	−464	3	−278	−282	−277	−339			−278
TcO_4^-	−819	184	−405	−421	−410	−518			−458
MnO_4^-	−7248	54485	−778	−832	−821	−1149			−939
OsO_4	−1295	1069	−521	−517	−503	−657			−505
RuO_4	−3330	8262	−740	−765	−733	−1037			≈ −820
FeO_4				−1224	−1172	−1957			

[a] SOS-IGLO approach; [b] Basis I: IGLO II basis for oxygen, RECP with (8s7p6d)/[6s5p3d] valence basis for metals, taken from Kaupp, Malkina and Malkin, 1997; [c] STO: Slater type basis of polarized triple zeta quality, taken from Schreckenbach and Ziegler, 1997a; [d] Basis II: EPR II basis for oxygen, RECP/valence basis for metals as in Basis I, taken from Adamo and Barone, 1998c; [e] taken from Kaupp, Malkina and Malkin, 1997.

these numbers have been collected from several sources which differ in the way the NMR shifts are being computed (GIAO and the SOS-DFPT IGLO variant), the way scalar relativistic effects are accounted for (RECP and the quasirelativistic approach), and the type of basis set (GTO and STO based). On the other hand, both HF and MP2 show a completely erratic behavior, which gets worse with increased deshielding. Interestingly, at the HF level the computationally predicted shieldings are too low, while MP2 errs no less dramatically in the opposite direction. A tendency of the MP2 methods to 'overcorrect' HF results can also be noted for many well-behaved compounds of lighter nuclei, but in the transition-metal species the errors can be spectacular: deviations from experiment of some 55000 ppm (!) in the case of MnO_4^- clearly indicate that meaningless numbers have been produced. These compounds are known to exhibit significant non-dynamical electron correlation effects pointing to the origin of this catastrophic behavior. The take-home message here is that in such cases Hartree-Fock and MP2 methods are simply useless. On the other hand, density functional theory seems to provide a general tool, applicable for the whole range of oxides studied in this particular example. Interestingly, if instead of GGA functionals their hybrid counterparts are employed, the agreement with experiment deteriorates. This is in contradiction with the usual observation that hybrid functionals perform better than GGA ones. To make things even more complicated, we will discuss cases below where the hybrid functionals behave as expected and yield chemical shifts which are significantly superior to GGA shifts. No general conclusion can be offered yet in this regard.

The computational prediction of not the ligand but the metal, particularly transition-metal chemical shifts poses an even more severe challenge to any method. Electron correlation cannot be neglected and at least for the heavier metals relativistic effects have a pronounced, direct influence on the results and need to be included. However, if instead of absolute shieldings relative chemical shifts are considered, the effects of relativity are attenuated. The reason for this is that relativistic effects are mostly due to the core electrons whose properties change only little when going from one molecular environment to another. Hence, they cancel to a large extent when relative shifts are considered. Only if the core tails of the valence orbitals start to become important can relativistic effects significantly affect chemical shifts. Qualitative trends in chemical shifts can therefore be reproduced with non-relativistic calculations for elements as heavy as 4d transition-metals. If we turn to the quantitative performance of various functionals, surprising differences to the corresponding ligand shifts are observed. In particular, the use of regular GGA functionals leads in most cases to only disappointing results while hybrid functionals perform significantly better. An extreme but still typical example is provided by the ^{57}Fe chemical shifts of the prototype organometallic compound, ferrocene. Using a fairly large basis set, pure GGA functionals severely underestimate the experimental chemical shift of 1532 ppm by some 900 ppm, while the B3LYP functional hits right on target at 1525 ppm as demonstrated by Bühl, 1997.

In conclusion, for most of the transition-metal chemical shifts studied up to now, hybrid functionals such as B3LYP perform much better than GGA functionals which tend to significantly underestimate the chemical shifts. However, as emphasized by Bühl et al., 1999, it is an entirely open question as to whether the good performance of B3LYP and related functionals is due to the underlying physics or just the result of a fortuitous error cancellation. We should point out though that Schreckenbach, 1999, offered an explanation for the dramatic effect obtained for ferrocene. He argues that the following factors brought about by mixing in exact exchange should be responsible: stabilization of the occupied orbitals and the concomitant larger energetic gaps between occupied and virtual orbitals, the more diffuse character of the unoccupied orbitals, and the inclusion of some coupling due to Hartree-Fock exchange in the UDFT equations. In any case, even though such functionals seem to help in these cases, they are certainly not a panacea since there are also examples – like ^{95}Mo chemical shifts – where the inclusion of exact exchange results in larger deviations from the experimental data. The, admittedly somewhat unsatisfactory, bottom line is that any serious investigation of metal chemical shifts must be preceded by a careful calibration of the available density functionals, since it is not possible to make a safe a priori prediction of how well a particular functional will perform.

Finally, we should note that a particularly important area of application where density functional techniques, in spite of the deficiencies noted above, are virtually without competition is provided by biochemically relevant molecules, such as enzymes or nucleic acids. The techniques discussed in this section are virtually the only quantum chemical methods which can be applied in this context due to their outstanding price/performance ratio. For example, the ^{13}C and ^{15}N chemical shifts in bacteriochlorophyll A have been studied by Facelli, 1998, and in another investigation the ^{57}Fe, ^{13}C and ^{17}O shifts in iron porphyrin derivatives gave important clues as to the structural details of these species, as shown by McMahon et al., 1998.

11.3 NMR Nuclear Spin-Spin Coupling Constants

The chemical shifts discussed in the previous section are an important, but not the only information carried by an NMR spectrum. In order to understand and interpret a complete NMR spectrum the nuclear spin-spin coupling constants need to be analyzed as well. However, the computational prediction of these quantities within approximate density functional theory has turned out to be difficult and only a couple of reports have appeared in the literature. In addition, only very few of the regularly available, commercial black-box program packages include modules which would allow the routine calculation of this property. We will therefore restrict ourselves to a brief overview of the current, still fairly experimental state of the art for the determination of (isotropic) spin-spin coupling constants in the context of density functional theory. One reason why these are difficult to address theoretically is that there are a total of four terms which contribute to this property, i. e., the diamagnetic and paramagnetic spin-orbit terms, the spin-dipole term and the Fermi-contact term, all of which pose stringent and at the same time different problems for their theoretical description. The most important contribution among the four is commonly the Fermi-contact term. Pictorially and very qualitatively speaking, the Fermi-contact term for spin-spin coupling arises from the interaction of a magnetic nucleus with the charge density at the position of this very same nucleus which induces a small polarization of the total spin density (even though the system is formally closed-shell). This distortion is then detected by a second nucleus through the same contact mechanism. In other words, the Fermi-contact term probes the charge density and its sensitivity with respect to spin polarization at the position of the two nuclei. In order to account for this phenomenon we need to describe the local density and its spin polarization at the corresponding nuclear positions as accurately as possible. This property is very sensitive to almost everything, in particular to the type of functional and the one-particle basis set used. Clearly, since the Fermi-contact interaction depends on the amplitude of the orbitals at the nuclei (which determines the charge density), neither ECP nor frozen core approaches as in the quasirelativistic *ADF* scheme are applicable. Rather, all-electron techniques in which the core electrons are explicitly accounted for are mandatory. The first report describing a practical implementation in the context of density functional theory appeared in 1994 by Malkin, Malkina, and Salahub. While these authors based their technique on GTO basis sets and GGA functionals as implemented in the *deMon* program, a complementary study using STO sets and the LDA approximation in the context of the *ADF* code was presented two years later by Dickson and Ziegler, 1996. Both schemes employed a mixed analytical/finite differences implementation in which the spin-dipole term was completely neglected. Hybrid functionals could not be used with either approach. An important step towards routine calculations of nuclear spin-spin coupling constants was achieved a few years later by Sychrovský, Gräfenstein, and Cremer, 2000, and by Helgaker, Watson, and Handy, 2000. The two groups presented almost simultaneously fully analytical implementations of nuclear spin-spin coupling constants which not only included all four contributions but also allowed the use of the popular and successful hybrid functionals.

Table 11-4. Nuclear Spin-Spin Coupling Constants [Hz].

Molecule	Coupling	LDA[a]	PWP[b]	BLYP[c]	B3LYP[c]	Exp.[c]
CH_4	$^2J_{HH}$	–6.5	–10.6	–9.8	–10.9	–12.6
	$^1J_{CH}$	121.6	118.4	124.0	123.5	125.3
C_2H_6	$^2J_{HH}$	–10.8	–7.8	–7.1	–8.2	–/–
	$^1J_{CH}$	119.8	120.9	127.7	127.5	124.9
	$^2J_{CH}$	–1.7	–2.5	–3.3	–4.1	–4.5
C_2H_4	$^2J_{HH}$	3.8	4.3	–/–	3.2[d]	2.5
	$^1J_{CH}$	140.6	152.0	–/–	154.2[d]	156.4
	$^2J_{CH}$	1.8	–0.7	–/–	–1.3[d]	–2.4
	$^1J_{CC}$	68.5	61.2	–/–	70.1[d]	67.6
	$^3J_{HH,trans}$	12.3	16.9	–/–	17.7[d]	19.1
	$^3J_{HH,cis}$	6.6	10.1	–/–	11.0[d]	11.6
C_2H_2	$^1J_{CH}$	232.8	249.1	256.1	254.4	248.7
	$^2J_{CH}$	45.9	49.1	52.6	51.5	49.3
	$^1J_{CC}$	204.7	184.3	197.3	201.7	171.5
	$^3J_{HH}$	2.6	9.0	9.4	10.2	9.5
CH_3F	$^1J_{CF}$	–297.4	–268.1	–252.1	–227.1	–161.9
	$^1J_{CH}$	141.3	–/–	144.3	144.9	149.1
	$^2J_{HH}$	–2.8	–/–	–6.5	–7.7	–9.6
	$^2J_{HF}$	33.2	–/–	49.5	50.8	46.4

[a] Taken from Dickson and Ziegler, 1996; [b] taken from Malkin, Malkina and Salahub, 1994; [c] taken from Sychrovský, Gräfenstein and Cremer, 2000; [d] taken from Helgaker, Watson and Handy, 2000. As pointed out by these authors, the corresponding values by Sychrovský, Gräfenstein and Cremer, 2000, are apparently in error due to an incorrect geometry.

In Table 11-4 we summarize some representative spin-spin coupling constants for simple organic molecules as obtained in the studies mentioned above. The somewhat unusual PWP functional used by Malkin, Malkina, and Salahub, 1994, is a GGA functional which combines the gradient-corrected exchange and correlation functionals due to Perdew/Wang and Perdew. All the calculations from Table 11-4 have peen performed using large basis sets since it is generally accepted that basis set requirements for calculating nuclear spin-spin coupling constants are pretty demanding and exceed those known for computing NMR chemical shifts. In particular, the fact that standard Gaussian functions are not able to reproduce the correct cusp condition at the location of the nuclei creates problems in GTO based implementations and standard basis sets are often inadequate. A possible remedy is to augment such sets with tight (i. e., large exponent) s-functions in order to improve the local density at the nucleus. Several authors have recommended the IGLO-III set although there are also examples where even more flexible basis sets are required. Similarly, the Slater type basis sets of Dickson and Ziegler, 1996, which do not share the problems of the GTO sets near the nucleus, were of doubly polarized triple-zeta quality (i. e., one set of d- and f-functions for first-row atoms).

The accuracy obtained with the GGA and hybrid functionals is satisfactory and deviations from experiment are generally of the order of 10–20 %, while the LDA spin-spin couplings are less useful. In their detailed comparisons which include many more molecules, Sychrovský, Gräfenstein, and Cremer, 2000, as well as Helgaker, Watson, and Handy, 2000, come to the conclusion that among the functionals investigated B3LYP provides the most accurate nuclear spin-spin coupling constants. However, as noted by Malkina, Salahub, and Malkin, 1996, the quality of the computationally predicted spin-spin coupling constants systematically deteriorates along the series H, C, N, O, F which was tentatively assigned to the increasing number of lone pairs as one passes from left to right in the Periodic Table. For example, the computed ^{13}C–^{19}F couplings in methylfluoride, CH_3F, of –268.1 (PWP), –252.1 (BLYP) and –227.1 (B3LYP) Hz dramatically overshoot the experimentally determined value of –161.9 Hz while the remaining computed and experimental couplings are within a few Hz. Applications of GGA functionals to study spin-spin coupling constants in larger organic molecules have been reported by Stahl et al., 1997, and Hricovíni et al., 1997. The results of applying the B3LYP hybrid functional to various ^{13}C-^{13}C and ^{13}C-^{1}H spin-spin coupling constants (considering only the dominant Fermi-contact term) in saccharides and related carbohydrates reported by Bose et al., 1998, Cloran, Carmichael, and Serianni, 1999a and 1999b, give encouraging results, too. As a general trend these authors note that experimental ^{13}C-^{1}H spin-spin coupling constants are typically underestimated by less than 10 %, while ^{13}C-^{13}C coupling constants seem to come out too large by a similar amount in the DFT calculations.

11.4 ESR *g*-Tensors

Let us now turn to ESR spectroscopy, which is used to explore the electronic structure of open shell compounds such as simple radicals. The theoretical determination of *g*-tensors in the context of density functional theory is also a very new field. The first implementation based on the GIAO scheme has been presented only recently by Schreckenbach and Ziegler, 1997b. The corresponding modules have not been included in most popular programs yet and the number of studies aimed at a systematic assessment of the quality of these approaches is therefore very small. Just as with the NMR spin-spin coupling constants it thus appears premature to present any general conclusions with regard to basis sets, functionals and resulting accuracy. Nevertheless, it seems that the trends observed for the NMR chemical shifts apply also to the *g*-tensor: density functional theory usually provides results of higher quality than HF based techniques, but basis sets of at least polarized triple zeta quality are required. Experimental trends are reproduced fairly well, both for the isotropic *g*-value as well as its individual tensor components. However, severe problems still exist in particular if heavier elements are involved. For these cases, relativistic effects such as spin-orbit coupling become more and more important. Without going into detail we mention that van Lenthe, Wormer, and van der Avoird, 1997, have presented an alternative approach which explicitly includes these relativistic effects through the ZORA technique which seems to be better suited for these cases than the Schreckenbach/Ziegler implementation. We close this section with a similar conclusion as the preceding one. Density functional theory seems to offer a promising avenue for the theoretical prediction of ESR *g*-tensors, but the current state of the art has not reached a stage where specific recommendations regarding functionals and basis sets would be possible. The future seems to be bright and full of fascinating challenges. One only has to consider the wide area of enzymatic reaction sequences where ESR techniques are heavily used to monitor the reaction. But, it is still some way to go before such applications will become routine.

11.5 Hyperfine Coupling Constants

Hyperfine couplings can be classified in two categories stemming from two physical mechanisms. The *anisotropic* contribution results from the interaction between the magnetic moments of the electrons and the nuclei. Since this interaction depends on the relative orientation of these magnetic moments, it is only detectable when the motion of the molecules is frozen such as in matrices and crystals. In solution the molecules tumble randomly and the anisotropic part of the hyperfine coupling vanishes. On the other hand, the *isotropic* contribution of hyperfine coupling constants are related to the interaction between the nuclear spin and the spin of the unpaired electron, which is due to a Fermi-contact mechanism very akin to that described earlier in the context of NMR spin-spin couplings, according to

$$A_I^{iso} = C \sum_{\mu}^{K} \sum_{\nu}^{K} P_{\mu\nu}^{\alpha-\beta} \int \eta_\mu \delta(\vec{r}_I) \eta_\nu d\vec{r} \,. \tag{11-4}$$

C contains several natural constants which in part depend on the nucleus I, $P_{\mu\nu}^{\alpha-\beta}$ is the difference between the density matrices of electrons with α and β spin (i. e., a measure of the spin density) and $\delta(\vec{r}_I)$ is the Dirac delta function which ensures that only the density at the position of the nucleus I is considered in this equation. Obviously, no orientation dependence exists and isotropic hyperfine coupling constants can also be measured in solution. These data contain important information about details of the geometries and allow insights into the electronic structure of radicals and in the following, we will deal exclusively with these isotropic values. The crucial requirement for obtaining accurate results for isotropic hyperfine coupling constants is the availability of reliable spin densities at the nuclear positions, which is known to be hard to achieve. Due to the similarity of the mechanisms underlying NMR spin-spin and ESR hyperfine couplings it is no surprise that the demands on the basis sets used are also comparable. Standard DZP or even TZP sets are generally too small. Eriksson et al., 1994, for example recommend that the IGLO-III basis sets also be used for this purpose while other studies have employed basis sets such as the EPR-II and EPR-III sets specifically designed by Barone and co-workers for the evaluation of EPR properties in density functional calculations (see, e. g., Barone, 1995). There are too few systematic studies on the determination of (isotropic) hyperfine couplings to allow for a solid evaluation of the correlation between expected accuracy and employed technique. However, it seems clear that density functional methods are capable of providing reasonable predictions for these properties, even though they are not able to rival the accu-

Table 11-5. Isotropic hyperfine structures [Gauss].

Molecule	Parameter	PW86[a]	B3LYP[b]	PBE1PBE[b]	Exp.[a,b]
CH_3	A_H	–20.8	–23.3	–26.1	–23, –25.1
	A_C	32.8	29.8	29.2	≈38, 28.4
CH_2O^+	A_H	133.3	130.3	134.2	133
	A_C	–31.8	–33.5	–34.6	–39
	A_O	–12.6	–15.4	–14.5	

[a] Taken from Eriksson et al., 1994; [b] taken from Adamo, Cossi and Barone, 1999.

racy obtained from highly correlated wave function based methods as shown in a comprehensive recent study by Gauld, Eriksson, and Radom, 1997. Again, the choice of the functional and/or the one-particle basis set to expand the Kohn-Sham orbitals is of crucial importance. This should not surprise us since the isotropic couplings depend on the spin density at the nuclear positions only and hence the performances of the different functionals are directly related to their abilities to generate good spin densities not on average but at these very positions. For example, Eriksson et al., 1994, demonstrated that the isotropic hyperfine couplings of the water cation computed from various GGA functionals vary significantly. Compared with the experimental coupling of –29.7 Gauss, the BP functional yields –8.6 Gauss, PW86 results in –24.6 Gauss and the more recent PW91 functional performs worst and produces +0.6 Gauss. As already alluded to in Section 11-3, the PW86 functional produces rather good spin densities which is also mirrored by its comparatively good performance in the present context, while other functionals such as the widely used BP are very poor. Similar conclusions regarding the strong dependence of the couplings on the functional forms and the disappointing performance of the BP functional have been reported by others, see, e. g. Barone, 1994. Consequently, Eriksson et al. recommend PW86 as the best suited GGA functional for the calculation of isotropic hyperfine coupling constants. Thus, while the LDA does not deliver trustworthy results and should not be used for computational prediction of hyperfine couplings, certain (but not all!) GGA functionals such as PW86 already generally provide results of acceptable accuracy. Just as in most other areas, hybrid schemes which include a certain amount of Hartree-Fock exchange usually represent a further improvement, as, for example, reported by Barone, 1994, or Adamo, Cossi, and Barone, 1999. A representative set of results is documented in Table 11-5, where isotropic hyperfine coupling constants are collected for the methyl radical and the formaldehyde cation radical obtained with various functionals.

The best results are obtained with the hybrid schemes, the standard B3LYP functional and the new, one-parameter PBE1PBE protocol. Of course, one must keep in mind that independent of the particular functional chosen, large and flexible basis sets must be used.

Promising accuracy is also reported for the isotropic hyperfine coupling constants involving metals. A case at hand is the study by Knight et al., 1999, of binary oxides such as ScO, YO, and LaO, where the unrestricted B3LYP model together with very large GTO basis sets provided a satisfactory agreement with the corresponding experimental hyperfine couplings. In addition, these authors showed that B3LYP performed considerably better than conventional restricted and unrestricted Hartree-Fock and even ROHF with configuration interaction with single excitations (ROHF-CIS) models. In a detailed evaluation of DFT methods for the prediction of transition-metal hyperfine coupling constants Munzarovà and Kaupp, 1999, reported the results obtained with eight different functionals (the GGA functionals BLYP, BP86, and BPW91, the corresponding three-parameter hybrids B3LYP and B3PW91, and three examples of 'half-and-half' hybrid functionals briefly mentioned in Section 6.6 which include 50 % exact exchange, namely BHLYP, BHP86, and BHPW91) and several basis sets in comparison with reliable experimental data and results from elaborate coupled cluster calculations. In terms of the basis set the authors note that for 3d transition-metals due to error compensation already a contracted GTO set of 9s7p4d quality yields hyperfine coupling constants comparable to results obtained with much larger basis sets. However, no generally valid hierarchy of functionals for the calculation of hyperfine coupling constants of transition-metal containing compounds could be established, since the performance of a given functional varies significantly for different classes of complexes. The subtleties of the electronic structures, the degree of spin contamination as well as other factors seem to be responsible for these variations. Nevertheless, for the majority of the systems studied, essentially all of the functionals showed deviations of only some 10-15 % from the experimental isotropic metal hyperfine coupling constants.

11.6 Summary

The accurate computational determination of magnetic properties still poses a challenge to density functional methods. The reasons for this are manifold. The theoretically most severe deficiency of today's implementations is the use of the standard functionals which depend only on the charge density. In the presence of a magnetic field the usual Hohenberg-Kohn theorem no longer applies and for a proper treatment functionals which depend not only on the charge density but also on the magnetic current density should in principle be employed. For the time being, no such $j(\vec{r})$ dependent functionals have been implemented in standard production codes and no quick solution to this problem seems to be in sight. Fortunately, from a pragmatic point of view, the use of regular charge density functionals seems to be only a minor flaw. The use of sophisticated functionals such as the modern hybrid variants together with large and flexible basis sets frequently affords results which are better than HF and which often rival or even outperform MP2 data. However, unlike for many other molecular properties, no clear-cut hierarchy of functionals in terms of accuracy has yet emerged. One should also keep in mind that all magnetic properties are very sensitive to the quality of the basis sets, the kind of functional selected, and in particular to the geometry of the system at hand. While using inaccurate geometries may render the whole calculation worthless, the comparison between theoretically predicted and experimentally obtained magnetic properties may on the other hand also be used as a probe for the correct structure of the target molecule. DFT-based NMR chemical shift calculations can and should use the techniques to deal with the gauge problem which have been developed earlier in the context of wave function based methods, namely IGLO or GIAO or related techniques. In conclusion, particularly if large molecules which may even contain transition-metal centers with a complex electronic structure are considered, density functional approaches are probably the only means available today to obtain reasonably accurate results for NMR and ESR properties.

12 Hydrogen Bonds and Weakly Bound Systems

The structures, energetics, vibrational frequencies etc. discussed in the preceding chapters concerned mostly covalently bound molecules with bond energies typically exceeding 50-100 kcal/mol. Let us now turn to more weakly bound systems with binding energies smaller by one or two orders of magnitude. An important category among such systems is the *hydrogen bond* or *hydrogen bridge*, i. e., the linkage A–H···B which involves hydrogen atoms in a moderately polar bond $A^{\delta-}$–$H^{\delta+}$ (the proton donor) and another atom or group $B^{\delta-}$ (the proton acceptor). Since 1939 the concept of hydrogen bonding has found wide acceptance due to the publication of Pauling's 'The Nature of the Chemical Bond' and it is commonly invoked to explain the properties of systems as important as water, carbohydrates, and nucleic acids, just to name a few. This type of bond is characterized by a complex formation energy larger than just dipolar and dispersion interaction energies and by an H···B bond length that is shorter than the sum of the van der Waals radii of H and B. Both A and B are usually atoms more electronegative than hydrogen like O, F, N, or Cl. If the proton accepting group B is oxygen or nitrogen, the existence of a hydrogen bond is indicated when the H···B distance is less than 2.5 Å and larger than the covalent N-H or O-H bond length (about 1.0 Å). Moreover, the presence of hydrogen bonds is evidenced by an elongation of the A-H bond, accompanied by a red shift of the corresponding infrared stretching frequency. This weakening of the donor A-H bond goes hand in hand with a strengthening of the H···B interaction and leads to binding situations which are intermediate between the two extremes A-H + B and A^- + $H\text{-}B^+$, the latter corresponding to the product of a simple acid-base proton transfer reaction. A decrease in electron density of the hydrogen atoms involved in a bridge is also indicated by a low field shift in ^{1}H-NMR experiments, which can be as high as 20 ppm for strong bonds involving ionic species. The strength of hydrogen bonds between neutral species in the gas-phase is usually of the order of 2-10 kcal/mol and hence intermediate between covalent bonds (usually exceeding 50 kcal/mol) and attractive van der Waals interactions (below 2 kcal/mol). Although rather weak in nature, hydrogen bonds often have a decisive influence on the chemical properties of substances. Hydrogen bonds stabilize the secondary and tertiary structures of proteins and are thought to play a major role in substrate recognition, binding and enzymatic catalysis. Owing to their major contributions to the molecular architecture, hydrogen bonds have been studied extensively, experimentally as well as theoretically, over the years.

The peculiarity of this unique bonding pattern is that the groups A–H and B are generally closed-shell (and in their electronic ground state) and it is not obvious at first sight how the hydrogen atom could be involved in more than one bond with its single 1s valence orbital. Although experimental work has been carried out for a vast variety of systems, many difficulties prevent the detailed understanding of the nature of this bond and electronic structure theory therefore plays an important role in this field of research. The development of energy partitioning schemes in particular provided theoretical means to qualitatively rationalize the underlying bonding mechanism. The scheme of Morokuma, 1977, for example, which has probably received the most wide-spread attention for over two decades now, splits the total interaction energy between two molecules into several parts such as an electrostatic component, a polarization component (originating from the polarization of one of the interacting molecules due to the presence of the other), a charge-transfer component and a few further ingredients. Applied to hydrogen bonding Morokuma's scheme has led to the understanding that electrostatic and charge-transfer interactions both constitute the most relevant components of the binding energy (Umeyama and Morokuma, 1977).[48] The detailed nature of the hydrogen bond, however, is still a matter of debate (see Dannenberg, Haskamp, and Masunov, 1999, and references cited) owing to the fact that most rationalizations depend strongly on the particular level of theory employed. What seems clear is that the interactions responsible for a hydrogen bond can originate from various physical effects. Both electrostatic and covalent contributions vary from species to species and thus a balanced description of the subtle interplay of electronic effects is needed for a successful theoretical description of hydrogen bonding.

Let us briefly outline some salient problems limiting the accuracy that a theoretical treatment of hydrogen bonded systems can reach. One contribution to the bonding stems from the interaction of molecular moments, like the dipole-dipole interaction. Considering the well known tendency of Hartree-Fock theory to overestimate dipole moments (cf. Chapter 10), it becomes obvious that this contribution to the binding energy will be exaggerated at this level. On the other hand, attractive dispersion interactions are completely missing in HF level calculations. Dispersion interactions are pure correlation effects and hence can only be recovered at more sophisticated, correlated levels of theory which, in turn, usually also yield better (i. e., smaller) dipole moments. Hence – as a consequence of error cancellation – both approaches may reproduce equally well, say, the experimental binding energy of a hydrogen bonded complex. Properties of other regions on the potential energy surface may be described quite differently by the two different approaches, though.

Another decisive point for the theoretical treatment of hydrogen bonded systems is the choice of basis set. The strength of hydrogen bonds computed by means of traditional ab initio theory requires highly flexible basis sets including diffuse functions and an explicit recovery of electron correlation effects. The fact that the basis sets employed are practically always incomplete is troublesome with respect to a balanced description of the molecular complex as well as its constituting fragments. There are two aspects of this problem from the perspective of post-HF theory. Quite a vexing one is the basis set superposition error (BSSE). It arises from the use of finite sized basis sets in the supermolecular approach, which is usually adopted to compute the interaction energy as the difference between the total energy of the A–H···B complex and the sum of the total energies of the non-interacting fragments A–H and B. Whereas the isolated fragments are just described in their own basis sets, in the interacting complex each of them will expand its respective wave function using

[48] Since the procedure used to compute some of these components violates the Pauli exclusion principle, the physical meaning of the interpretations emerging from the Morokuma scheme has sometimes been questioned (see, e. g., Chakravorty and Davidson, 1993). However, although its physical basis is certainly not 'rock-solid', this scheme (as well as others, see Reed et al., 1986, Glendening and Streitwieser, 1994, Remer and Jensen, 2000) splits the interaction energy into physically intuitive components and allows for a rationalization of quantum chemical results. For a rigorous theoretical approach to electron density partitioning see Bader, 1994.

virtual orbitals of the other. This will lead to a spurious lowering of the total energy of the complex with respect to its fragments and thus to an artificial overestimation of the complexation energy. An approximate and highly popular way to estimate the BSSE a posteriori is the counterpoise correction of Boys and Bernardi, 1970,[49] in which the fragment energies are computed using the basis functions of the entire complex but considering the atoms of one respective fragment only. In this way the fragment energies will be lowered since their basis set expansion is becoming more complete. Notwithstanding some – quite intelligible – critical opinions about the applicability of the counterpoise correction, it appears that its use is now widely accepted provided that adequate basis sets are used (for an in-depth discussion and original references, see van Duijneveldt et al., 1994).

One further problem is caused by the incomplete recovery of the correlation energy and the very large basis set requirements in correlated calculations. Correlation effects are larger in the interacting complex than in the fragments – hence the incomplete coverage of correlation effects will lead to an underestimation of the interaction energy. This effect is usually the more pronounced the smaller (less complete) the basis sets are. This error and the BSSE act in opposite directions and could cancel (a rather optimistic standpoint taken in some studies), but the actual influence of both errors on computed binding energies is hard to predict.

All this illustrates the problems when attempting to properly describe hydrogen bonding by means of electronic structure calculations and to rationalize the underlying physics (for reviews see, Scheiner, 1991 and 1997, Guo et al., 1997, Del Bene, 1998, Lii, 1998, and Rappé and Bernstein, 2000). Such problems also exist in principle for covalently bonded systems but they are usually ignored since the binding energies are one or two orders of magnitude larger than the uncertainties caused by the errors mentioned above. Whenever calculations aim at a similar relative accuracy, say, 2 % of the binding energy of weakly interacting molecules, the computational expenses are enormous and a systematic improvement of the basis set quality and the correlation treatment is necessary in order to allow for an estimate of remaining errors. Of course, such an approach is inevitably constrained to the smallest molecular systems. For some cases traditional wave function based approaches have been pushed to their limits and highly accurate computational results exist, which eventually challenge experimental accuracy. For very high level calculations on the water dimer, for example, see Klopper, van Duijneveldt-van de Rijdt and van Duijneveldt, 2000, van Duijneveldt-van de Rijdt and van Duijneveldt, 1999, or Schütz et al., 1997, as well as references cited therein.

Finally, we need to step back a little in order to draw the readers attention to a more fundamental problem inherent to any theoretical treatment of hydrogen bonding situations, be it based on traditional wave function based or density functional theory. The topic is the evaluation of frequency shifts for situations in which the choice of coordinate system is of particular concern. Let us begin by classifying hydrogen bonds according to their strength or, correspondingly, according to the A–H stretching frequency. We will use Figure 12-1 to illustrate the issue for the O–H···O type of bonds.

[49] A priori approaches have also been developed, see, Valiron, Vibók, and Mayer, 1993, as well as Paizs and Suhai, 1998, and references cited therein.

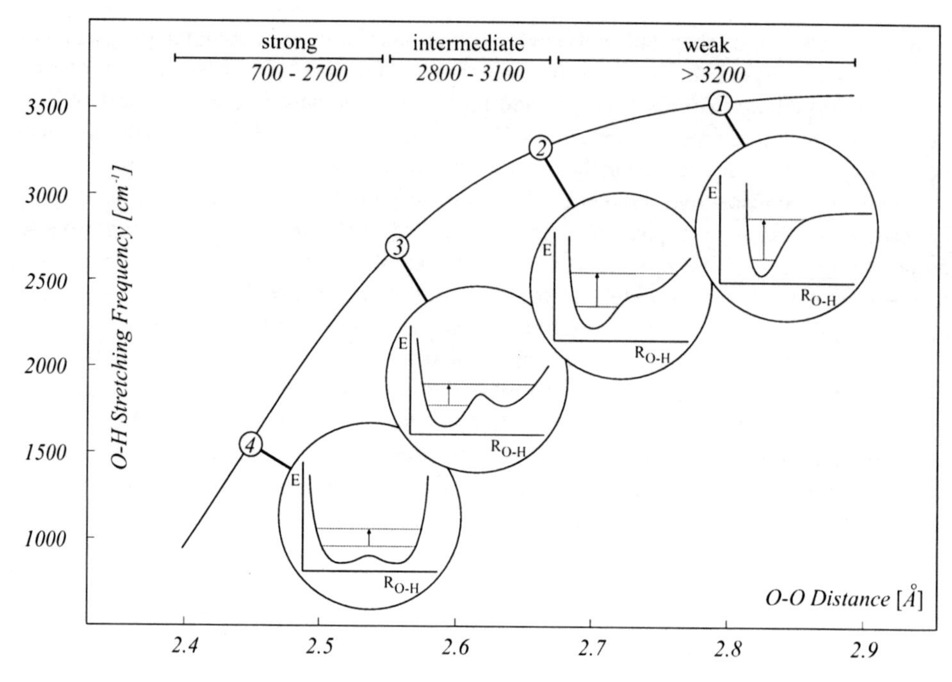

Figure 12-1. Classification of O-H···O hydrogen bonding situations of different strengths and schematic representation of corresponding potential energy curves along the O-H stretching coordinate.

The graph shows qualitatively the correlation between O-H stretching frequency and the O-O separation in increasingly stronger hydrogen bonds from right to left. Borrowing from Novak, 1974, we distinguish three binding situations, weak, intermediate, and strong, which are of interest in the forthcoming discussion, and which are classified according to the shape of the potential energy surface (PES) in the region of the O-H stretch. The O–H stretching potential in a weak OH···O bond is very similar to that in the isolated water molecule (sketched as ① in Figure 12-1). This is the standard situation for a quantum chemical treatment and the methodological issues determining the accuracy of computed frequencies discussed in Chapter 8 are perfectly valid. The harmonic approximation for the evaluation of force constants usually gives sufficiently accurate results or, if not, one has to correct for the anharmonicity of the potential energy surface in one dimension. In the present context that is along the proton coordinate in the O–H bond (see Bleiber and Sauer, 1995). Proceeding to situation ② with a hydrogen bond of intermediate strength, the O-O distance becomes shorter, the PES is broader and the vibrational levels are more closely spaced, in line with a stronger frequency shift. This is due to the increased H···O interaction, which starts to influence the PES considerably. The potential along the OH coordinate has a shoulder or, in the stronger binding situation ③, a shallow second minimum. The theoretical recovery of anharmonic frequencies on such a PES is possible only by explicit consideration of two dimensions, e. g., the O-H and the O-O coordinate (see Sokolov and Savel'ev, 1977, Sokolov, Vener, and Savel'ev, 1990, as well as Del Bene and Jordan, 1998). In even stronger bonds (situation ④) the PES is commonly very flat in the O···H···O region and the barrier separating the two minima might even disappear taking the zero point vibrational level into account. A meaningful description of the OH stretching frequency requires at least a two dimensional treatment (see Ojamäe, Shavitt, and Singer, 1995). This collection of problems connected to the PES of other than weak hydrogen bonds illustrates what sort of difficulty one may encounter in studies on seemingly easy systems. Unlike the relatively simple search for stationary points in order to determine binding energies, the evaluation of anharmonic frequencies via explicit treatment of higher dimensional potential energy surfaces is anything but trivial. For strong hydrogen bonding situations the results obtained from studies of harmonic frequency shifts can therefore be far from relevant. To the best of our knowledge, no attempts have been made to explore the performance of density functional theory under such circumstances. The theory of weak hydrogen bonds, on the other hand, is well developed and the available literature is full of successful comparisons with experimental data and high level wave function based theory. This is the area we will mainly concentrate on in the following.

Facing huge computational demands even for the smallest species, theoretical research in this field is eagerly in need of much more efficient methods in order to assess larger, scientifically more relevant species. It is no wonder, therefore, that the excellent performance of modern density functional methods in other areas of chemical research encouraged many to test this methodology on hydrogen bonded complexes. In the following section we will elaborate on the performance of density functional theory in this highly demanding field by discussing one of the archetype examples: the water dimer. Later in this chapter we will also present examples for a variety of other hydrogen bonds in order to see how well the different binding situations are described by modern functionals. As to computed frequency shifts, we will restrict our presentation to rather weak intermolecular hydrogen bonds. Since all effects characterizing the physics of hydrogen bonding are present in these species, we can fully assess the capabilities of density functional implementations by comparison to experiment or high level wave function based theory. Strong bonding situations and most intramolecular hydrogen bridges are intrinsically multidimensional problems in the bridging region and anharmonicity effects become particularly severe. The theoretical treatment of vibrational frequencies of such systems requires expert knowledge, a presentation of which is definitely outside the scope of this book. We will conclude this chapter with a brief discussion on the shortcomings of present-day functionals when it comes to coping with dispersion interactions, which, as we will see, constitute the limiting factor for the applicability of density functional theory to weakly interacting systems.

12.1 The Water Dimer – A Worked Example

The water dimer is probably the most intensively studied intermolecular hydrogen bonded system of all. Hence, ample theoretical and experimental data is available for this system,

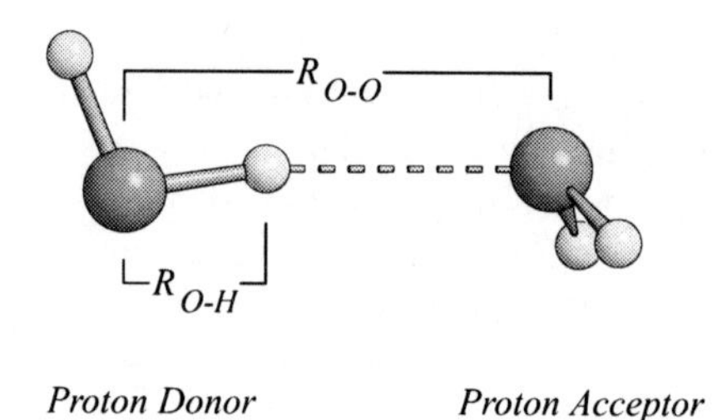

Figure 12-2. The water dimer.

which is therefore well suited for demonstrating the performance of theoretical methods.[50] There are three essential properties of the water dimer which have been determined experimentally: the oxygen-oxygen intermolecular distance, the binding energy, and the O-H stretching frequency shift of the donor molecule. As to the equilibrium geometry, the C_s-symmetric linear trans-structure (depicted in Figure 12-2) has been established by microwave spectroscopy.

Comparisons of the computed $R_{O\text{-}O}$ with experiment are somewhat complicated by the fact that the region of the PES associated with the intermolecular coordinates is quite flat and subject to large vibrational anharmonicity effects. The experimentally measured, vibrationally averaged R_0 value should therefore be corrected before a comparison is made with the computed R_e distance. For the water dimer the experimentally determined R_0 for the oxygen-oxygen distance is 2.976 Å and a R_e value of 2.952 Å has been estimated (see van Duijneveldt-van de Rijdt and van Duijneveldt, 1992). Clearly, this coordinate is highly sensitive to the level of theory applied and the accuracy with which it is reproduced by theoretical methods thus constitutes a good benchmark. Hartree-Fock theory gives an intermolecular distance which is too long, and the deviations reach 0.1 Å if improper basis sets are chosen. As evident from the data compiled in Table 12-1, increasing the basis set quality expands the O–O distance and further deteriorates the agreement with experiment. MP2, in turn, does a fair job and nearly halves the deviations seen at the HF level. But this lowest level of correlation treatment underestimates $R_{O\text{-}O}$, even if quite large basis sets are used. The remaining discrepancy has been attributed to the neglect of the BSSE in the geometry optimization procedure and indeed, an improved distance of 2.917 Å results from a counterpoise corrected MP2/aug-cc-pVQZ evaluation of the equilibrium structure (Hobza, Bludský, and Suhai, 1999). The more sophisticated correlation treatment in the CCSD(T) approach affords still smaller deviations than MP2 if small basis sets are used, but the gain in accuracy is not so impressive with the larger basis sets, despite the much higher computational costs. Also this approach underestimates the bond length owing to BSSE effects as shown by Schütz et al., 1997, and a counterpoise corrected CCSD(T) value of 2.925 Å was obtained by this group. The shortcomings of even these highly demanding geometry optimizations demonstrate the huge demands of correlated post-HF approaches with re-

[50] A recent discussion of the water dimer in the DFT domain has been given by Guo et al., 1997.

Table 12-1. Deviation in the computed $R_{O\text{-}O}$ distance of the water dimer [Å] from the 'experimental' R_e value of 2.952 Å.

Basis Set	HF	MP2	CCSD(T)	SVWN	BLYP	SLYP	BVWN	B3LYP
6-31++G(d,p)	0.035	–0.038	–0.030	–0.256	–0.040	–0.301	0.061	–0.066
6-311++G(d,p)	0.082	–0.042	–0.028	–0.244	–0.025	–0.291	0.082	–0.052
aug-cc-pVDZ	0.080	–0.034[a]	–0.033[b]	–[c]	–0.012	–0.290	0.117	–0.041
aug-cc-pVTZ	0.086	–0.044[a]	–0.057[b]	–0.241	–0.004	–0.288	0.124	–0.034
aug-cc-pVQZ	0.086	–0.057[b]	–	–0.238	–0.003	–0.286	0.119	–0.033

[a] Taken from Kim and Jordan, 1994. [b] Only $R_{O\text{-}O}$ optimized, Halkier et al., 1997. [c] No C_s-*trans* minimum structure.

spect to the basis set size: it seems that even the highly flexible aug-cc-pV5Z basis – comprising no less than 574 contracted basis functions for the water dimer – is not yet close to the basis set limit. Even considering today's computational standards, such calculations cannot be performed in production mode and, even worse, the unfavorable scaling of correlated post-HF methods with the basis set size is strongly discouraging for anyone thinking of larger systems.

Comparing a few example density functional models with these results, we first note that SVWN severely underestimates the O–O bond distance and the deviations from experiment are more than three times larger compared to the HF level. Quite disturbingly, in combination with Dunning's aug-cc-pVDZ basis sets the linear trans-structure is not even a minimum but is characterized by three imaginary frequencies as a chemically irrelevant higher order saddle point. This seems to be a particular shortcoming of the SVWN/aug-cc-pVDZ combination as the larger triple and quadruple bases render this structure correctly a minimum. The poor LDA performance is in line with early findings by others, documenting the severe underestimation of intermolecular distances in hydrogen bonded complexes at this level (see Sim et al., 1992, Kieninger and Suhai, 1994, Kaschner and Seifert, 1994, Guo et al., 1997). The inclusion of gradient corrections to exchange and correlation yields major improvements for the predicted oxygen-oxygen separation. But the agreement of the BLYP results with experiment is in fact baffling, in particular in combination with the correlation-consistent basis sets. We have not come to trust the quality of this functional with respect to structure predictions for it has shown a general trend to overestimate bond distances (recall the discussion in Chapter 8). Curious about the influence of the particular functional components, we applied also the BVWN and SLYP functionals to this system (i. e., adding gradient corrections only to the exchange or correlation part, respectively). What we obtained is, putting it mildly, irritating: compared to the SVWN results, inclusion of the LYP gradient-corrected correlation functional shortens $R_{O\text{-}O}$ constantly by some 0.05 Å, regardless of the basis set, exaggerating the faulty bond compression at the LDA level. Becke's 88 gradient correction to exchange, on the other hand, overshoots this distance more than HF does. As a consequence the individual components of the BLYP functional give intermolecular bond lengths which differ by 0.4 Å! This rather unsatisfactory finding clearly indicates that the excellent performance of the BLYP functional is entirely due to fortuitous error cancellation. From the last column in Table 12-1 we see that the B3LYP functional, which – skeptic personalities now might put it this way – admixes a little overestimation from HF with a little more underestimation from LDA to the well cancelled B88 and LYP errors, performs not quite as well as the pure GGA and gives a slightly underestimated $R_{O\text{-}O}$. But still, the agreement with experiment compares well with MP2 or CCSD(T) results. And it is encouraging to note that Simon, Duran, and Dannenberg, 1999, showed in a recent contribution that application of a (modified) counterpoise procedure to the geometry optimization scheme yields a much improved agreement with the 'experimental' R_e value for the B3LYP functional, even using very moderately sized basis sets. On the corrected B3LYP potential surface obtained with a polarized double-zeta basis set augmented with diffuse functions (D95++(d,p)) these researchers located a minimum at R_e = 2.912 Å, compared to an uncorrected value of 2.880 Å. Considering the basis set dependence of deviations seen above, still better values seem obtainable with higher quality basis sets.

In conclusion, these dreary findings cast some doubt on the reliability of DFT results for hydrogen bonded systems. However, although the interatomic oxygen-oxygen separation is a highly sensitive measure for judging the quality of theoretical results, it is not the only one. But before we concentrate on other properties of the dimer, let us look at some experimentally known properties of the water monomer and see how well DFT does here. Clearly, the description of the water dimer is closely connected to the question of how well the water molecule itself is tackled. From the data given in Table 12-2 it is evident that the oxygen-hydrogen bond distance is reproduced best at the MP2 level of theory. Just as we would expect from the conclusions drawn in Chapter 8, we find an underestimation of $R_{O\text{-}H}$ by HF and an overestimation by the BLYP functional. The constituent SLYP and BVWN functionals both exaggerate the O–H distance, the former more than the latter, and BLYP gives an error right in the middle. B3LYP, also quite foreseeably, compensates the overestimation of BLYP and by including the underestimation of HF gives a deviation from the experimental value similar to MP2. Related to the geometric deviations are those of the (unscaled) harmonic vibrational frequencies and most interesting in the present context are the O-H stretching vibrations. For ν_s and ν_{as}, the symmetric and antisymmetric stretch, respectively, the errors generally ensue from the trends observed for the bond length: bonds which are too short result in frequencies which are substantially too high for the HF method, which shows the largest deviations. Bonds which are too long, in turn, cause frequencies which are too low in the case of SVWN and BLYP (as well as SLYP and BVWN). The MP2 approach, which furnishes Table 12-2 with a highly accurate $R_{O\text{-}H}$, also shows the best agreement with experiment for the frequencies. The deviations of B3LYP are right in the range of what we have noted as typical for this functional in Chapter 8, and the systematic underestimation by some 30–40 cm^{-1} could profit from scaling.[51] As components of the interaction energy, the quality of computed dipole moments and polarizabilities is also of interest. It is a well

[51] We also noted already that BLYP results show a better agreement with directly observed, anharmonic frequencies. This is also the case here: the BLYP deviations from the fundamental experimental frequencies (ν_s = 3657 cm^{-1}, ν_{as} = 3756 cm^{-1}) are 2 and 0 cm^{-1}, respectively.

Table 12-2. Experimental values and deviation from experiment of the R_{O-H} bond distance, the symmetric (ν_s) and the antisymmetric (ν_{as}) stretching frequency [cm^{-1}], the dipole moment [D], and the mean polarizability [$Å^3$] of the water molecule. The aug-cc-pVTZ basis set is used throughout.

Property	Exp.	HF	MP2[a]	SVWN	BLYP	SLYP	BVWN	B3LYP
R_{O-H}	0.957[a]	–0.016	+0.004	+0.013	+0.015	+0.019	+0.010	+0.005
ν_s	3832[a]	+288	–9	–106	–177	–155	–132	–33
ν_{as}	3943[a]	+279	+5	–107	–186	–156	–142	–42
μ [D]	1.854[a]	+0.084	+0.006	+0.005	–0.051	+0.007	–0.052	–0.006
$\Delta(\langle\alpha\rangle)$[b]	1.427[a]	–0.207	–0.004	+0.109	+0.143	+0.179	+0.075	+0.026

[a] Taken from Kim and Jordan, 1994. [b] Mean polarizability computed as $\langle\alpha\rangle = 1/3\,(\alpha_{xx} + \alpha_{yy} + \alpha_{zz})$, see Section 10.3.

known deficit of the HF approach that the dipole moment of the water molecule is overestimated by some 5 % as a consequence of the neglect of electron correlation, while MP2 reproduces the experimental value with pleasing accuracy. Following the trends noted in Section 10-2, the B3LYP functional performs as good as MP2 does and also the LDA does a good job. The BLYP deviation of 3 % from the experimental value is smaller than the Hartree-Fock error and, as apparent from the data collected in Table 12-2, caused by the B88 gradient correction to exchange. Thus, the good B3LYP dipole moment could once more be seen as a compensation effect of errors present in the building blocks, i. e., HF and BLYP. The mean polarizability of the water molecule is underestimated by 15 % at the HF level, but well reproduced by MP2. The LDA overestimates this quantity by some 8 % and the BLYP error amounts to 10 %. Now, it is the LYP gradient-corrected correlation functional which is obviously causing the deviation present in the BLYP results. The hybrid B3LYP again performs rather well, reproducing the mean polarizability within 2 % of the experimental value, but slightly overestimating this property. Hence, notwithstanding the sometimes faulty performance of the constituting components, the B3LYP hybrid functional yields an excellent description of physical properties of the water monomer and lags only marginally behind the more demanding computations at the MP2 level of theory.

Let us now return to the water dimer and focus on the computed binding energy. Intermolecular dissociation energies (corrected for zero point vibrational and thermal effects) of 5.4 ± 0.7 kcal/mol and 5.4 ± 0.2 kcal/mol have been obtained in experimental studies (see the references cited by Kim and Jordan, 1994), with the former value being more often quoted. The data collected in Table 12-3 underlines what we have said above about the effort one has to spend when using traditional wave function based methods in order to obtain reliable results. A look at the binding energies obtained in the Hartree-Fock approximation shows that at the lowest level of calculation presented, HF/cc-pVDZ, the two water molecules are bound by 5.7 kcal/mol, in good agreement with experiment. The counterpoise correction at this level seems at first sight not to do any good – the corrected binding energy drops down to 3.6 kcal/mol. Saturating the basis set, however, shows that it is in fact the BSSE estimation procedure, which works well: the converged (essentially BSSE-free) interaction energy computed at the HF/cc-pV5Z level (3.5 kcal/mol) confirms the former

Table 12-3. Interaction energy (ΔE) and counterpoise-corrected interaction energies (ΔE_{CP}) of the water dimer at different levels of theory [kcal/mol]. The CCSD(T)/aug-VTZ geometry has been used for single point energy calculations on the dimer. The BSSE is given in parentheses. Data taken from Halkier et al., 1997. An experimental value of 5.4 ± 0.7 kcal/mol has been reported in this study.

Basis Set	HF			MP2			CCSD(T)		
	ΔE	ΔE_{CP}		ΔE	ΔE_{CP}		ΔE	ΔE_{CP}	
cc-pVDZ	5.7	3.6	(2.1)	7.3	3.9	(3.4)	7.0	3.7	(3.3)
cc-pVTZ	4.3	3.5	(0.8)	6.0	4.4	(1.6)	5.9	4.3	(1.6)
cc-pVQZ	3.9	3.5	(0.4)	5.5	4.7	(0.8)	5.4	4.7	(0.7)
cc-pV5Z	3.6	3.5	(0.1)	5.1	4.8	(0.3)	5.1	4.8	(0.3)
aug-cc-pVDZ	3.8	3.5	(0.3)	5.2	4.3	(0.9)	5.2	4.3	(0.9)
aug-cc-pVTZ	3.6	3.5	(0.1)	5.1	4.6	(0.5)	5.2	4.7	(0.5)
aug-cc-pVQZ	3.6	3.5	(0.1)	5.0	4.8	(0.2)	5.1	4.9	(0.2)
aug-cc-pV5Z	3.5	3.5	(0.0)	5.0	4.8	(0.2)	–	–	–

counterpoise corrected estimate. Adding a single set of diffuse functions leads to a significant decrease of the BSSE – already the fairly small aug-cc-pVDZ set shows a BSSE of only 0.3 kcal/mol and the basis seems to be converged with respect to both computed binding energy as well as BSSE, from the aug-cc-pVTZ level on. Use of the cc-pVDZ basis set leads to a significantly larger BSSE at the MP2 level. The computed binding energy of 7.3 kcal/mol is corrected to 3.9 kcal/mol by the counterpoise procedure but both values miss the experimental interaction energy by 1–2 kcal/mol. MP2 results fall within the experimental error range using the cc-pVTZ and better basis sets. The augmented basis sets almost halve the BSSE with respect to their counterparts without diffuse functions and the uncorrected binding energies seem in fact almost converged from the augmented double-zeta basis on. This underlines the need to use at least one set of diffuse functions, as commonly advised in theoretical studies on hydrogen bonds. It is further interesting to note that the uncorrected results approach the basis set limit from above whereas the counterpoise procedure yields lower limits of the correct binding energy. Such a behavior is of course highly desirable for it allows the correct binding energy to be extrapolated and constitutes a way of specifying the remaining error at a given level of theory. Essentially the same observations can be made for the CCSD(T) method, which gives marginally better results than MP2 but at substantially higher costs. Again, as already alluded to above, the unfavorable scaling of the latter method with the system size renders it a tool for benchmarking of small systems rather than a production method to give results at justifiable costs.

Now that we have considered the traditional approaches in some detail – how does density functional theory deal with the problem? The LDA data presented in Table 12-4 confirms what can be speculated from the findings above: the underestimated bond lengths are paralleled by a significantly overestimated interaction energy. As was the case for the intermolecular O–O distance, the BLYP functional shows a reasonable agreement with the experimental binding energy, although one of its component functionals, SLYP, overesti-

Table 12-4. Computed interaction energy (ΔE) and counterpoise-corrected interaction energies (ΔE_{CP}) of the water dimer [kcal/mol]. The BSSE is given in parentheses.

Basis Set	SVWN		BLYP		SLYP		BVWN	
	ΔE	ΔE_{CP}	ΔE	ΔE_{CP}	ΔE	ΔE_{CP}	ΔE	ΔE_{CP}
6-31++G(d,p)	10.8	9.8 (1.0)	5.6	4.8 (0.8)	13.2	12.3 (0.9)	4.1	3.4 (0.7)
6-311++G(d,p)	10.5	9.5 (1.0)	5.4	4.6 (0.8)	12.9	11.9 (1.0)	4.0	3.3 (0.7)
6-311++G(3df,2p)	9.3	9.1 (0.2)	4.5	4.2 (0.3)	11.7	11.5 (0.2)	3.1	2.9 (0.2)
aug-cc-pVDZ	–[a]		4.3	4.1 (0.2)	11.6	11.6 (0.0)	2.9	2.8 (0.1)
aug-cc-pVTZ	9.0	9.1 (–0.1)	4.2	4.2 (0.0)	11.4	11.6 (–0.2)	2.9	2.8 (0.1)
aug-cc-pVQZ	9.0	9.1 (–0.1)	4.2	4.2 (0.0)	11.3	11.6 (–0.2)	2.9	2.9 (0.0)

[a] No C_s-*trans* minimum structure.

mates the binding energy drastically. Again, the inclusion of gradient corrections to the correlation functional gives worse results than SVWN and overestimates ΔE by some 2–3 kcal/mol more than LDA does already. This resembles the situation found for atomization energies, where the inclusion of the LYP correlation functional also lead to increased binding energies, larger even than those of the pure LDA (see Chapter 9). The BVWN functional, on the other hand, underestimates the binding energy by 2–3 kcal/mol.

In Table 12-5 we compare the binding energies computed using several hybrid functionals and basis sets, attempting to approach the basis set limit for each functional in a systematic (but not necessarily cost effective) way. At first we note the reasonable performance of all functionals. The converged results, however, indicate a slight tendency to underestimate the experimental value by about 1–2 kcal/mol. This trend is slightly more emphasized for

Table 12-5. Computed interaction energy (ΔE) and counterpoise-corrected interaction energies (ΔE_{CP}) of the water dimer [kcal/mol]. The BSSE is given in parentheses.

Basis Set	B3LYP			B3PW91			*m*PW1PW91		
	ΔE	ΔE_{CP}		ΔE	ΔE_{CP}		ΔE	ΔE_{CP}	
6-31++G(d,p)	6.0	4.6	(1.4)	5.5	4.7	(0.8)	6.2	5.3	(0.9)
6-311++G(d,p)	5.8	5.1	(0.7)	5.3	4.5	(0.8)	5.9	5.1	(0.8)
6-311++G(3df,2p)	4.8	4.6	(0.2)	4.3	4.0	(0.3)	4.9	4.6	(0.3)
cc-pVDZ	–[a]			7.3	3.9	(3.4)	7.9	4.5	(3.4)
cc-pVTZ	6.1	4.5	(1.6)	5.3	4.0	(1.3)	5.9	4.6	(1.3)
cc-pVQZ	5.3	4.6	(0.7)	4.6	4.0	(0.6)	5.2	4.6	(0.6)
cc-pV5Z	4.8	4.6	(0.2)	4.2	4.0	(0.2)	4.8	4.6	(0.2)
aug-cc-pVDZ	4.7	4.5	(0.2)	4.2	4.0	(0.2)	4.8	4.6	(0.2)
aug-cc-pVTZ	4.6	4.6	(0.0)	4.0	4.0	(0.0)	4.6	4.6	(0.0)
aug-cc-pVQZ	4.6	4.6	(0.0)	4.0	4.0	(0.0)	4.6	4.6	(0.0)

[a] No C_s-*trans* minimum structure.

the B3PW91 functional. We also tested the *m*PW1PW functional of Adamo and Barone, 1998b. Its exchange part has explicitly been designed with an improved description of the low density regions in mind, which dominate the interactions between weakly bound systems. Indeed, with respect to the B3PW91 hybrid, an improved representation of the binding energy results, now providing the very same quality as the B3LYP functional. In view of the large error compensation effects of the functional ingredients of the B3LYP hybrid, however, we refrain from a further discussion of subtle binding energy differences at this point. Instead we note another interesting finding, namely that the counterpoise-corrected results reveal a much less pronounced BSSE for any of the DFT methods than found for the traditional calculations. It is striking that Dunning's augmented correlation-consistent bases hardly show any BSSE from double-zeta quality on in connection with the hybrid functionals, whereas the non-augmented sets are slightly more prone to this error even up to pentuple-zeta quality. The diffuse set of functions in the augmented series is apparently an essential part needed in the basis and is more important than higher angular momentum functions – for DFT calculations.[52] This is not the case for the MP2 and CCSD(T) calculations, where the augmented bases show a smaller but still significant BSSE and thus both, diffuse and higher angular momentum functions are indispensable to approach the basis set limit. The Pople-type basis sets show essentially the same binding energies and BSSEs as the correlation-consistent bases of comparable size. The counterpoise procedure apparently works very well, which allows for a reasonable estimate of the converged binding energies at moderate cost. This is in line with findings of Paizs and Suhai, 1998, who applied a counterpoise procedure as well as an a priori correction scheme to DFT calculations and showed that both procedures give results very close to each other. From the results shown above we see that for a given basis set, however, different functionals show differences in the estimated BSSE by up to 0.6 kcal/mol. This is an indication that the standard basis sets taken from the world of traditional wave function based methods are not equally well suited for different functionals (we have come to the same conclusion already in Chapter 9).

Overall, the accuracy reached by the hybrid functionals tested is quite pleasing, in particular if we keep in mind the favorable scaling of computational demands and the obviously much lower basis set requirements. The binding energy of the water molecule computed with the B3LYP and the *m*PW1PW functional miss the lower experimental error bar by only 0.1 kcal/mol and approach quite closely the best available conventional wave function based data.

Finally, we test the ability of a variety of functionals to predict the characteristic frequency shift $\Delta\nu_{OH}$, which the donor O–H stretching mode experiences upon forming a hydrogen bridge. Following Sauer et al., 1994, we compare in Table 12-6 the harmonic donor ν_{OH} stretching mode of the dimer with the arithmetic mean $\overline{\nu} = (\nu_s + \nu_{as})/2$ of the symmetric and the asymmetric harmonic stretching modes of the free monomer in order to account for the strong coupling of these two modes in the latter species. In addition, the elongation $\Delta R_{O\text{-}H}$ of the O–H bond involved in the interaction is given in Table 12-6.

[52] In a different context, very similar conclusions about the BSSE in density functional applications were obtained by Dargel et al., 1998.

Table 12-6. Harmonic frequency shifts [cm^{-1}] of the donor O-H stretching mode and elongation of the O-H bond [Å] in the water dimer computed at several levels of theory (aug-cc-pVTZ basis set).

	HF	MP2	MP4[a]	SVWN	BLYP	SLYP	BVWN	B3LYP
$\Delta\nu_{OH}$	–100	–169	–121	–360	–181	–401	–134	–174
$\Delta R_{O\text{-}H}$	0.004	0.008	0.007	0.010	0.008	0.021	0.005	0.008

[a] Best available *ab initio* data: MP4/VTZ(2df,2p), Bleiber and Sauer, 1995.

Experimentally observed frequency shifts range from –105 cm^{-1} to –206 cm^{-1} (see references cited by Bleiber and Sauer, 1995) and depend strongly on the type of experiment applied. Furthermore, they include potentially strong anharmonicity effects which are not easy to account for. Thus, it seems more adequate to compare the results of the various DFT methods to accurate post-HF results. Firstly we note the excellent agreement between the BLYP and B3LYP results with the MP2 data. The computed $\Delta R_{O\text{-}H}$ show the expected behavior: the reasonable agreement in the frequency shifts for MP2 on one hand and BLYP and B3LYP on the other is reflected in a common elongation by 0.008 Å of this bond for all these methods.[53] The observed overbinding for SVWN and even more so for SLYP goes hand in hand with a much overestimated shift in frequencies and bond lengths by these methods. Reduced values are obtained for both shifts for the BVWN functional, which underestimates the binding energy. Different from what we have seen with the geometric data above, the importance of higher order correlation effects is apparently more pronounced for the computation of frequencies: the best available MP4-level data shows a somewhat lower shift in frequency and $R_{O\text{-}H}$. However, we have learned already to expect errors around 50 cm^{-1} when applying density functional theory to the evaluation of harmonic frequencies. The inclusion of scaling factors should lead to a further decrease in errors, but we also suspect that the deviations between computed and experimental geometries are a non-negligible source of error for the computed frequencies. Indeed, Hobza, Bludský, and Suhai, 1999, computed counterpoise corrected harmonic frequencies at the MP2/aug-cc-pVDZ level of theory, and found a red shift 20 cm^{-1} lower than that resulting from uncorrected MP2/aug-cc-pVDZ calculations.

To conclude this section, we add some data which appeared in a recent study of Tuma, Boese, and Handy, 1999, who tested several of the more recently developed DFT methods for their ability to reproduce the properties of hydrogen bonded systems, including the water dimer. Table 12-7 shows selected results for the water dimer properties. Employing a TZ2P-quality basis set, these authors obtained results of very similar quality for the hybrid schemes B3LYP, B97-1, and PBE1PBE and the BLYP gradient-corrected functional. Somewhat larger deviations in the frequency shifts are seen for the gradient-corrected PBE scheme. The HCTH GGA functional gives the largest deviations for the geometry and binding energy, whereas a newly developed improved version of this functional, called HCTH38,

[53] A similarly good agreement between B3LYP and MP2 calculations has been noted also for much smaller basis sets, see Del Bene, Person, and Szczepaniak, 1995.

Table 12-7. Selected computed properties for the water dimer (taken from Tuma, Boese, and Handy, 1999).

Property	B3LYP	B97-1	PBE1PBE	BLYP	PBE	HTCH	HTCH38
$\Delta R_{O\text{-}O}$[Å][a]	–0.047	–0.044	–0.077	–0.009	–0.072	0.109	–0.003
ΔE_{CP}[kcal/mol]	4.8	5.2	5.2	4.5	5.4	2.9	4.6
$\Delta \nu_{OH}[cm^{-1}]$	–175	–175	–199	–187	–227	–143	–181

[a] Deviation from the experimental R_e of 2.952 Å.

yields an excellent $R_{O\text{-}O}$ value together with a binding energy and frequency shift very similar to B3LYP and B97-1.

12.2 Larger Water Clusters

The reasonable success in the description of the water dimer by gradient-corrected and hybrid density functional methods has led to investigations on larger clusters of water molecules and the properties of the water trimer have been the subject of a variety of theoretical and experimental studies. The water trimer is one of the simplest (and therefore one of the best studied) species for which cooperative effects can be investigated, i. e., the effect on the nature of the first hydrogen bond when a second is formed between one of the first two partners and a third water molecule. The global minimum of this species is sketched in Figure 12-3 and has been established as a cyclic structure with a nearly planar six-membered ring and three exocyclic O-H bonds.

It was only recently that the binding energy and the structural parameters were accurately determined by means of state-of-the-art wave function based theory (Nielsen, Seidl, and Janssen, 1999) and this data now serves as a benchmark for the accuracy of alternative theoretical methods. The oxygen-oxygen distance obtained at the MP2/aug-cc-pVQZ level of theory amounts to 2.78 Å, and a value of 10.4 kcal/mol has been extrapolated for the infinite basis set MP2 classical binding energy at this geometry, including zero-point vibrational energies obtained at the MP2/aug-cc-pVTZ level. DFT calculations performed by

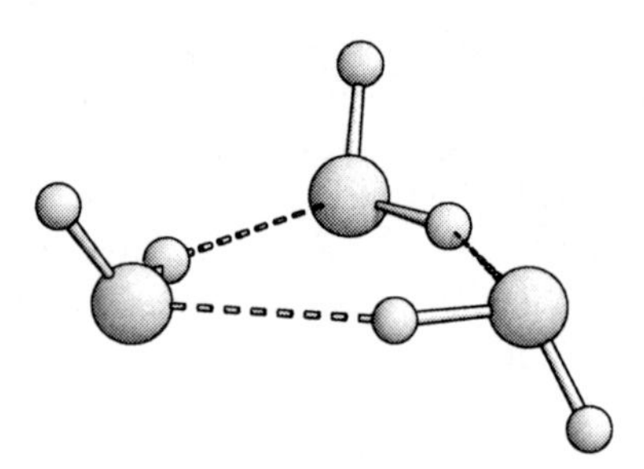

Figure 12-3. Global minimum configuration of the water trimer.

Gonzáles et al., 1996, employed BLYP as representative of the GGA family, as well as the B3LYP, B3P86, and B3PW91 hybrid functionals in combination with the 6-31+G(d,p) basis for geometry optimization and the 6-311++G(3df,2p) basis set for single point energy calculations. The deviations from the results of Nielsen et al. for the optimized (three slightly inequivalent) R_{O-O} distances follow precisely the trends described above for the water dimer: with oxygen-oxygen separations between 2.79 and 2.80 Å, the BLYP functional gave the best structural parameters, followed by B3LYP (2.77–2.78 Å) and B3PW91 (2.76–2.77 Å). The largest deviation was found for the B3P86 hybrid (R_{O-O} = 2.72–2.73 Å). The binding energy computed at the B3P86 level (10.1 kcal/mol) agreed best with the value of Nielsen, Seidl, and Janssen, followed by B3LYP (8.9 kcal/mol), BLYP (8.1 kcal/mol), and B3PW91 (7.6 kcal/mol). These energies were not corrected for the BSSE, but from what we found for the water dimer, we can assume that this data obtained with the 6-311++G(3df,2p) basis approaches the basis set limit to within about 1 kcal/mol. However, a slight underestimation of the binding energies is apparent in this data, as also noted for the water dimer.

Larger water clusters consisting of up to 8 monomers have been studied by Estrin et al., 1996, who applied the PWP86 and the BP86 functionals. Their results on the water dimer indicate a tendency of the former GGA functional to give a somewhat larger binding energy than the latter (ΔE_{CP} = 5.8 kcal/mol and 4.4 kcal/mol, respectively. 'CP' indicates that the value has been corrected for BSSE via the counterpoise correction) but both values are in reasonable agreement with the best available conventional wave function based data. For the water trimer, these authors obtained a slightly higher binding energy (ΔE_{CP} = 12.7 kcal/mol) than Gónzales et al., also overestimating the value of Nielsen, Seidl, and Janssen. Estrin et al. concluded, that the methods applied were somewhat inferior to published B3LYP data, but that the approach taken should allow for meaningful predictions on larger water clusters, of which they investigated the tetramer, the pentamer, two hexamer, and three different octamer forms. Marked differences between these and earlier results of Laasonen et al., 1993, for the same set of species[54] have been attributed to limitations inherent to the plane-wave approach used by the latter authors. The results of Estrin et al. compare at least qualitatively to wave function based data on small water clusters of up to six monomers (Xantheas and Dunning, 1993) corroborating the applicability of this DFT approach, although a general trend to overestimate binding energies, to underestimate intermolecular distances, and to overestimate O–H bond lengths seems to persist at the GGA level. Kim et al., 1999, extended these studies and investigated a variety of structural isomers of the water heptamer. The structures, binding energies, and vibrational spectra obtained at the B3LYP/6-311++G(d,p) level of theory show a reasonable correlation with selected MP2 data and experiment. In a pragmatic approach, Lee, Chen, and Fitzgerald, 1994, used the BP86 and BLYP functionals (which yield qualitatively correct binding energies of smaller clusters if no BSSE is taken into account) for an assessment of structural and energetic properties of water clusters consisting of up to 20 monomers. In this work, particularly stable bonding patterns of planar four-membered rings were identified, and the existence of

[54] Later, these studies have been extended to clusters as large as 20 water monomers (see Laasonen and Klein, 1994).

magic numbers (4, 8, 12) for particularly stable clusters was predicted. Furthermore, by empirically fitting a simple 1/n function to these results, the authors were able to extrapolate the stability of larger water clusters with an accuracy better than 1 kcal/mol up to the limit of ice at 0 K. Such predictive capabilities further strengthen the role of DFT as an efficient practical tool in this contemporary field of research.

12.3 Other Hydrogen Bonded Systems

Closely related to water clusters are hydrogen-bonded aggregates of alcohol molecules and mixed clusters of water and alcohols, which have attracted some attention in recent years. In a variety of studies the (at least qualitatively) correct description of thermochemical and physical properties of such species by gradient-corrected or hybrid functionals has been repeatedly documented by comparison to experiment or correlated ab initio data.[55] All these studies are centered around hydrogen bonding involving the moderately strong OH···O bridges in neutral clusters. The strengths of these hydrogen bonds seem in general to be very well described, albeit slightly underestimated by current DFT methodology. Let us now see how DFT covers other binding situations ranging from very weak interaction energies to the strongest hydrogen bonds.

In one of the earliest of all comparative DFT studies on hydrogen bonded systems, Latajka and Bouteiller, 1994, studied the hydrogen fluoride dimer and applied a number of different functionals in combination with a variety of polarized Pople-type basis sets. While the quality of the computed geometry and vibrational frequency of the monomer follows essentially what has been said above for the water monomer, these authors observed a significant underestimation of available experimental polarizabilities. This has been attributed to deficiencies in the 6-311++G(d,p) basis set used and better values have been obtained by increasing the number of polarization functions. As found for the water dimer, all gradient-corrected and hybrid functionals tested were well suited to assess the structural and energetic properties of the FH dimer, and particularly low deviations from experimental data were found for the BLYP functional. As for the water dimer, the computationally predicted vibrational frequencies and the red shift of F-H stretching modes showed somewhat larger deviations. Later, Maerker et al., 1997, investigated the performance of DFT for the description of various properties of $(FH)_n$ clusters (n = 1-6) in a highly elaborate study and the results corroborate the applicability of DFT approaches for this type of system. For the bigger clusters, however, somewhat larger deviations were observed than for the dimer.[56]

---

[55] Related studies include: Mó, Yáñez, and Elguero, 1997 (minima and interconnecting transition structures on the methanol trimer potential energy surface), González, Mó, and Yáñez, 1999 (the ethanol dimer and trimer), Jursic, 1999 (the mixed water-methanol dimer), González et al., 1998a (mixed water/methanol clusters), González, Mó, and Yáñez, 1996, (the H_2O_2 dimer and the $H_2O_2 \cdots H_2O$ complex), Rablen, Lockman and Jorgensen, 1998 (complexes of small organic species with water), Mó and Yáñez, 1998 (tropolone-$(H_2O)_2$ clusters).

[56] The interested reader is referred to the original publication as a survey of the present day knowledge of properties of larger hydrogen fluoride clusters, a good bibliography, and further interesting theoretical results.

Hirata and Iwata, 1998, extended these studies to linear oligomers and an infinite linear hydrogen fluoride polymer. Also this study substantiates the applicability of the BLYP and B3LYP functionals, for which reasonable agreement with experiment has been found with respect to structure, binding energies and vibrational frequencies of the species explored.

Novoa and Sosa, 1995, investigated the complexes $(FH)_2$, $(H_2O)_2$, $C_2H_2\cdots H_2O$, $CH_4\cdots H_2O$, and $(NH_3)_2$ and compared the results of DFT and Møller-Plesset perturbation theory up to fourth order. While an excellent agreement between BLYP and B3LYP on the one hand and correlated wave function based methods on the other was documented for geometries and counterpoise-corrected binding energies, the constituting SVWN, BVWN and SLYP functionals performed even worse for some of the species than noted above for the water dimer. For example, the O–C distance in the $H_2O\cdots CH_4$ complex, for which MP2, BLYP, and B3LYP agree quite reasonably (3.77 Å, 3.94 Å, and 3.89 Å, respectively), is computed as 5.91 Å (BVWN), 3.23 Å (SLYP), 3.35 Å (SVWN), and 4.11 Å (HF). These findings again underline the high degree of error cancellation which is operative in gradient-corrected and hybrid functionals. That this error compensation is indeed successful is impressively demonstrated by the computed binding energies, where the BLYP, B3LYP and MP4 results agree to within 1 kcal/mol, even for weakly bound complexes. For the $H_2O\cdots CH_4$ complex, which is bound by 0.4 kcal/mol at the MP4 level of theory, however, this accuracy is not sufficient for any qualitative let alone quantitative results (actually, this complex is unbound by 0.3 kcal/mol at the BLYP level).

Civalleri, Garrone, and Ugliengo, 1997, have studied $(FH)_2$ as well as binary adducts of hydrogen fluoride with NH_3 and CO, the former as an example of a hydrogen bond stronger than that in $(FH)_2$, the latter representative for weaker interactions. This work also shows a good conformity between all tested DFT methods (with the anticipated exception of the local SVWN functional) and available wave function based approaches for the computed binding energy of $(FH)_2$ and the $FH\cdots NH_3$ complex. The largest relative deviations were found for the FH···OC heterodimer, but all density functional methods correctly reproduce the greater stability of the FH···CO isomer. In contrast, at the Hartree-Fock level of theory, this isomer is incorrectly described as being equally stable as the FH···OC isomer, probably in relation to the reversed dipole moment of CO at that level. Computed harmonic frequency shifts for the F–H stretch were found to be significantly different for BLYP and B3LYP, and both methods overestimate the best available values for this quantity. For anharmonic frequencies, which have been considered for FH···CO and $FH\cdots NH_3$ in this study, a fair agreement between B3LYP, MP2 and experiment has been noted, whereas BLYP gave grossly underestimated frequencies. This should not surprise us too much in the light of Chapter 8, where we found an exceptionally good agreement between observed fundamentals and harmonics computed by BLYP. Any correction for anharmonicity effects necessarily leads to lower, and therefore underestimated, frequencies.

For the hydrogen bonded complexes $(FH)_2$, $(HCl)_2$, $(H_2O)_2$, FH···CO, FH···OC, $FH\cdots NH_3$, $ClH\cdots NH_3$, $OH_2\cdots NH_3$, and $H_3O^+\cdots H_2O$, with binding energies ranging from 1.7 kcal/mol (FH···CO) to 32.9 kcal/mol ($H_3O^+\cdots H_2O$), Tuma, Boese, and Handy, 1999, compared results of several density functional methods with high level conventional wave function based data. With the exception of the HCTH functional, all functionals were found to over-

Table 12-8. Deviation of counterpoise-corrected interaction energies [kcal/mol] for several hydrogen bonded systems from best available computed data (TZ2P basis set quality, all data from Tuma, Boese, and Handy, 1999).

System	Reference	B3LYP	B97-1	PBE1PBE	BLYP	PBE	HTCH	HTCH38
$H_3O^+\cdots H_2O$	32.9	2.9	3.2	4.1	3.1	5.1	0.4	2.9
$FH\cdots NH_3$	12.6	0.4	0.7	1.2	0.2	1.7	–2.2	0.1
$H_2O\cdots H_2O$	5.0	–0.2	0.1	0.2	–0.6	0.3	–2.1	–0.4
$FH\cdots OC$	1.7	–0.3	0.0	–0.1	–0.6	–0.2	–1.2	–0.3

estimate the stability of hydrogen bonds stronger than that of water and to underestimate that of less strongly bound species (for selected examples see Table 12-8). The HCTH functional consistently underestimates the hydrogen bond stabilities somewhat more than the other functionals do. This method, in turn, outperforms every other functional regarding geometry changes and harmonic frequency shifts, which were also considered in this study.

In a study of Topol, Burt, and Rashin, 1995, enthalpies and entropies of dimerization (including counterpoise corrections, zero point vibrational energy effects and thermal corrections to 298 K) for water, methanol, ethanol, formic acid, acetic acid, and trifluoroacetic acid were computed using the BP86 GGA functional in combination with (LDA optimized) basis sets of polarized double- and triple-zeta quality and compared to experimental data. For the weakly bound water ($\Delta H = 4.1$ kcal/mol), methanol ($\Delta H = 3.2$ kcal/mol), and ethanol ($\Delta H = 3.3$ kcal/mol) dimers, the experimental values were slightly underestimated but the deviations did not exceed 1 kcal/mol. A similar accuracy was observed for the rather strongly bound carboxylic dimers of formic acid ($\Delta H = 16.6$ kcal/mol) whereas slightly larger overestimations of up to 2.7 kcal/mol were found for the most stable conformers of the acetic acid ($\Delta H = 17.3$ kcal/mol) and trifluoroacetic acid dimers ($\Delta H = 16.4$ kcal/mol). In an investigation on the stability of structural isomers of the formic acid tetramer, Stein and Sauer, 1997, obtained a pleasing agreement between MP2 and B3LYP results, although the latter method showed a tendency to exaggerate the binding energies as well.

The most strongly bound neutral hydrogen bonded species known, the phosphinic acid dimer, has been studied by González et al., 1998b. Its binding energy estimated at the B3LYP/6-311+G(3df,2p) level (23.2 kcal/mol) agrees nicely with experiment (23.9 ± 6 kcal/mol). Also the computed vibrational properties are consistent with available experimental data. A still stronger hydrogen bond is present in the anionic hydrogen diformiate complex $HCOO^-\cdots HOOCH$, which represents a model system relevant for enzymatic catalysis. Süle and Nagy, 1996, applied several DFT methods to evaluate the geometry and binding energy of this species. All functionals applied (BP86, BLYP, B3LYP, and B3P86) have been found to underestimate the experimental interaction energy of 36.8 kcal/mol by up to 8 kcal/mol when using the 6-311++G(d,p) basis set. At the B3P86/6-311++G(3df,2p) level, this error is reduced to 5 kcal/mol. In combination with a polarized triple-zeta basis set built from Slater-type orbitals, the BP86 functional implemented within the *ADF* code gave precisely the same result – however, the use of more diffuse basis functions might improve the computed thermochemistry of this anionic system. Kumar et al., 1998, also studied this species (as well as the neutral formic acid dimer, see also Smallwood and McAllister, 1997) and found good agreement between B3LYP and MP2 thermochemical results, but they obtained an even lower binding energy owing to the use of very limited basis sets and geometric constraints in the calculations. These authors also investigated solvation effects on the hydrogen bond strength by explicit inclusion of one and two water molecules or by a self-consistent reaction field model in the calculations. Also the influence of different substituents (replacing X = H in XCOOH) on the interaction energy has been considered. Furthermore, miscellaneous properties of strong hydrogen bonds have been investigated by means of DFT methods in a recent series of papers (see Kumar and McAllister, 1998, and references cited therein).

Let us now turn to very weakly bound systems. A systematic comparison between results from wave function theory, experiment and several DFT methods for the weakly bound $OC\cdots H_2O$ and $CO\cdots H_2O$ complexes has been undertaken by Lundell and Lataijka, 1997. These species (bound by ΔE_{CP} = 1.3 and 0.8 kcal/mol, respectively, at the CCSD(T)/6-311++G(2d,2p) level) seem to mark the borderline of meaningful applications of modern approximate density functional theory. Using the 6-311++G(2d,2p) basis set, the DFT optimized structures for both species showed slightly overestimated intramolecular bond lengths and slightly too short intermolecular distances as compared to MP2 and CCSD(T) calculations or to experiment. Also in this study the BLYP functional keeps up its leading performance for the prediction of intermolecular distances, whereas some functionals without gradient corrections showed an alarming overestimation by up to 1 Å. The correct description of the $H\cdots O$ distance in the less stable $CO\cdots H_2O$ isomer, however, is apparently more demanding also for the hybrid functionals, which showed deviations from CCSD(T) results of almost 0.4 Å for this bond. The counterpoise-corrected binding energies for the $OC\cdots H_2O$ complex are slightly underestimated by all GGA schemes and the hybrid functional results scatter around the conventional ab initio values with deviations of less than 1 kcal/mol. The relative energy of the other isomer, being 0.6 kcal/mol higher in energy at the CCSD(T) level, is also obtained within 1 kcal/mol by all GGA and hybrid functionals. While the intramolecular vibrational properties were described sufficiently well by most gradient-corrected and especially by hybrid functionals, larger relative deviations from MP2 results have been noted for the intermolecular modes, overestimating the red shifts by up to 20 cm^{-1}, which corresponds to a deviation of more than 200 %. Particularly problematic were the modes of the less stable $CO\cdots H_2O$ isomer. While these deviations are in fact lower than those observed in other applications described above in absolute terms, the large relative errors render the predictive power of such results ambiguous owing to the small size of observable effects. The authors related these deficiencies to the predominance of dispersion energy contributions to the bonding, which are not covered by the functionals investigated (see below), but constitute 60-80 % of the binding energies of the two species as shown by Lundell, 1995. Nevertheless, in the recent literature DFT methods have been applied with some success to weak hydrogen bonds involving π-electron systems (Chandra and Nguyen, 1998). In this context we note that modifications of the original B3LYP parameterization have been suggested for an improved assessment of vibrational frequency shifts and successfully applied to the $(FH)_2$ complex (Dkhissi, Alikhani, and Bouteiller, 1997). Also an empirical modification of the adjustable β-parameter in the B88 functional used in combination with exact exchange has been proposed and led to improved intermolecular distances and frequency shifts for the $(H_2O)_2$ and the $(FH)_2$ system. Non-uniform results, however, were obtained for other weakly interacting species (García et al., 1997).

In conclusion, it seems that the thermochemical, structural and vibrational properties of a broad variety of hydrogen bonded species in very different binding situations can be described reasonably well by gradient-corrected and hybrid density functional methods. The applicability of DFT methods to hydrogen bonded systems has been confirmed by benchmarking the results with experimental and higher level theoretical data. As of today, a wealth of papers has appeared,[57] which exploit the performance of DFT methods as a practical means to supplement and guide experimental work. Representative examples comprise studies on molecular clusters (Hagemeister, Gruenloh, and Zwier, 1998, and Pribble, Hagemeister, and Zwier, 1997), correlation between proton NMR chemical shift and hydrogen bond strength (Kumar and McAllister, 1998, Garcia-Viloca et al., 1998) molecular dynamics (Wei and Salahub, 1994 and 1997, Termath and Sauer, 1997, Haase, Sauer, and Hutter, 1997, Cheng 1998), molecular adsorption in zeolites (Krossner and Sauer, 1996, Sauer, 1998, Zygmunt et al., 1998), binding and vibrational properties of nucleic acid bases (Šponer and Hobza, 1998, Santamaria et al., 1999), cooperative hydrogen bonds in enzyme catalysis (Guo and Salahub, 1998) and many more. While the interactions in strongly bound species are generally overestimated, weaker hydrogen bonds are often found to be underestimated in stability. Many functionals reach an accuracy in the description of binding energies in the order of 1–2 kcal/mol provided that sufficiently flexible basis sets of at least polarized triple-zeta quality are used. Even though this is what is called chemical accuracy, it is not sufficient to achieve predictive power for very weakly bound species. Finally, we mention a study by Milet et al., 1999. These authors corroborate the above conclusions that DFT is able to deliver reasonably reliable results for minimum structures of such complexes. But they also point out that density functionals have more severe problems when it comes to probing other regions of the PES. For example, the angular dependence of the water dimer energy is described significantly worse by the GGA and hybrid functionals used than the properties of the minimum structure.

12.4 The Dispersion Energy Problem

In reviewing the performance of density functional theory applied to hydrogen bonded complexes of moderate strength, we repeatedly noted a systematic underestimation of the interaction energies for many types of functionals, usually below 2 kcal/mol. This has been related by some researchers to the inability of modern functionals to describe those contributions to intermolecular binding energies which stem from dispersion forces. Dispersion

[57] Searching the Chemical Abstracts database for the combined keywords 'DFT' and 'hydrogen bond' reveals 201 entries for the time between January 1997 and August 1999.

forces, also referred to as London forces, are long-range attractive forces which act between separated molecules even in the absence of charges or permanent electric moments. These forces, which are purely quantum mechanical in nature, arise from an interplay between electrons belonging to the densities of two otherwise non-interacting molecules or atoms. Owing to their like electric charges, the molecular electron densities of two different systems repel each other if they come too close together. But at intermediate distances the motion of electrons in one unit induces slight perturbations in the otherwise evenly distributed electron densities of the neighboring molecule. This correlation of electronic motion leads to a temporary dipole moment. The induced dipole moment, in turn, induces a charge polarization in the first molecule, creating an attractive force between the two systems. In the asymptotic limit, this induced dipole-induced dipole attraction decays with the inverse sixth power of the intermolecular distance. The actual presence of interactions from higher order electric moments leads also to other terms like induced quadrupole-dipole, quadrupole-quadrupole interactions, etc., which vary as $1/r^8$, $1/r^{10}$ and so on. It is solely this type of interaction which is responsible for the minute binding forces between rare gas atoms and is the only reason why, for example, He even liquefies at very low temperatures. This effect is entirely due to electron correlation and the Hartree-Fock model is therefore not applicable to such situations.

Clearly, the so far unknown exact density functional must account for such electron correlation effects. However, in present implementations only the exchange-correlation energy of a given, local molecular electron density is considered and remains unaffected by the density of another, distant system if no overlap is present. The exchange-correlation potential $V_{XC}(\vec{r})$ at a point $\vec{r}$ is determined by the density (and its gradients and perhaps other local information) exactly at this point. In other words, two unshared electron distributions do not contribute by any means to an energy lowering in functional forms which depend only on a local electron density. In order to describe London interactions, a fully nonlocal functional must be applied and a local density functional is in principle not capable of describing this long-range, nonlocal correlation effect. Accordingly, some standard functionals, while correctly describing the short-range repulsion, were found to completely fail in the description of the attractive branches in the potentials of van der Waals complexes like He_2, Ne_2 or Ar_2. Numerical applications show in fact minima on the potential energy surfaces of such systems, although too deep and at the wrong positions, but these usually vanish after correcting for basis set superposition errors. The presence of actual minima has been attributed to overlapping densities, which decay exponentially in r, and not to a physically correct description of true dispersion interactions dominated by the long range fluctuating dipole ($1/r^6$) term (Kristyán and Pulay, 1994). Pérez-Jordá and Becke, 1995, investigated the performance of the SVWN and BP86 functionals as well as Becke's two hybrid approaches for the description of the He_2, Ne_2, Ar_2, HeNe, HeAr, and NeAr rare gas dimers. Also this group found a strong overbinding for the LDA with minima located at too short distances, and only repulsive interactions for the GGA and the related hybrid functional. Interestingly, the half-and-half approach not including any gradient corrections provided a quite reasonable description of the potential shapes, but this approach gave minima which were too shallow. From the latter finding it appears that some DFT models might in fact give reasonable results for related systems. But as so many times before, this is not the right answer for the right reason, i. e., a proper description of the physics of dispersion forces, but merely a consequence of error cancellation. Others have outlined the strengths and weaknesses of density functional theory associated with the description of weakly interacting systems dominated by charge-transfer interactions and dispersion forces (for helpful entry points into the recent literature consult Hobza, Šponer, and Reschel, 1995, Kang, 1996, Jeong and Han, 1996, Ruiz, Salahub, and Vela, 1996, Meijer and Sprik, 1996, Wesolowski et al., 1997, and Lundell and Latajka, 1997).

While calculations at the GGA level with the B88 exchange functional plus some correlation functional have been shown to give purely repulsive interactions for van der Waals complexes, other functionals yield relatively strong binding interactions. Work along these lines, although rather sparse, has been put forward in the literature. For instance, Patton and Pederson, 1997, tested two standard gradient-corrected functionals in combination with mostly converged basis sets and demonstrated a reasonable performance for the description of a variety of rare gas dimers. While the LDA gave grossly overestimated atomization energies, the PWPW91 GGA functional led to reduced errors. For the gradient-corrected PBE protocol a reasonable agreement with experimental energies, bond distances, and even vibrational frequencies was obtained for the lighter He_2 and Ne_2 diatomics. Dimers consisting of heavier rare gas atoms, however, were found to be too weakly bound at this level, and better agreement was found with the PWPW91 functional. Zhang, Pan, and Yang, 1997, published a comparative study on the performance of seven different gradient-corrected exchange functionals in combination with the PW91 correlation functional on the same six rare gas diatomics which Peréz and Becke have explored. The former authors emphasized the particular influence that the choice of exchange functional has on the outcome of calculations on these van der Waals systems. Hence, it might well be that inclusion of data for weakly interacting systems into the data base used for the construction (empirical fitting) of new functionals might lead to progress in this field. Notwithstanding all criticism with respect to the lack of the underlying physics, such pragmatism has evidently helped a lot in the past. Thus, we conclude this chapter by noting that as of today density functional theory can be used successfully, albeit only with great care, for rather weakly bound systems (see also the discussion in Adamo, di Matteo, and Barone, 1999). However, for the most part contemporary density functional theory does not seem to offer sound and reliable predictive capabilities when it comes to describing systems dominated by very weak van der Waals forces. It will be interesting to see to what extent the many attempts to develop new functionals (see, Lundqvist et al., 1995, Andersson, Langreth, and Lundqvist, 1996, Dobson and Dinte, 1996, Osinga et al., 1997, Kohn, Meir, and Makarov, 1998, Dobson, 1998, Lein, Dobson, and Gross, 1999) or to improve existing ones (e. g., Adamo and Barone, 1998b) will lead finally to an adequate and satisfying description also of the energetically low energy end of chemical interactions.

13 Chemical Reactivity: Exploring Potential Energy Surfaces

The rates of chemical reactions are in many cases limited because of the presence of energy barriers between reactants and products, and the barrier heights typically determine the branching ratio of products. A rigorous understanding of all elementary steps along the reaction coordinates leading from reactants to products is a prerequisite for the development of guidelines for rationalizing or predicting the corresponding chemical transformations. However, the complete characterization of the actual reaction mechanisms, i. e., explicit information on the structural and energetic details of all intermediates and transition structures relevant in the course of a particular reaction, by experimental means alone has been possible in only a very limited number of cases.[58] Quantum chemical calculations offer in principle a complementary source of information. Among the most prominent applications of modern electronic structure theory is therefore the localization and characterization of stationary points on those parts of a potential energy surface, which are associated with a chemical reaction. In particular the ability to directly model the transition structures connected to the activation barriers is a most appealing feature. However, even for a qualitatively correct picture of a reaction path electron correlation effects need to be taken into account in the framework of conventional ab initio molecular-orbital theory. If quantitative accuracy is the target, a sophisticated treatment of correlation effects is usually required in order to ensure a balanced description of minima and transition structures. Hartree-Fock theory in most instances fails miserably. Typically, this approach leads to an overestimation of reaction barriers. One important reason for the poor performance of the HF method is that the stretching of bonds – which is a key feature of transition structures involving the shift of individual or groups of atoms – beyond a certain point leads to a break-down of the one-determinantal wave function. The HF model cannot cope with such situations and, in principle, a multi-determinantal description of the wave function combined with a sufficient recovery of dynamical electron correlation is needed in order to achieve chemically meaningful accuracy. Even though such methods have been developed and successfully applied in the past, approaches like CASPT2, MRCI, or ACPF, which are based on CASSCF zeroth order wave functions, are much too expensive to play a role as a standard tool and their application is limited to small, chemically less relevant cases only. Single reference approximations like MP2 or CCSD(T) have shown to provide sufficiently accurate results in many cases, however, it is often not clear from the outset whether they are appropriate or not. In addition, these methods also suffer from an unfavorable scaling with molecular size, as has been already critically noted in several places in this book. This problem is particularly pronounced due to the fact that – depending on the particular system under study – the use of very large basis sets is mandatory in order to obtain converged results. The compu-

[58] Note, however, that the 1999 Nobel Prize for Chemistry was awarded to A. Zewail for his 'studies of the transition states of chemical reactions using femtosecond spectroscopy' (Academy's citation, October 12, 1999).

tational efficiency of density functional approaches combined with the inclusion of nondynamical and dynamical correlation effects inherent to the functionals used has therefore made approximate DFT a promising competitor to the conventional methods. In recent years density functionals have been applied to a great variety of chemical reactions ranging from organic and inorganic systems through organometallic catalysis in the gas-phase, in homogeneous and heterogeneous environments, to bioinorganic reactions and models for enzymatic catalysis (for a collection of recent examples see, e. g., Truhlar and Morokuma, 1999). Early systematic investigations involving simple organic and organometallic reactions include the work of Andzelm, Sosa, and Eades, 1993, Stanton and Merz, 1994, and Baker, Muir, and Andzelm, 1995; for reviews see, e. g., Seifert and Krüger, 1995, Springborg, 1997, Ziegler, 1997, and Salahub et al., 1999. These studies revealed that the LDA gives extremely unreliable results, and should not be used, whereas a better agreement with post-HF and experimental reference results can be afforded by application of gradient-corrected approaches, although they were shown to have a pronounced bias to underestimate barriers.

In this final chapter we do not attempt to cover all the recent applications of density functional theory to chemical reactivity. Such an endeavor seems hard to accomplish even in an entire book, let alone a single chapter – plus, a review would be outdated by the time of publishing given the vast number of research papers appearing in the literature month by month. Rather, we concentrate on a few systematic studies on prototype reactions in which the performance of density functional methods is compared to high level post-HF computational results or to reliable experimental data. The examples have been chosen to demonstrate that the performance of approximate DFT for different classes of reactions frequently differs significantly and that no general rule of thumb is available. Rather, each reaction may offer its particular surprises. Therefore, the importance of a careful calibration of the theoretical methods prior to their application to the actual, uncharted territory of interest cannot be overstressed. Still, certain transformations are characterized by typical problems and some of them are also the subject of the following sections.

In particular, reactions involving transition-metals have attracted a lot of interest recently because of the connection to catalytic and enzymatic processes. Unfortunately, the proper computational description of such reactions is one of the great challenges of today's theoretical chemistry and the question for the general applicability of density functional methods in the field is an area of active research. We chose to provide a single but – as we think – representative example to illustrate the difficulties one has to face in theoretical studies of transition-metal reactivity.

13.1 First Example: Pericyclic Reactions

The first pair of examples we would like to discuss occurs in a field which lends itself naturally to be conquered by theory. Indeed, the past three decades have seen the exploration of mechanistic details of pericyclic reactions as one of the major success stories of computational chemistry. Rooted in qualitative molecular orbital theory, the key concept of *conservation of orbital symmetry* (Woodward and Hoffmann, 1970) has opened the way to the detailed mechanistic understanding of electrocyclic reactions, cycloadditions, sigmatropic shifts, cheletropic reactions, and the like. In earlier times, however, numerical applications using semiempirical methods or Hartree-Fock theory were insufficient to allow for unequivocal conclusions about mechanistic conceptions – a vividly written documentation of related controversies has been presented by Houk, Gonzalez, and Li, 1995. Activation energies for pericyclic reactions computed at the HF level are substantially too high, usually by about 50-100 %, due to the neglect of electron correlation contributions. Inclusion of dynamic correlation effects in most cases ameliorates the situation and usually affords a much better agreement with experiment. Another severe problem occurs if a balanced description of alternate pathways, like concerted and stepwise reaction mechanisms is envisaged. A computational discrimination between these alternatives is complicated by the fact that the latter involve biradical species, which can be extraordinarily troublesome for single-reference methods, such as HF or Møller-Plesset perturbation theory, but also difficult for density functional methods (see, e. g., Goddard and Orlova, 1999, also for possible remedies). As in many other areas, the advent of approximate density functional theory has had a major impact on this area of research, once its general applicability in the field seemed established. Modern functionals have been shown to yield a very reasonable description of the potential energy surfaces connected to the chemistry of pericyclic reactions, even in demanding electronic situations such as in radical cation species, which pose a fundamental challenge to the traditional approaches (see, e. g., Wiest, 1999). An overview of this theoretically well-covered field of research can be found in recent reviews (Wiest and Houk, 1996, Bertran et al., 1998, Wiest, 1998, and references therein). Here, we choose to introduce the reader to two prominent examples, the $[_{\pi}4_{c}]$ electrocyclic ring opening of cyclobutene and the prototype of a Diels-Alder reaction, the $[_{\pi}4_{s}+_{\pi}2_{s}]$ cycloaddition of ethylene to butadiene.

13.1.1 Electrocyclic Ring Opening of Cyclobutene

The conrotatory thermal electrocyclic ring opening of cyclobutene **1** has been extensively studied using a great variety of theoretical methods. In line with the Woodward-Hoffmann rules, all theoretical methods applied to this problem compute a C_2-symmetric transition structure **2** with a twisted carbon framework. This transition structure connects cyclobutene with the *gauche*-1,3-butadiene **3**,[59] which subsequently rotates along the central C-C bond via transition structure **4** to yield the global minimum along the reaction coordinate, i. e., *trans*-1,3-butadiene **5**. The reaction path emerging from experimental and high-level theoretical work is pictured in Figure 13-1.

From a first inspection of the experimental and theoretical energetic data for stationary points along the reaction path compiled in Table 13-1, we see that Hartree-Fock theory

[59] The planar *cis*-butadiene is not a minimum but rather represents the transition structure for the degenerate isomerization of the two identical *gauche* isomers.

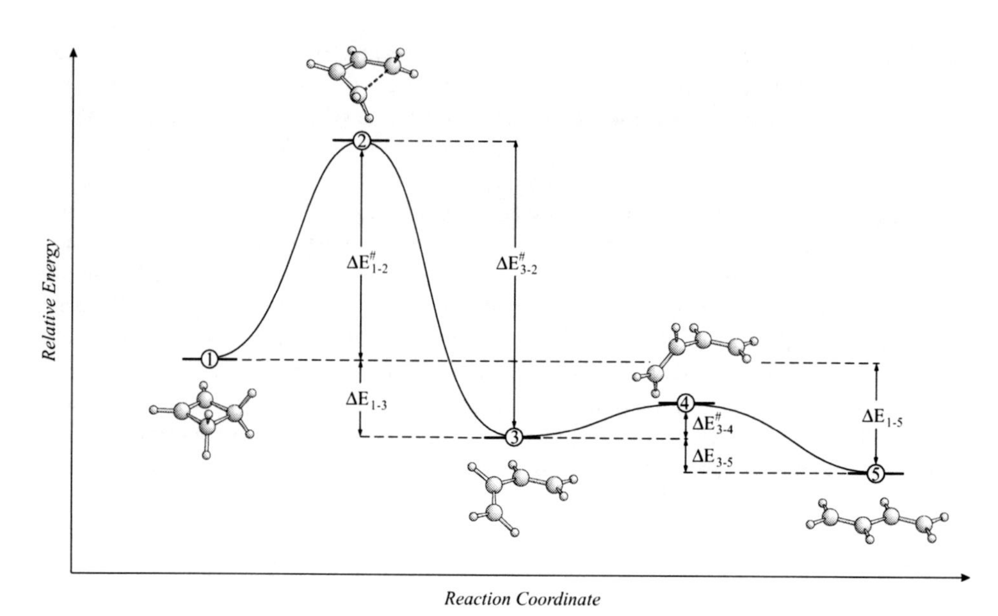

Figure 13-1. Reaction path for the electrocyclic ring opening of cyclobutene.

Table 13-1. Computed reaction barriers and isomer stabilities [kcal/mol] for the electrocyclic ring opening of cyclobutene (relative to cyclobutene **1**, including zero-point vibrational contributions). Except for G2, the results were obtained using the 6-311+G(d,p) basis set.

Barrier	Exp.[a]	G2	HF	SVWN	SLYP	BVWN	BLYP	B3LYP
$\Delta E^{\#}_{1\text{-}2}$	33	33	43	33	33	28	27	32
$\Delta E^{\#}_{3\text{-}2}$	41	43	56	36	34	44	41	44
$\Delta E_{3\text{-}5}$	−3	−3	−3	−4	−3	−4	−4	−3
$\Delta E^{\#}_{3\text{-}4}$	3	3	2	4	5	3	4	3
$\Delta E_{1\text{-}3}$	−8	−10	−13	−3	−1	−16	−14	−12
$\Delta E_{1\text{-}5}$	−11	−13	−16	−6	−4	−20	−18	−16

[a] Compiled from the experimental references cited in Murcko, Castejon, and, Wiberg, 1996, and Wiest, 1998.

overshoots the activation barrier $\Delta E^{\#}_{1\text{-}2}$ for the initial ring opening step by 10 kcal/mol. A closer inspection of the restricted HF wave function reveals a RHF→UHF instability,[60] i. e., there is an unrestricted Hartree-Fock solution with a lower energy than the restricted, closed-shell RHF determinant. This is a strong indication of the necessity to include electron correlation in order to properly describe the bond rupture in transition structure **2**. Indeed, if electron correlation effects are accounted for at the rather sophisticated G2 level, the ex-

[60] The stability of a Slater determinant can be checked by means of the *stable* keyword in *Gaussian* 98. For an extension to DFT, see Bauernschmitt and Ahlrichs, 1996a.

perimental barrier is reproduced perfectly. If we look at the relative energies resulting from the LDA treatment, we note an impressive agreement between SVWN and experiment as to the height of the activation barrier – in contrast with common expectations, the LDA does not underestimate the barrier. Further, quite interestingly, the inclusion of gradient corrections to correlation does not seem to have any influence on the computed $\Delta E^{\#}_{1\text{-}2}$: the SLYP barrier height also matches the experimental value. Switching on gradient corrections to exchange in the BVWN functional, however, leads to an underestimation of the activation barrier by 5 kcal/mol. Also the BLYP GGA functional underestimates the barrier by 6 kcal/mol, and only the B3LYP hybrid functional catches up with the excellent LDA quality and accurately describes $\Delta E^{\#}_{1\text{-}2}$. Note that unlike in the HF scheme, irrespective of the actual functional applied all non-interacting Slater determinants generated from the approximate KS orbitals are stable with respect to symmetry breaking – starting from an unrestricted set of guess orbitals of broken symmetry for transition structure **2**, the determinants in all cases collapse to the restricted solutions. This is an encouraging documentation of what we have noted already before pertaining to the dissociation potential of H_2 in Section 5.3.5: density functional theory is significantly more robust with respect to symmetry breaking for stretched bonds than Hartree-Fock theory. That is, the onset of the point where restricted and unrestricted calculations differ from each other is shifted to larger bond distances for density functional calculations. In the present context this means that, ignoring all other sources of error, this class of methods lends itself intrinsically better to a description of stretched bonds in transition structures than Hartree-Fock and related wave function based concepts. One would be tempted to relate the good performance of the density functionals to this pleasing feature, if there were not the disturbing deviations upon inclusion of gradient corrections to exchange. Remember that these were rather decisive for an accurate evaluation of atomization energies as noted in Chapter 9. So, is there something wrong with the feeling we developed for the hierarchy of density functional methods in the preceding chapters? Certainly not! In fact, a different picture emerges, if we look at the same reaction from a different point of view, viz., as an electrocyclic ring closure of *gauche*-butadiene **3**, and compare the calculated energy difference $\Delta E^{\#}_{3\text{-}2}$ computed by the various methods. Now we find a situation which corresponds more closely to what we would have expected on the basis of our previous findings. While HF overestimates the activation barrier of the reverse reaction even more than before and G2 theory also deviates by +2 kcal/mol, SVWN underestimates the barrier by 5 kcal/mol. This is ameliorated a little at the SLYP level and a much larger barrier is obtained by application of the BVWN functional, which is now in quite good agreement with the experimental data. Inclusion of the gradient corrections to correlation in the BLYP functional brings about a perfect match with experiment. Admixture of exact HF-exchange leads to a slight overestimation of the barrier, but still, the experiment is well reproduced. The relative stabilities of *gauche*- and *trans*-butadiene expressed in $\Delta E_{3\text{-}5}$ are in most pleasing agreement with experiment at all levels of theory: the deviations do not exceed 1 kcal/mol in any case. Likewise, the relative energy of the transition structure **4**, which connects the *gauche*- (**3**) and *trans*-forms (**5**) of butadiene, is accurately described and within 2 kcal/mol of the reference values at all levels, with the BVWN and B3LYP functionals performing best. Significant deviations are seen for $\Delta E_{1\text{-}3}$ and $\Delta E_{1\text{-}5}$, and it is hence the description of the strained cyclobutene species **1**, which is causing the trouble. Its electronic structure is obviously very different from the other species such that errors cancel only incompletely. While the G2 technique places ΔE_{1-3} and ΔE_{1-5} within 2 kcal/mol of the experimental results – which is just the accuracy this method has been designed for – the DFT methods perform much worse. The SVWN, and even more so the SLYP functional, underestimate the energy difference between cyclobutene and the two butadiene minimum conformations, whereas BVWN, BLYP, and B3LYP overestimate these energy differences. In conclusion, among the functionals tested, B3LYP provides the best overall description of the entire reaction sequence with a maximum deviation of 5 kcal/mol.

13.1.2 Cycloaddition of Ethylene to Butadiene

The Diels-Alder reaction is a most useful synthetic tool in organic chemistry, and the parent [4+2] cycloaddition of ethylene to butadiene has been well studied by experimentalists and theorists alike and constitutes a good test case for our current presentation. It has been established that the reaction proceeds through a synchronous concerted transition state, which means that the two new bonds are being formed not only in one single step but also in a synchronous way (Goldstein, Beno, and Houk, 1996). The alternative stepwise mechanism involving biradical species is, however, energetically not far from the concerted one. A variety of theoretical methods has been applied to Diels-Alder reactions and it has been shown that an accurate description of this reaction type is in need of a rather high level of electron correlation as far as classical electronic structure theory is concerned. The problems range from the overestimation of barriers at the HF level to non-converging results at different levels of Møller-Plesset perturbation theory and do not end unless non-dynamical correlation is adequately accounted for in a balanced way to study the radical pathway alternative (for related literature see Bertran et al., 1998, and cited references). Density functional theory has been successfully used in this field ever since its first implementation in standard quantum chemical codes. It has furnished organic chemistry with much mechanistic comprehension about this type of reaction and has been used – usually in combination with limited basis sets due to the size of the molecules involved – as an easy-to-use tool for synthetic organic chemists (for recent examples see Sodupe et al., 1997b, Venturini et al., 1997, Chen, Houk, and Foote, 1998, Tietze, Pfeiffer, and Schuffenhauer, 1998, and Goldstein, Beno, and Houk, 1999).

The course of the reaction is sketched in Figure 13-2 and Table 13-2 contains the activation and reaction energies computed at different levels of theory. G2[61] and experimental results agree to within 2 kcal/mol so that this data serves well as a benchmark. We note the large deviations from these results at the Hartree-Fock level which we have got so used to by now: the reaction barrier is overestimated by no less than 100 %, rendering this level of theory completely useless for gaining chemical insight. A most dramatic underestimation

[61] Related, yet simpler extrapolation schemes have been applied to evaluate the activation barrier (Froese et al., 1997) and the entire sequence (Barone and Arnaud, 1997).

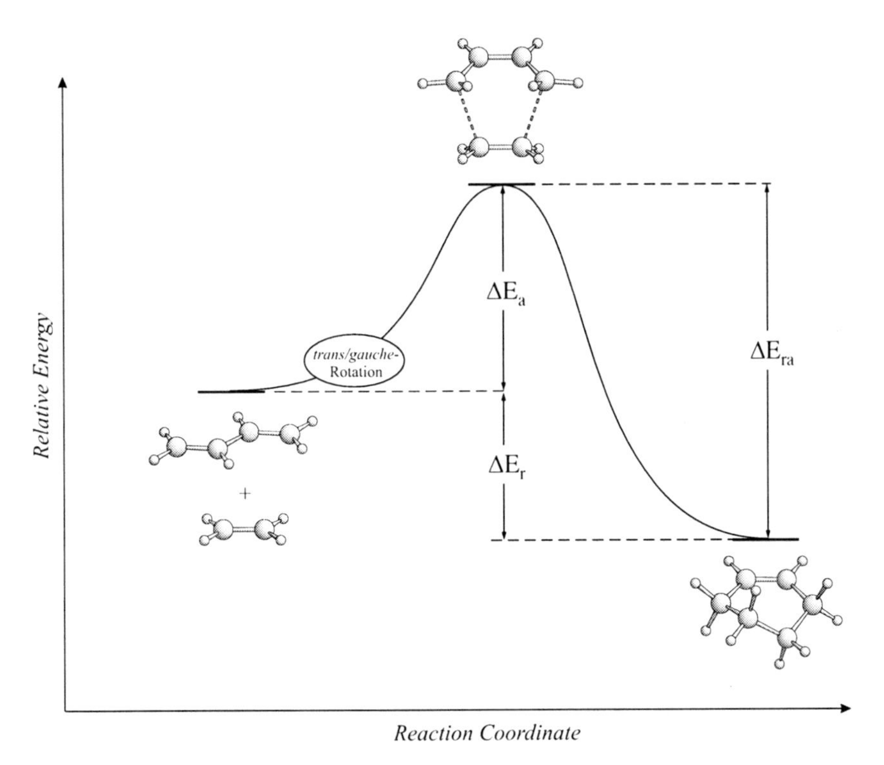

Figure 13-2. Reaction path for the cycloaddition of ethylene to butadiene.

Table 13-2. Computed activation (ΔE_a) and reaction energies (ΔE_r) for the concerted gas-phase cycloaddition of ethylene to *trans*-butadiene [kcal/mol]. The HF and DFT calculations were performed with the 6-311+G(d,p) basis set and include zero-point vibrational contributions.

	Exp.[a]	G2	HF	SVWN	SLYP	BVWN	BLYP	B3LYP
ΔE_a	27 ± 2	25	51	5	–2	33	26	28
ΔE_r	–38	–38	–30	–59	–67	–14	–22	–29

[a] Taken from references cited in Wiest, 1998.

of the barrier is seen at the LDA level and the SLYP functional places the barrier below even the relative energy of the reactants. However, the energy difference between the cyclohexene product and the transition structure (ΔE_{ra}) obtained with these two methods (64 and 69 kcal/mol, respectively) is close to the reference data, which is indicative of an incomplete error cancellation on the part of the separated reactants. The BVWN functional has just the reverse problem: correcting exchange with the gradient-corrected functional due to Becke results in a substantial improvement for the computed barrier height, but now the stability of the product is significantly underestimated. This leads to an underestimation of the reverse barrier by this functional by about the same amount as SVWN and SLYP err in the forward direction. It is apparent that gradient corrections to exchange are more important for a balanced description of separated reactants and the transition structure, whereas gradient corrections to correlation do well for the relative energies of transition structure and product. The combination of both in the BLYP functional leads to an improved description for the entire reaction sequence, but the product stability remains underestimated by 16 kcal/mol – an unacceptably large error for chemical purposes. The B3LYP hybrid slightly improves the barrier height, but still a large deviation of 9 kcal/mol is seen for the relative energy of cyclohexene. This latter finding is rather surprising as this reaction has not been found problematic for the B3LYP hybrid functional in previous published work (see, e. g., Barone and Arnaud, 1996 and 1997). The only difference between these investigations and the present results is the use of smaller basis sets in the previous studies. So, it seems worthwhile to have a look at the basis set dependence. Let us therefore focus on results from calculations with basis sets of varying size.

From the energetics compiled in Table 13-3 we clearly see that a systematic improvement of the basis set quality causes a severe deterioration of the description of the relative stability of the cyclohexene product. The height of the activation barrier is less affected and the disagreement with the G2 data increases only slightly when going from smaller to larger basis sets. It appears that even the B3LYP hybrid functional is not well suited to consistently describe the dramatic changes in molecular electron density when fragmentation processes are considered, even if all species involved are closed-shell. Hence, it is rather plausible that the many honors density functional theory has earned in this field in the past are due to massive error compensation effects arising from the use of small basis sets. Admittedly, this error compensation has been highly effective and constant for a broad variety of systems studied, and has led to extraordinarily good agreement with experimental data (see, e. g., Beno, Houk, and Singleton, 1996). So the B3LYP/6-31G(d) combination is without doubt a cost efficient and therefore valuable procedure with predictive capabilities.[62] But, in contrast to what we have concluded in other instances, there is no way to improve the results if we encounter problems and we have to be very careful what we can

Table 13-3. Basis set dependence of activation (ΔE_a) and reaction energies (ΔE_r) computed using the B3LYP functional for the concerted gas-phase cycloaddition of ethylene to *trans*-butadiene [kcal/mol]. All calculations include zero-point vibrational contributions evaluated at the B3LYP/6-311+G(d,p) level.

	6-31G(d)	6-31+G(d)	6-31++G(d,p)	6-311++G(3d,2p)	cc-pVTZ
ΔE_a	25	27	27	28	28
ΔE_r	–37	–33	–32	–29	–28

[62] As noted by Adamo, di Matteo, and Barone, 1999, similarly good results can be expected from the traditional schools of theory only at the CCSD(T)/TZ2P level, but this computational approach is prohibitively expensive for most chemically relevant systems to be studied.

and cannot believe. There might be situations which small basis sets are not able to cope and the error compensation does not work. And worse, there is probably no way to identify such situations from the outset. In any event, the bottom line is that one way to identify potential shortcomings of density functionals used to study chemical reactivity is to check the basis set influence on the results. In view of these facts let us reiterate the big caveat that will accompany us through this chapter: one should not apply current approximate density functionals to mechanistic problems without careful prior benchmarking of the employed methodology for thermochemical data of related well-characterized systems!

13.2 Second Example: The S_N2 Reaction at Saturated Carbon

Another fundamental reaction in organic chemistry is the bimolecular nucleophilic substitution (S_N2), also referred to as the Walden inversion (see, e. g., Shaik, Schlegel, and Wolfe, 1992). One experimentally and theoretically particularly well characterized example is the gas-phase S_N2 reaction of $Cl^- + CH_3Cl \rightarrow ClCH_3 + Cl^-$. Starting from the separated chloride anion and methylchloride, an ion-molecule complex $[Cl{\cdots}CH_3Cl]^-$ is formed. From this complex, the reaction proceeds through a trigonal bipyramidal D_{3h} symmetric transition structure, in which one of the two identical carbon-chlorine bonds is formed to the same extent as the other one is broken. The product side of this identity reaction is of course the mirror image of the first half of the reaction and indistinguishable from the reactant side. Obviously, the practical interest in this sequence is limited, but besides the fundamental implications it suffices to establish the performance of density functional methods in the present context. Figure 13-3 sketches the energetic regime along the reaction coordinate

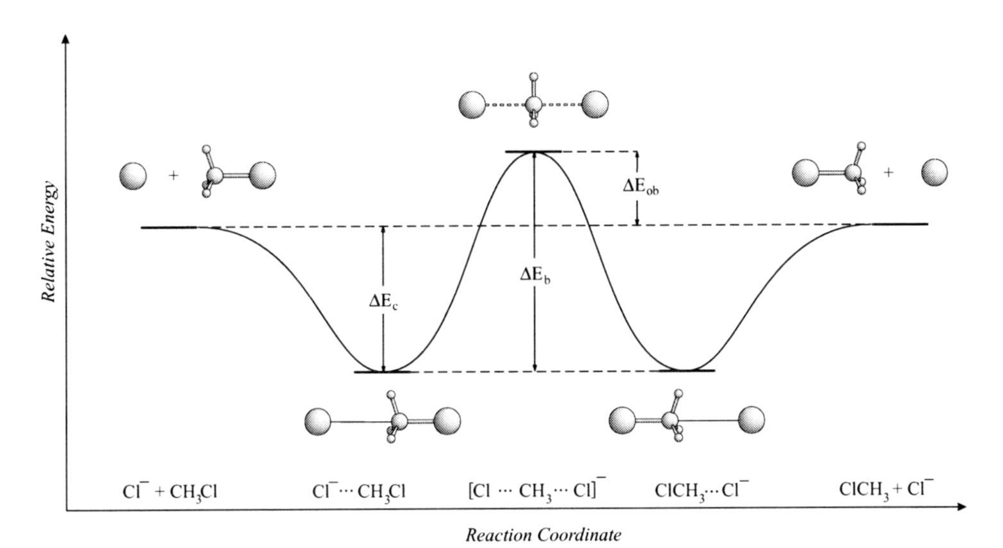

Figure 13-3. Reaction path for the gas-phase S_N2 reaction of $Cl^- + CH_3Cl$.

Table 13-4. Complexation energy (ΔE_c) and barrier heights (ΔE_{ob} and ΔE_b, see text) for the gas phase bimolecular S_N2 identity reaction $Cl^- + CH_3Cl \rightarrow ClCH_3 + Cl^-$ [kcal/mol]. HF and DFT calculations were done with the 6-311+G(d,p) basis set and include zero-point vibrational contributions.

	Exp.[a]	G2[b]	HF	SVWN	SLYP	BVWN	BLYP	B3LYP
ΔE_c	-12 ± 2	−11	−9	−15	−17	−9	−10	−10
ΔE_{ob}	$3 / 1 \pm 1$[c]	3	7	−8	−11	−4	−6	−2
ΔE_b	13 ± 2	13	16	6	6	5	4	8

[a] Compiled from references cited in Glukhovtsev et al., 1996; [b] taken from Glukhovtsev, Pross, and Radom, 1995; [c] two experimental values given.

and Table 13-4 provides experimental and accurate computational reference data, together with Hartree-Fock and a few density functional results.

If we compare the experimental ΔE_c, the (negative) complexation energy of the ion-molecule complex, with the computed data we see that the Hartree-Fock level underestimates this quantity, while G2 reproduces the experimental value very well. The SVWN functional overbinds the complex by 4 kcal/mol compared to the G2 value, but lies in fact just 1 kcal/mol outside the experimental error range – given the well known overbinding tendency of the LDA, one might take this as an indication that the true value lies rather close to the upper bound of the experimental uncertainty. As is common, the inclusion of the LYP gradient corrections increases the overbinding and the SLYP functional overestimates the G2 benchmark data by 6 kcal/mol. The SVWN overbinding is largely compensated for by inclusion of gradient corrections to exchange and BVWN gives a reasonable complexation energy, matching the Hartree-Fock value. The BLYP GGA functional and the B3LYP hybrid give the same result, missing the experimental target by 2 kcal/mol. The overall reaction barrier, ΔE_{ob}, is the difference between the relative energies of the entrance channel and the transition structure. HF overestimates the barrier relative to G2 and experiment by some 4-6 kcal/mol. The qualitative picture changes with application of density functional theory: all functionals place the central barrier below the relative energy of the entrance channel and give negative values for ΔE_{ob}. The LDA approach and even more so the SLYP functional both underestimate the overall barrier by 11 and 14 kcal/mol, respectively. Again, BVWN does a better job, but still underestimates the G2 value[63] by 7 kcal/mol. Additional inclusion of the LYP correlation functional in the BLYP scheme leads to a deterioration of the results by 2 kcal/mol. The admixture of exact exchange within the B3LYP hybrid gives the best result with a deviation from experimental and G2 results in the order of 5 kcal/mol. The intrinsic reaction barrier, that is the energy difference ΔE_b between the ion-molecule complex and the transition structure, is described with similar deviations. These results are representative for other studies on this and related systems: standard GGA functionals underestimate the activation barriers more than the corresponding hybrid functionals (Deng, Branchadell, and Ziegler, 1994, and Glukhovtsev et al., 1996).

[63] Large scale CCSD(T) calculations by Botschwina, 1998, confirm the G2 result for ΔE_{ob}.

Table 13-5. Computed complexation energies (ΔE_c) and barrier heights (ΔE_{ob} and ΔE_b, see text) [kcal/mol] for the gas phase bimolecular S_N2 identity reaction $Cl^- + CH_3Cl \rightarrow ClCH_3 + Cl^-$ from different sources (single point energy calculations using the 6-311++G(3df,3pd) basis on top of 6-311+G(d,p) geometries).

	B3LYP[a]	B3LYP[b]	B1LYP[c]	PBE1PBE[b]	LGLYP[c]	LG1LYP[c]
ΔE_c	–9	–9	–10	–10	–12	–11
ΔE_{ob}	–1	–1	–1	1	–1	–2
ΔE_b	8	8	9	10	11	9

	BPW91[d]	B3PW91[d]	*m*PWPW91[d]	*m*PW1PW91[d]	*m*PW3PW91[d]
ΔE_c	–9	–9	–11	–10	–10
ΔE_{ob}	–3	0	–4	0	0
ΔE_b	6	10	6	11	10

[a] 6-31+G(d,p) basis: Glukhovtsev et al., 1996; [b] Adamo and Barone, 1999; [c] Adamo and Barone, 1998a; [d] Adamo and Barone, 1998b.

Slightly better results have been obtained, however, with the recently developed *m*PW and PBE functionals as can be seen from the data compiled in Table 13-5. The improvements do not exceed 2 kcal/mol though. It is interesting to note that the application of smaller (6-31+G(d)) or larger basis sets (6-311++G(3df,3pd)) does not change the picture much, in contrast to the strong influence of the basis set quality we have noted before.

13.3 Third Example: Proton Transfer and Hydrogen Abstraction Reactions

The transfer of protons between atoms is of utmost importance in chemical and biological transformations. Theoretical research in this field is entangled in a combination of problems associated with the highly demanding aspects of hydrogen bridges and those difficulties associated with a consistent description of different regions on potential energy surfaces encompassing the formation and cleavage of bonds. The breaking of covalent bonds to hydrogen atoms is another important subject related to combustion chemistry. In this field, the abstraction of hydrogen atoms by radical species is of great interest but, as we will see, highly involved as far as theory is concerned. The present section provides two examples, which are well investigated and covered by literature: the intramolecular proton transfer process in malonaldehyde enol, and the simple, seemingly trivial, hydrogen exchange reaction $H + H_2 \rightarrow H_2 + H$.

13.3.1 Proton Transfer in Malonaldehyde Enol

The enol form of malonaldehyde is favored over the tautomeric aldehyde due to the presence of an intramolecular hydrogen bond. It constitutes one of the smallest model systems

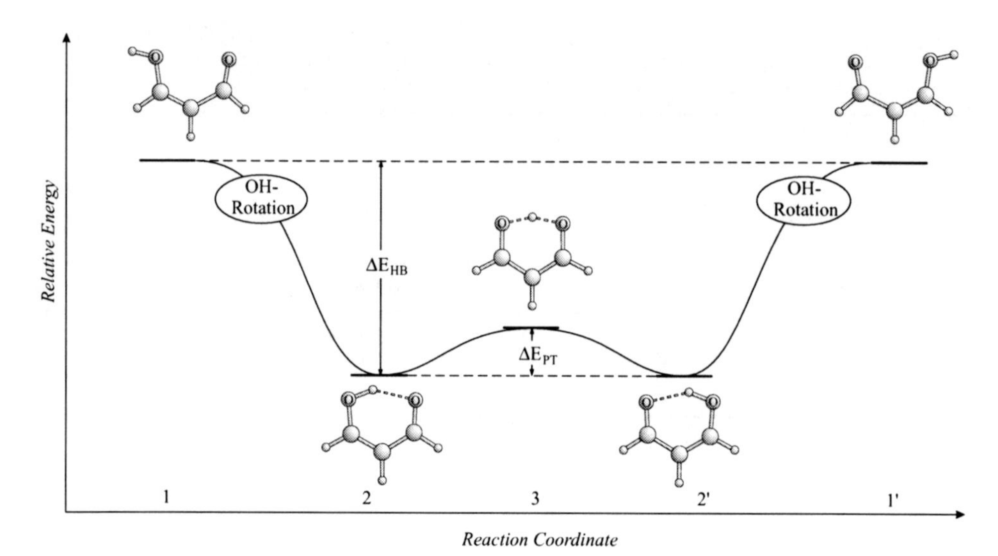

Figure 13-4. Energy profile for the proton transfer in malonaldehyde enol.

for hydrogen bonded species subject to an intramolecular proton transfer – an important process in enzymatic reactions. As an extension to Chapter 12, where we have evaluated the capabilities and limitations of contemporary density functionals to describe species with intermolecular hydrogen bonds, we now look a little closer at the intramolecular binding and the proton transfer process in malonaldehyde from a thermochemical point of view. The asymmetric nature of the hydrogen bond in this species has been established by crystallographic, microwave, and NMR experiments (see Perrin and Kim, 1998, and references therein) and the shift of the proton from one oxygen atom to the other – concomitant with a rearrangement of the double bond framework – proceeds via a C_{2v} symmetric transition state. This double minimum situation is depicted in Figure 13-4.

As we have outlined in the introduction to Chapter 12, the flat potential energy surface in the hydrogen bonding region and its double minimum nature obscures the validity of the harmonic approximation commonly used to evaluate vibrational frequencies and thermochemical properties. For the related situation in the protonated water dimer, Valeev and Schaefer, 1998, have argued that strong anharmonicity effects of low frequency modes can cause substantial errors for computed zero-point vibrational energies. Here, we are mainly interested in demonstrating the general ability of density functionals to describe potential energy surfaces. Thus, we are not seeking final answers to the chemical problem – in order to do so there are many severe puzzles to be solved, starting from the choice of a multidimensional coordinate system (e. g., Carrington and Miller, 1986) and ranging through a description of proton tunneling (see Shida, Barbara, and Almlöf, 1989). Rather, we will address the basic question: is present day density functional theory able to provide a good enough potential energy surface to justify its use in a subsequent, physically sound treat-

Table 13-6. Computed total energy differences [kcal/mol] and distances [Å] for the proton transfer in malonaldehyde enol (6-311++G(d,p) basis).

Method	ΔE_{HB}	ΔE_{PT}	Min (**2**)		TS (**3**)	
			$R_{O\cdots O}$	$R_{O\cdots H}$	$R_{O\cdots O}$	$R_{O\cdots H}$
HF	0.3	10.8	2.70	1.91	2.32	1.19
SVWN	20.3	0.0	2.37	1.29	2.36	1.21
SLYP	22.5	–[a]	–[a]	–[a]	2.37	1.21
BVWN	11.5	3.4	2.65	1.76	2.40	1.22
BLYP	13.0	2.2	2.59	1.67	2.40	1.23
B3LYP	12.9	3.2	2.58	1.69	2.37	1.21
MP2	12.1	3.3	2.58	1.69	2.36	1.20
CCSD(T)	11.5	4.4	2.58[b]	1.69[b]	2.36[b]	1.20[b]
CCSD(T)[c]	12.4	4.0	2.58[b]	1.69[b]	2.36[b]	1.20[b]

[a] No asymmetric C_s structure; [b] MP2/6-311++G(d,p) structure; [c] 6-311++G(2df,2pd) basis.

ment of the chemical problem? Does it make sense to use it at all? Let us try to find an answer from the results summarized in Table 13-6.

We can see at a glimpse that for HF theory as well as for SVWN or SLYP the answer to both questions is 'No'. The energy difference ΔE_{HB} between conformers **1** and **2**, which represents the strength of the intramolecular hydrogen bridge, is severely underestimated by the former and dramatically overestimated by the latter methods. Accordingly, the O···O and the O···H distances are computed too long at the HF, and too short at the SVWN level,[64] both compared to the MP2 reference geometry. Using the SLYP functional the C_S minimum structure does not even exist but rather collapses into a C_{2v} minimum – a dramatic failure. On the contrary, as so often is the case, the BVWN functional improves the agreement with the CCSD(T) energetic reference data. The structural parameters differ in the order of 0.01 Å from the optimized MP2 structure, though. Better agreement is found for the BLYP functional, which also describes the relative energies of stationary points investigated on this PES reasonably well, although the barrier height is underestimated by 2 kcal/mol. A slightly better performance results from the B3LYP treatment of the problem. Both energetics and structural parameters agree nicely with the accurate conventional ab initio results. Overall, the B3LYP functional yields structures and relative energies of very similar quality as MP2 and approaches the accuracy of the higher level benchmarks to within 1 kcal/mol, with a slight bias to underestimate the proton transfer barrier (see also Zhang, Bell, and Truong, 1994, Barone, Orlandini, and Adamo, 1994b, and Barone and Adamo, 1996 and 1997b). While this accuracy is certainly at the borderline of being useful for studies of such flat potential energy surfaces as that present in malonaldehyde enol, we conclude that the hybrid functional level can provide proton transfer potential energy surfaces of a quality comparable to MP2, but at substantially lower costs. Therefore, this level of theory should lend itself as an efficient tool to a further assessment of proton transfer

[64] Similar findings have been reported already in the literature (Sim et al., 1992).

processes. Similar conclusions have been drawn by others, who tested various recently developed functionals on related problems (Sirois et al., 1997, Adamo and Barone, 1998a and 1998b, Sadhukhan et al., 1999).

13.3.2 A Hydrogen Abstraction Reaction

One of the simplest chemical reactions involving a barrier, $H_2 + H \rightarrow [H\cdots H\cdots H]^{\neq} \rightarrow H + H_2$, has been investigated in some detail in a number of publications. The theoretical description of this hydrogen abstraction sequence turns out to be quite involved for post-Hartree-Fock methods and is anything but a trivial task for density functional theory approaches. Table 13-7 shows results reported by Johnson et al., 1994, and Csonka and Johnson, 1998, for computed classical barrier heights (without consideration of zero-point vibrational corrections or tunneling effects) obtained with various methods. The CCSD(T) result of 9.9 kcal/mol is probably very accurate and serves as a reference (the experimental barrier, which of course includes zero-point energy contributions, amounts to 9.7 kcal/mol).

As usual, we see the uncorrelated HF level overestimating the barrier, this time by 8 kcal/mol. MP2 does better, but it is not able to fully recover the correlation energy so that the CCSD(T) benchmark data is still overestimated by 3 kcal/mol. Turning to the density functionals, the SVWN and SLYP implementations produce only chemical nonsense: the H_3 radical is computed as a stable minimum, lying energetically below the reactants, rather than as a saddle point. The BVWN functional changes the situation giving at least a qualitatively correct picture with a barrier separating the two asymptotic channels $H + H_2$. Yet the barrier height is severely underestimated by some 7 kcal/mol or over 50 %. This wholly unsatisfactory picture does not change significantly if the BLYP and B3LYP functionals are employed.

Durant, 1996, has found the incorporation of a larger amount of exact exchange into the hybrid functional formulation as one way to improve the performance with respect to this reaction: the BHLYP functional delivers a barrier height of 6 kcal/mol. Accordingly, Chermette, Razafinjanahary, and Carrion, 1997, have reoptimized the empirical mixing parameters inherent to the B3LYP hybrid functional formulation with respect to the description of hydrogen-only systems. These authors stressed that the results were found particularly sensitive to the amount of exact exchange admixture, which they readjusted from 0.20 in the original B3LYP composition to 0.559 in their so-called B3(H) functional.

Table 13-7. Computed classical barrier heights ΔE [kcal/mol] for the reaction $H_2 + H \rightarrow [H\cdots H\cdots H]^{\neq} \rightarrow H_2 + H$ (6-311++G(,3pd) basis set); data compiled from Johnson et al., 1994, and Csonka and Johnson, 1998.

	HF	MP2	CCSD(T)	SVWN	SLYP	BVWN	BLYP	B3LYP
ΔE	17.6	13.2	9.9	–2.8	–3.5	3.7	2.9	4.1
ΔE_{SIC} [a]	–	–	–	6.6	6.0	13.2	12.6	11.1

[a] Corrected for self-interaction contributions.

Applied to the reaction barrier, this functional indeed gave an improved value of 7 kcal/mol.

Johnson et al., 1994, and subsequently Csonka and Johnson, 1998, went deeper into the problem and investigated the influence of the self-interaction contributions (see Section 6.7) on this reaction. Some of their results obtained after applying a self-interaction correction are given in Table 13-7. Indeed we see a dramatic effect on all functionals applied: all barriers are shifted by 7 – 9.5 kcal/mol to higher energies with the smallest effect for the B3LYP hybrid. Its self-interaction corrected barrier amounts to 11.1 kcal/mol, which agrees much better with the CCSD(T) value. The data given in Table 13-7 shows that the self-interaction error causes a significant and unphysical lowering of the activation barrier, which is, apparently, least pronounced for the B3LYP hybrid functional. We have noted already that inclusion of an even larger portion of exact exchange in the BHLYP hybrid functional improves the computed energy barrier. Without further reference, this could now be explained in two ways. One point of view is that the larger amount of exact exchange – that is, exchange which is free of the self-interaction error by construction – reduces this particular shortcoming to some extent. On the other hand, seen from an even more naive standpoint, the increased barrier is only a reflection of the fact that we are mixing in a certain amount of HF quality into the functional form. We have noted above that the pure Hartree-Fock level severely overestimates the barrier, hence the more that HF character is included in the functional the more the barrier is shifted to higher relative energies. While both views allow for a – albeit hand-waving – rationalization of the performance of the respective functionals, the explicit analysis of self-interaction errors by Csonka and Johnson, 1998, reveals that the situation is more complicated. These authors studied the individual components which contribute to the overall self-interaction error for all three species involved in the reaction – see Table 13-8.

For the BLYP functional there is a pronounced self-interaction error for the H_3 transition structure, which is much larger than the errors for the H radical or the H_2 molecule taken together. The correlation part has no self-interaction problem, since the LYP correlation energy has been designed as perfectly self-interaction free, and hence the error in the computed barrier height is entirely due to the incomplete cancellation of the self-interaction error in the Coulomb/exchange part. So far, everything is according to expectation. For the B3LYP hybrid, however, the breakdown of the self-interaction error into its components

Table 13-8. Self-interaction error components for Coulomb and exchange energies ($E_J + E_X$) as well as for the correlation energy (E_C), and the resulting sum for the H atom, the H_2 molecule, and the H_3 transition structure [kcal/mol]. Data taken from Csonka and Johnson, 1998.

Method	H			H_2			H_3		
	$E_J + E_X$	E_C	SIC	$E_J + E_X$	E_C	SIC	$E_J + E_X$	E_C	SIC
BLYP	0.86	0.00	−0.86	0.38	0.00	−0.38	−8.51	0.00	8.51
B3LYP	2.96	−2.62	−0.34	5.20	−5.45	0.24	0.92	−7.93	7.01

shows a startling situation: the errors for $E_J + E_X$ are three to thirteen times larger than those found for the pure Becke exchange functional for H and H_2. Furthermore, for the H_3 transition structure the exchange related self-interaction error is about nine times smaller than for BLYP. There, $E_J + E_X$ was about –9 kcal/mol, but this changes to +0.9 kcal/mol when the B3LYP functional is used. Hence, the self-interaction error trends in the hybrid functional treatment of this reaction are just opposite to that of the pure GGA! But even more baffling is the finding that the correlation energy error is no longer zero but gets significant! This error stems from the admixture of the VWN correlation functional into the B3LYP hybrid form (see equation (6-28)) and scales quite constantly with the size of the number of atoms in the system. It is acting in the opposite direction and in fact compensates almost perfectly for the self-interaction error of the Coulomb-exchange energy for H and H_2 such that the total self-interaction errors are well below 1 kcal/mol. This is not so for H_3, where we have a small error from exchange of +0.92 kcal/mol which contrasts with a huge correlation self-interaction error of –7.93 kcal/mol. Thus, a total error of 7.01 kcal/mol results. The large underestimation of the barrier height is thus due to the imperfect cancellation of the individual error components.

This hydrogen abstraction reaction has been found particularly problematic also for other GGA and hybrid functionals, so it is clear that this and related reactions constitute severe problems for currently available functionals (see Durant, 1996, Sadhukhan et al., 1999). Of particular interest, however, is the finding that functionals that depend also on the non-interacting kinetic energy density – like LAP and B95 – show an improved description of the barrier (BLAP3: 7.3 kcal/mol, B1B95: 7.6 kcal/mol, see Salahub et al., 1999).[65] In conclusion, although one important source of error for the failure of density functional methods to describe this hydrogen abstraction reaction has been pinned down, the procedure to correct for self-interaction contributions is not commonly available and not easily applicable to standard Kohn-Sham procedures (see Csonka and Johnson, 1998, for an easy-to-read outline of procedures and further references). Therefore, one should expect that gradient-corrected functionals tend to severely underestimate the barriers to radical hydrogen abstraction reactions. To some lesser extent this also applies to hybrid functionals. Further examples corroborating these statements can be found for similar reactions in Johnson, 1995, or Baker, Muir, and Andzelm, 1995. It may nevertheless be that related reactions can be very well described by standard functionals provided that a more fortuitous error compensation occurs for the particular species under study. Better results than reported here have in fact been described but the performance changes unpredictably from system to system and a general bias to underestimate the activation barriers for such processes appears to persist (see, e. g., Bernardi and Bottoni, 1997, Nguyen, Creve, and Vanquickenborne, 1996, Jursic 1996. See also Ventura, 1997, for a review).

[65] We find also other recently developed exchange functionals in combination with LYP and the 6-311++G(,3dp) basis marginally better than the B3LYP hybrid: B1LYP: 4.8 kcal/mol, LG1LYP: 5.1 kcal/mol, *m*PW1LYP: 4.4 kcal/mol. Note that these hybrid functionals do not contain correlation contributions from the VWN functional.

13.4 Fourth Example: H_2 Activation by FeO^+ in the Gas-Phase

In this last example we extend the investigation of reaction mechanisms to transition-metal chemistry. As repeatedly mentioned in the previous chapters, the presence of transition-metals leads to additional difficulties for an appropriate computational description, see, e. g., Koch and Hertwig, 1998. However, it is particularly this difficult terrain where DFT methods have acquired a good reputation as offering the best value for money. This also applies to mechanistic details as verified by the large and continuously growing number of investigations uncovering organometallic and related reaction mechanisms that appear in the literature. Rather than attempting to survey all these developments, we will use one particular example in the following for discussing some of the problems and peculiarities inherent to the computational description of transition-metal reactivity. The example chosen is the activation of molecular hydrogen by an iron oxide cation, FeO^+. While at first glance this may seem to be a fairly exotic model reaction, it will turn out to be highly instructive in the following. Before we start with some background information, let us add that this reaction involves open-shell and coordinatively unsaturated transition-metal complexes. Because of the complicated electronic structure inherent in such species, the demands on computational techniques are particularly high. The performance of density functional methods established for this example represents therefore a worst case scenario and the errors should be upper bounds. Reactions involving larger, coordinatively saturated, closed-shell transition-metal complexes which obey the 18-electron rule are expected to be less problematic and associated with smaller computational uncertainties. Indeed, an ample number of contributions testifies to the suitability of DFT methods in the regime of organometallic chemistry (for representative examples see Cui et al., 1995, Dapprich et al., 1996, Siegbahn, 1996b, Deng and Ziegler, 1997, Frankcombe et al., 1997, Siegbahn and Crabtree, 1997, Niu and Hall, 1997, Cui, Musaev, and Morokuma, 1998a and 1998b, Torrent, Deng, and Ziegler, 1998, Basch et al., 1999, Niu, Thomson, and Hall, 1999, Amara et al., 1999, Pavlov, Blomberg, and Siegbahn, 1999, Deubel and Frenking, 1999, Kragten, van Santen, and Lerou, 1999, Petitjean, Pattou, and Ruiz-López, 1999, just to mention a few). In many cases density functional approaches are actually the only feasible way to tackle these questions, because typically rather large model systems are required to assign enough chemical relevance to the calculations. The need to choose large model systems complicates both theoretical and experimental work and due to the lack of accurate data to compare to, most such theoretical studies are rather qualitative in nature and are not suited as examples in the present context. The chemical insight gained in many instances is nevertheless highly impressive and density functional calculations at moderately advanced levels are often the only means to rationalize experimental findings.

The situation is somewhat better for the gas-phase chemistry of isolated transition-metal ions or complexes, and this area of research has received a lot of attention in the past. On the experimental side, comprehensive mass-spectrometric techniques allow for an explicit measurement of thermochemical and kinetic parameters of reactants, intermediates, and products occurring along the reaction pathways. These data can be obtained without the influence of ligands, counter ions, solvents etc. which would be a highly complicated enterprise – if possible at all – for experiments in homogeneous or heterogeneous environments. For similar reasons, reactions among isolated species in the gas-phase are particularly well suited for quantum chemical investigations, which can supplement the experimental results with direct information on stationary points occurring along the reaction coordinate. Therefore, the synergy between experiment and theory has been particularly useful in this field (Schröder et al., 1997).

We now turn to our specific example. In the last decade, bond activation processes mediated by bare metal oxide cations have been extensively studied reflecting the important role that such reactions play in various areas ranging from heterogeneous catalysis to biochemical transformations.[66] As a rather advanced example, we present here the results from state-of-the-art theoretical studies on the activation of molecular hydrogen by iron oxide cation, FeO^+ ($^6\Sigma^+$) + $H_2 \rightarrow Fe^+$ (^6D) + H_2O. Let us start with an illustration of the chemical problem: the most intriguing features of this reaction are (a) the unexpectedly low reactivity found in gas-phase experiments. Although the overall reaction is highly exothermic ($\Delta H = -36$ kcal/mol) and spin-conserving (both reactants and products are high-spin with $S = 5/2$, i. e., sextet multiplicity) only one in 100-1000 collisions is reactive under the conditions of mass-spectrometric measurements. This contrasts with the general finding that FeO^+ efficiently reacts with alkanes, alkenes, and aromatic systems. (b) A very low kinetic isotope effect ($k_H/k_D \approx 1$-1.5) and (c) an inverse temperature dependence, i. e., an even lower reactivity at elevated temperatures has been found for the reaction. While finding (a) is in line with the assumption of a high barrier in the course of the reaction, the latter two findings do not readily agree with a central barrier. Experimental work alone has not been able to resolve these apparent contradictions and a combination of density functional theory and multireference post-HF calculations has been used to unravel the intrinsic mechanistic details of this reaction, first by Fiedler et al., 1994, and subsequently by Filatov and Shaik, 1998b, and Irigoras, Fowler, and Ugalde, 1999.

The first step in computational studies like this should always be to calibrate the chosen density functional methods against the known energetic properties of related systems in order to define the level of accuracy that can be expected. Our first interest lies in the high-spin/low-spin state energy differences, as the formation and cleavage of covalent bonds to the transition-metal always involves electron population changes among the respective d and s orbitals. We begin by comparing computed and experimentally determined energies to excite the Fe^+ ion from its $^6D(d^6s^1)$ ground state into the first excited ^4F (d^7) state. From the data in Table 13-9 we see already significant deviations (up to half an eV) from the experimental value for all density functionals, all erroneously predicting a quartet ground state of Fe^+. Part of this error is due to the neglect of differential relativistic effects, but the well-known preference of density functional methods for d^n over $d^{n-1}s^1$ configurations (recall the discussion in Section 9.2) contributes as well and inflicts a substantial error on the $Fe^+ + H_2O$ exit channel. Of the density functional methods tested, the B3LYP functional performs best, followed by BP86 and FT97. There have been two studies with the B3LYP

[66] For a presentation of the field and experimental references the reader is referred to the review of Schröder and Schwarz, 1995.

Table 13-9. High-spin/low-spin excitation energies ΔE [kcal/mol] for Fe^+ ($^6D \rightarrow ^4F$), $[Fe(H_2O)]^+$ ($^6A_1 \rightarrow ^4B_2$), and FeO^+ ($^6\Sigma^+ \rightarrow ^4\Phi$).

Species	B3LYP[a]	BP86[a]	FT97[a]	B3LYP[b]	CCSD(T)[b]	CASPT2[c]	Exp.
ΔE (Fe^+)	–2.4	–4.3	–6.5	–4.2	5.4		5.9
ΔE ($[Fe(H_2O)]^+$)	–5.4	–8.2	–11.2	–8.9	2.7		
ΔE (FeO^+)	8.0	12.7	15.5	7.3	12.5	19.1	

[a] Wachters basis for Fe, Dunning TZ2P basis for H and O, Filatov and Shaik, 1998b; [b] modified Ahlrichs TZVP basis for Fe, Pople 6-311++G(2df,2p) for H and O, Irigoras, Fowler, and Ugalde, 1999; [c] ANO [8s7p6d4f2g] basis for Fe, [3s2p1d] and [5s4p3d2f] for H and O, respectively, Fiedler et al., 1994.

functional employing different basis sets, due to Wachters (for Fe)/Dunning (for H, O) and Ahlrichs (for Fe)/Pople (for H, O), respectively. Both sets, which we will abbreviate as WD and AP, respectively, are of approximately polarized triple-zeta quality. The deviations between the corresponding results give a feeling for the errors inherent in the choice of the basis set. For the atomic excitation energy the basis set effect amounts to 2 kcal/mol, with the Wachters basis set performing better.

Very similar deviations from the CCSD(T) benchmark data[67] are seen for the computed spin-state splittings in the $[Fe(H_2O)]^+$ complex. Apparently, the error in the atomic state splitting also affects that of the complex. The basis set effect amounts to 3.5 kcal/mol with the WD basis set combination again performing better. The computed high-spin/low-spin state splittings for the FeO^+ molecule reported in Table 13-9 compare fairly well with the CCSD(T) data, with B3LYP exhibiting the largest deviation. If compared to the CASPT2 results the differences are larger and show a preference of the density functionals to favor low-spin states. Note also that the CCSD(T) and CASPT2 calculations differ by 6.6 kcal/mol from each other. The rather complicated electronic structure of FeO^+ is probably best described at the multireference CASPT2 level of theory, but due to the lack of more precise data a final recommendation as to which of these approaches is to be preferred cannot be given. At least we find that all levels consistently predict a high-spin $^6\Sigma^+$ ground state for FeO^+ while the uncertainty of the DFT results is as high as some 10 kcal/mol.

The binding energies of FeO^+ and the $[Fe(H_2O)]^+$ complex are known experimentally and are compared to the corresponding computed results in the first two entries in Table 13-10. In both cases the BP86 functional exhibits a dramatic overbinding of about 20 kcal/mol. Oddly, the FT97 GGA functional shows an equally large deviation for the strongly bound oxide, whereas it perfectly reproduces the experimental value for the water-iron binding energy. The B3LYP functional provides an excellent description of the FeO^+ binding energy and modestly overestimates that of the weakly bound $[Fe(H_2O)]^+$ species by

[67] All CCSD(T) data reported in the following have been obtained using B3LYP geometries and the CASPT2 calculations are based on BP86 structures. This exemplifies another important use of density functional theory in the field: as detailed in Chapter 8, modern functionals usually yield reliable geometries. Thus, performing energy calculations with correlated, computationally expensive conventional ab initio methods on top of DFT optimized geometries represents an economic yet accurate strategy.

Table 13-10. Computed binding energies of FeO^+ and $[Fe(H_2O)]^+$ (D_0 with respect to atomic and molecular sextet states) and overall reaction energies for three examples [kcal/mol].

	B3LYP[a]	BP86[a]	FT97[a]	B3LYP[b]	CCSD(T)[b]	Exp.[c]
FeO^+	80.5[d]	108.4[d]	108.0[d]	–	–	81.4 ± 1.4
$[Fe(H_2O)]^+$	34.9	50.1	30.7	32.5	30.6	30.7 ± 1.2
$FeO^+ (^6\Sigma^+) + H_2 \rightarrow Fe^+(^6D) + H_2O$	38.5	3.8	7.7	33.2	37.2	36.0 ± 1.4
$Fe^+(^6D) + H_2O \rightarrow FeOH^+(^5A') + H$	–33.0	–9.8	–21.5	–26.2	–33.1	–30.4 ± 2.8
$Fe^+(^6D) + H_2O \rightarrow FeH\ (^5\Delta) + OH$	–	–	–	–56.7	–65.3	–69.2 ± 1.4

[a] Wachters basis for Fe, Dunning TZ2P basis for H and O, Filatov and Shaik, 1998b; [b] modified Ahlrichs TZVP basis for Fe, Pople 6-311++G(2df,2p) for H and O, Irigoras, Fowler, and Ugalde, 1999; [c] experimental references reported in Irigoras, Fowler, and Ugalde, 1999; [d] D_e values.

about 4 and 2 kcal/mol with the WD and AP basis sets, respectively. As a final check of the computational methods, we compare the energetics of three reactions for which experimental data is available. The thermochemistry of $FeO^+ + H_2 \rightarrow Fe^+ + H_2O$ is described very well at the B3LYP level, the WD basis set combination leading to an overestimation, the AP combination to an underestimation of the exothermicity of this reaction, both values lying only slightly outside the experimental uncertainty. FT97 significantly underestimates the reference value and this is even more pronounced with BP86. Similar findings apply to the thermochemical description of the hydrogen abstraction from H_2O by Fe^+ to give $FeOH^+$ and a hydrogen radical, although the errors for the two GGA functionals are less severe. A remarkable difference of 6.8 kcal/mol is observed between the B3LYP results obtained with the WD and the AP basis set combinations. The former is in excellent agreement with the CCSD(T) data, while both B3LYP energies are slightly outside the experimental error range, but on opposite sides. The final entry in Table 13-10 also shows a critical deviation between the B3LYP results and experiment: the hybrid functional underestimates the endothermicity of this reaction by 12.4 kcal/mol while the CCSD(T) result underestimates it by 3.9 kcal/mol.

From these few observations we can already conclude that the most consistent density functional description of the energetics is provided by the B3LYP hybrid with errors of the order of 5 kcal/mol for binding energies and errors about twice as large for the high-spin/low-spin excitation energies. Weakly bound metal complexes are more problematic to describe than strongly bound species, a problem which is quite commonly found for this functional. The use of two different basis sets has revealed that an additional uncertainty of 2 to 3 kcal/mol is present in particular for computed binding energies, but deviations can in fact be as large as 7 kcal/mol. The overall accuracy of the B3LYP functional derived from these benchmark studies is hence in the order of ±5 kcal/mol. Let us stress this particular point in other words: errors as large as half an eV or some 10 kcal/mol can occur for the relative energetics of species involved in the reaction as summarized in Table 13-10. The FT97 and BP86 functionals show much larger errors and they probably can not be recommended for this kind of application.

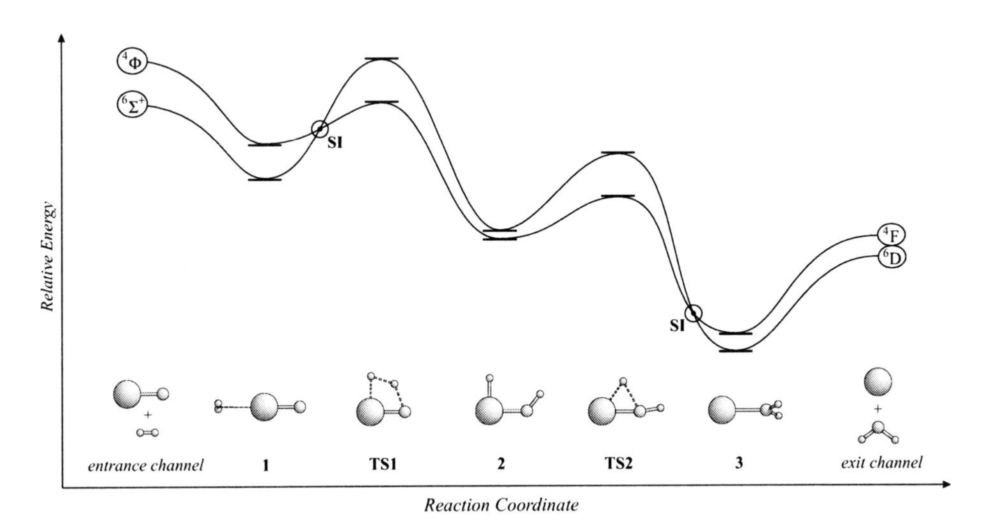

Figure 13-5. Energy profile for the gas-phase reaction $H_2 + FeO^+ \rightarrow Fe + H_2O$.

Table 13-11. Computed energies [kcal/mol] of stationary points for the activation of H_2 by FeO^+ (D_0 relative to separated FeO^+ ($^6\Sigma^+$) + H_2).

	B3LYP[a]	BP86[a]	FT97[a]	B3LYP[b]	CCSD(T)[b]	CASPT2[c]
$FeO^+(^4\Phi) + H_2$	8	13	16	7	13	19
^{6}1	–15	–16	–12	–13	–12	-5
^{4}1	-7	–16	–12	-6	0	3
^{6}TS1	8	9	12	10	13	19
^{4}TS1	1	0	–1	1	8	6
^{6}2	–38	–33	–26	–34	–31	–14
^{4}2	–41	–41	–34	–38	–30	–25
^{6}TS2	–13	–7	4	–11	–7	
^{4}TS2	–34	–31	–21	–30	–22	
^{6}3	–73	–54	–38	–66	–70	–67
^{4}3	–79	–62	–50	–75	–67	
$Fe^+(^6D) + H_2O$	–39	–4	-8	–33	–37	–36
$Fe^+(^4D) + H_2O$	–41	-8	–14	–37	–32	

[a] Wachters basis for Fe, Dunning TZ2P basis for H and O, Filatov and Shaik, 1998b; [b] modified Ahlrichs TZVP basis for Fe, Pople 6-311++G(2df,2p) for H and O, B3LYP geometries, Irigoras, Fowler, and Ugalde, 1999; [c] ANO [8s7p6d4f2g] basis for Fe, [3s2p1d] and [5s4p3d2f] for H and O, respectively, BP86 geometries, Fiedler et al., 1994.

Let us now turn to the description of the reaction pathways. Figure 13-5 schematically depicts the shapes of the corresponding potential energy curves for the sextet and quartet spin-states and Table 13-11 contains the thermochemical information obtained at different levels of theory.

The reaction starts on the high-spin surface by forming a planar ion-molecule encounter complex $[(H_2)FeO]^+$, **^{6}1**, from the separated reactants (the superscript indicates the multiplicity). According to the CASPT2 results the relative energy of this complex is merely 5 kcal/mol below the entrance channel, whereas the CCSD(T) calculations suggest a stabilization energy of 12 kcal/mol. The density functional methods compute relative energies for this complex between –12 to –16 kcal/mol, in much better agreement with the CCSD(T) than with the CASPT2 value. The corresponding complex **^{4}1** on the low-spin surface is found to be less stable than its sextet counterpart by 8 kcal/mol (CASPT2) and 12 kcal/mol (CCSD(T)), both values are satisfactorily reproduced by the B3LYP functional. The FT97 and BP86 functionals both describe the high-spin and the low-spin species as equally stable. Following the formation of this complex, the reaction proceeds on the sextet potential energy surface through an energetically rather high-lying multicentered H_2-insertion transition state **^{6}TS1**, which is found to be 19 kcal/mol above the sextet entrance channel at the CASPT2 level of theory. The relative energy of this structure is computed to be much lower at all other levels, with B3LYP using the WD basis set giving the lowest estimate of 8 kcal/mol. While the discrepancy for the computed absolute barrier height is rather large between CASPT2 and all other methods, we can see a highly consistent description among all methods if we look at the barrier computed with respect to the relative energy of the encounter complex, **^{6}1**: now all methods agree to within 1 kcal/mol with the CASPT2 value of 24 kcal/mol. As we have already outlined earlier in this chapter, it is apparent that a balanced description of fragmentation (or, in this case, association) processes is particularly demanding for density functional methods. It is clear from the results in Table 13-11 that this holds true also for post-HF methods, albeit to a lesser extent. Proceeding further on the sextet surface, this transition state collapses into the intermediate $HFeOH^+$, **^{6}2**, with a relative energy of –14 kcal/mol (CASPT2) or –31 kcal/mol (CCSD(T)). Note the large discrepancy between the two post-HF results. The B3LYP and BP86 energies agree much better with the CCSD(T) result, and only the FT97 functional assigns a substantially higher energy to this complex. A second transition state along the reaction coordinate, **^{6}TS2**, is placed 7 kcal/mol below the separated reactants at the CCSD(T) level and connects **^{6}2** with the cationic iron water product complex. This second transition structure is much lower in energy than **^{6}TS1**. Actually, all computed results agree on that: irrespective of the basis set used, both B3LYP calculations give a lower barrier, whereas the BP86 results agree with the reference value perfectly. Only the FT97 functional gives a relative barrier 4 kcal/mol above the entrance channel. Seen from another angle, the B3LYP calculations place **^{6}TS2** 21 kcal/mol below **^{6}TS1**, in excellent agreement with a CCSD(T) difference of –20 kcal/mol. The two GGA functionals give significantly smaller energy differences of 16 (BP86) and 8 kcal/mol (FT97). A very consistent picture also emerges if we look at the barrier **^{6}TS2** relative to **^{6}2**: all methods agree with the CCSD(T) value of 24 kcal/mol within 2 kcal/mol except for FT97, which gives a larger barrier of 30 kcal/mol.

Following the transition state **^{6}TS2** downhill, we end up with the cationic iron water complex **^{6}3**, which is found –70 or –67 kcal/mol more stable than the reactants at the CCSD(T) or CASPT2 levels, respectively. The B3LYP numbers are similar, yielding –73

(WD) and –66 (AP) kcal/mol. The BP86 and FT97 functionals err significantly by about 30 kcal/mol.

If we step back into the region of the initial hydrogen activation and consider the energy of the corresponding transition structure with a quartet multiplicity, ^{4}**TS1**, a much lower barrier than on the high-spin surface results (6 kcal/mol for CASPT2 or 8 kcal/mol for CCSD(T)). All density functional methods give an even lower value of around ±1 kcal/mol. The subsequent minimum ^{4}**2** is at all levels of theory more stable than ^{6}**2** except for CCSD(T) which favors the latter by 1 kcal/mol. Given the large deviations between the CASPT2 and CCSD(T) energies in this region of the potential surface none of the methods applied is able to provide a conclusive answer as to the favored spin-state. The quartet transition structure ^{4}**TS2** is, however, definitely more stable than the corresponding sextet structure ^{6}**TS2**. While CCSD(T) favors the low-spin transition state by 15 kcal/mol, the two B3LYP calculations do so by 21 and 19 kcal/mol, respectively and BP86 (24 kcal/mol) and FT97 (25 kcal/mol) emphasize the quartet stability even more. ^{4}**TS2** is also well below the relative energy of ^{4}**TS1** as is quite consistently shown by the B3LYP, BP86 and CCSD(T) results, which give an energy difference between these species of –35 to –30 kcal/mol; use of the FT97 functional results in a smaller difference of –22 kcal/mol. The subsequently formed cationic iron complex with water is found to be more stable as a quartet by all density functionals, whereas CCSD(T) favors the sextet, and the same trends are observed for the exit channel $Fe^+ + H_2O$. The preference for low-spin states of density functionals is obvious and has been identified above as a consequence of shortcomings in the description of the atomic splittings.

With respect to the reaction mechanism the following conclusions emerge: the highest energy barrier along the reaction coordinate is that of the initial H_2 activation ^{6}**TS1**; all other barriers are lower in energy by an amount which is most probably outside the error range of the B3LYP hybrid functional and the post-HF calculations. The presence of a high lying transition structure for the initial activation of the H_2 molecule could readily explain the experimental observation of an exothermic, but highly inefficient reaction. However, a barrier height of +19 kcal/mol computed at the CASPT2 level and also the +13 to +8 kcal/mol obtained with the other methods, should in fact prevent the reaction from taking place at all under thermal conditions.[68] The finding of a much lower barrier on the quartet surface, however, provides a more favorable pathway. In particular the occurrence of a barrier very close to the energy of the entrance channel, as computed by the density functional approaches, provides an appealing possibility to interpret the experimental observation of a very inefficient reaction. In order for the reaction to proceed through quartet spin states, a crossover from high-spin to low-spin, mediated by spin-orbit coupling has to take place. This junction region where the sextet and quartet potential surfaces cross is located between the encounter complex and the transition state – marked as SI (*spin inversion*) in Figure 13-5. Even though the crossing point is probably below the energy of the quartet

[68] A detailed elucidation of the electronic structure of the species involved in the course of the reaction is well beyond the aim of this section so we rather refer the reader to the presentations by Fiedler et al., 1994 and Filatov and Shaik, 1998b.

transition state **^{4}TS1** and thus most likely below the energy of the entrance channel, a rather low crossing probability has been identified by Danovich and Shaik, 1997, providing an explanation for the low reactivity. Experimentally we know that a second crossing between high-spin and low-spin surfaces has to occur, for the sextet exit channel is unambiguously lower in energy than the quartet asymptote. Indeed, the CCSD(T) calculations are in accord with this reasoning, with the crossing taking place between **^{4}TS2** and **^{6}3**: while **^{4}TS2** is between 15 and 25 kcal/mol more stable than **^{6}TS2**, the high-spin product complex **^{6}3** is more stable by 3 kcal/mol than its quartet counterpart. All density functionals, on the other hand, artificially favor the low spin complex **^{4}3** as well as the quartet exit channel and are evidently unable to provide a fully satisfying answer. The erroneous stabilization of the low-spin state in the DFT regime has been frequently noted. However, if we simply correct the sextet/quartet gap of the exit channel by the atomic error, all functionals (except for the FT97 functional) prefer the sextet by about 2 kcal/mol, close to the CCSD(T) result. The conclusions for the multiplicity of transition state **TS2** would remain unaltered, and we obtain the same qualitative picture at the density functional level as deduced from the CCSD(T) calculations with respect to the second surface-crossing region. While these assumptions seem very reasonable, the accuracy of the applied theoretical models is simply not good enough for unambiguous conclusions. In summary, theoretical findings provide quite unexpected, but important insights into the origin of the experimental observations: the apparent spin conservation for the overall reaction originates in a double crossing of the high-spin and low-spin surfaces along the reaction coordinate. The evolving mechanism, which has been coined 'two-state reactivity', provides an appealing way of interpreting experimental findings for a wide range of transition-metal mediated reactions (see, e. g., Shaik et al, 1995 and 1998).

So much for this interesting chemistry. But what can we deduce for the applicability of density functional theory to such complicated and multifaceted electronic problems? Strictly speaking, we are forced to conclude from the presentation of results above, that a consistent description of energy differences below 5 kcal/mol is out of reach for present day density functional theory and even errors of 10 kcal/mol are not uncommon (see, e. g., Brönstrup et al., 2001). Part of this uncertainty is also due to the rather larger basis set effects. The results obtained with the WD and the AP bases are in some instances quite different and it is not generally clear which one is to be preferred. This underlines the need to develop basis sets particularly designed for use in connection with density functional methods in general (see, e. g., Porezag and Pederson, 1999) and hybrid functionals in particular. For the two gradient-corrected functionals applied, we have seen much larger errors in relative energies than those found for the B3LYP hybrid, which provided a much better overall agreement with experimental or computational reference data. In addition, a much more satisfying consistency of results is obtained with the hybrid functional if relative energies of similarly bound species are considered – a finding that not only applies for transition-metal reactions. The largest errors are usually connected to the fragments in the exit or entrance channels; if the energies of the various stationary points within a reaction sequence are considered, much smaller errors are obtained. However, we have also seen discrepancies as large as 17 kcal/mol between advanced post-HF methods like CCSD(T) and CASPT2, which puts things into a different perspective: it is by no means clear if these methods really perform better than the B3LYP functional. This brings us back to the recommendation mentioned already above: the importance of benchmarking the performance of a particular density functional method on a set of experimentally well established thermochemical facts related to the reaction at hand. Only from a carefully performed calibration can we get insight into the problems relevant to a new system under study. Experience with the application of density functional theory on first and second-row transition-metal reactivity has revealed that one has to be very careful interpreting hybrid density functional results on energy differences below 10 kcal/mol. Errors can be large in some cases and perfectly cancel out in others – of course, we never know unless we explicitly check with more accurate data. However, in particular the application of the B3LYP hybrid functional has provided valuable insights into mechanistic scenarios of transition-metal mediated reactions, which by far exceed the insights from experimental means alone. If applied with care, these techniques offer new avenues for the investigation of transition-metal chemistry and reactivity. For representative further work related to the reactivity of bare transition-metal ions, the reader is referred to Holthausen and Koch, 1996a and 1996b, Holthausen et al., 1996 and 1997, Hertwig et al., 1997, Hoyau and Ohanessian, 1997, Abahkin, Burt, and Russo, 1997, Wittborn et al., 1997, Westerberg and Blomberg, 1998, Blomberg et al., 1999, Yoshizawa, Shiota, and Yamabe, 1998 and 1999, Luna et al., 1997, 1998a, and 1998b. For related work on the reactivity of third-row transition-metal ions, see, e. g., Pavlov et al., 1997, Hertwig and Koch, 1999, and Sändig and Koch, 1997 and 1998.

Bibliography

Abashkin, Y. G., Burt, S. K., Russo, N., 1997, "Density Functional Study of the Mechanisms and the Potential Energy Surfaces of $MCH_2^+ + H_2$ Reactions. The Case of Cobalt and Rhodium (M = Co, Rh)", *J. Phys. Chem. A*, **101**, 8085.

Adamo, C., Barone, V., 1997, "Toward Reliable Adiabatic Connection Models Free from Adjustable Parameters", *Chem. Phys. Lett.*, **274**, 242.

Adamo, C., Barone, V., 1998a, "Implementation and Validation of the Lacks-Gordon Exchange Functional in Conventional Density Functional and Adiabatic Connection Methods", *J. Comput. Chem.*, **19**, 418.

Adamo, C., Barone, V., 1998b, "Exchange Functionals With Improved Long-Range Behavior and Adiabatic Connection Methods Without Adjustable Parameters: The *m*PW and *m*PW1PW Models", *J. Chem. Phys.*, **108**, 664.

Adamo, C., Barone, V., 1998c, "Toward Chemical Accuracy in the Computation of NMR Shieldings: The PBE0 Model", *Chem. Phys. Lett.*, **298**, 113.

Adamo, C., Barone, V., 1999, "Toward Reliable Density Functional Methods Without Adjustable Parameters: The PBE0 Method", *J. Chem. Phys.*, **110**, 6158.

Adamo, C., Cossi, M., Barone, V., 1999, "An Accurate Density Functional Method for the Study of Magnetic Properties: The PBE0 Model", *J. Mol. Struct. (Theochem)*, **493**, 145.

Adamo, C., Cossi, M., Scalmani, G., Barone, V., 1999, "Accurate Static Polarizabilities by Density Functional Theory: Assessment of the PBE0 Model", *Chem. Phys. Lett.*, **307**, 265.

Adamo, C., Scuseria, G. E., Barone, V., 1999, "Accurate Excitation Energies from Time-Dependent Density Functional Theory: Assessing the PBE0 Model", *J. Chem. Phys.*, **111**, 2889.

Adamo, C., di Matteo, A., Barone, V., 1999, "From Classical Density Functionals to Adiabatic Connection Methods. The State of the Art", *Adv. Quantum Chem.*, **36**, 45.

Adamo, C., Ernzerhof, M., Scuseria, G. E., 2000, "The meta-GGA Functional: Thermochemistry with a Kinetic Energy Density Dependent Exchange-Correlation Functional", *J. Chem. Phys.*, **112**, 2643.

Adamson, R. D., Dombroski, J. P., Gill, P. M. W., 1996, "Chemistry Without Coulomb Tails", *Chem. Phys. Lett.*, **254**, 329.

Adamson, R. D., Gill, P. M. W., Pople, J. A., 1998, "Empirical Density Functionals", *Chem. Phys. Lett.*, **284**, 6.

Alkorta, I., Elguero, J., 1998, "Ab Initio Hybrid DFT-GIAO Calculations of the Shielding Produced by Carbon-Carbon Bonds and Aromatic Rings in ^{1}H NMR Spectroscopy", *New J. Chem.*, 381.

Allen, W. D., Császár, A. G., 1993, "On the Ab Initio Determination of Higher-Order Force Constants at Nonstationary Geometries", *J. Chem. Phys.*, **98**, 2983.

Allen, M. J., Tozer, D. J., 2000, "Kohn-Sham Calculations Using Hybrid Exchange-Correlation Functionals with Asymptotically Corrected Potentials", *J. Chem. Phys.*, **113**, 5185.

Almlöf, J. E., Zheng, Y. C., 1997, "A Grid-Free Implementation of Density Functional Theory" in *Recent Advances in Density Functional Methods Part II*, Chong, D. P. (ed.), World Scientific, Singapore.

Altmann, J. A., Handy, N. C., Ingamells, V. E., 1996, "A Study of the Performance of Numerical Basis Sets in DFT Calculations on Sulfur-Containing Molecules", *Int. J. Quant. Chem.*, **57**, 533.

Amara, P., Volbeda, A., Fontecilla-Camps, J. C., Field, M. J., 1999, "A Hybrid Density Functional Theory/Molecular Mechanics Study of Nickel-Iron Hydrogenase: Investigation of the Active Site Redox States", *J. Am. Chem. Soc.*, **121**, 4468.

Amos, R. D., 1987, "Molecular Property Derivatives" in *Ab Initio Methods in Quantum Chemistry – Part I*, Lawley, K. P. (ed.), Wiley, Chichester.

Amos, R. D., Murray, C. W., Handy, N. C., 1993, "Structures and Vibrational Frequencies of FOOF and FONO Using Density Functional Theory", *Chem. Phys. Lett.*, **202**, 489.

Andersson, Y., Langreth, D. C., Lundqvist, B. I, 1996, "Van der Waals Interactions in Density Functional Theory", *Phys. Rev. Lett.*, **76**, 102.

Andrews, L., Zhou, M., Chertihin, V., Bauschlicher, C. W., Jr., 1999, "Reactions of Laser-Ablated Y and La Atoms, Cations, and Electrons With O_2. Infrared Spectra and Density Functional Calculations of the MO, MO^+, MO_2, MO_2^+, and MO_2^- Species in Solid Argon", *J. Phys. Chem. A*, **103**, 6525.

Andzelm, J., Wimmer, E., 1992, "Density Functional Gaussian-Type-Orbital Approach to Molecular Geometries, Vibrations, and Reaction Energies", *J. Chem. Phys.*, **96**, 1280.

Andzelm, J., Sosa, C., Eades, R. A., 1993, "Theoretical Study of Chemical Reactions Using Density Functional Methods With Nonlocal Corrections", *J. Phys. Chem.*, **97**, 4664.

Armentrout, P. B., Kickel, B. L., 1996, "Gas-Phase Thermochemistry of Transition Metal Ligand Systems: Reassessment of Values and Periodic Trends" in *Organometallic Ion Chemistry*, Freiser, B. S. (ed.), Kluwer, Amsterdam.

Atkins, P. W., Friedman, R. S., 1997, *Molecular Quantum Mechanics*, 3rd edition, Oxford University Press, Oxford.

Bachrach, S. M., 1995, "Population Analysis and Electron Densities from Quantum Mechanics" *Rev. Comput. Chem.*, **5**, 171.

Bader, R. W. F., 1994, *Atoms in Molecules. A Quantum Theory*, Clarendon Press, Oxford.

Baerends, E. J., Ellis, D. E., Ros, P., 1973, "Self-Consistent Molecular Hartree-Fock-Slater Calculations: I. The Computational Procedure", *Chem. Phys.*, **2**, 41.

Baerends, E. J., Gritsenko, O. V., 1997, "A Quantum Chemical View of Density Functional Theory", *J. Phys. Chem. A*, **101**, 5383.

Baerends, E. J., Branchadell, V., Sodupe, M., 1997, "Atomic Reference Energies for Density Functional Calculations", *Chem. Phys. Lett.*, **265**, 481.

Baerends, E. J., 2000, "Perspective on 'Self-Consistent Equations Including Exchange and Correlation Effects' ", *Theor. Chem. Acc.*, **103**, 265.

Baker, J., Scheiner, A., Andzelm, J., 1993, "Spin Contamination in Density Functional Theory", *Chem. Phys. Lett.*, **216**, 380.

Baker, J., Andzelm, J., Scheiner, A., Delley, B., 1994, "The Effect of Grid Quality and Weight Derivatives in Density Functional Calculations", *J. Chem. Phys.*, **101**, 8894.

Baker, J., Muir, M., Andzelm, A., 1995, "A Study of Some Organic Reactions Using Density Functional Theory", *J. Chem. Phys.*, **102**, 2063.

Bally, T., Sastry, G. N., 1997, "Incorrect Dissociation Behavior of Radical Cations in Density Functional Calculations", *J. Phys. Chem. A*, **101**, 7923.

Banerjee, A., Harbola, M. K., 1999, "Density-Functional-Theory Calculations of the Total Energies, Ionization Potentials and Optical Response Properties with the van Leeuwen-Baerends Potential", *Phys. Rev. A*, **60**, 3599.

Barone, V., 1994, "Inclusion of Hartree-Fock Exchange in Density Functional Methods. Hyperfine Structure of Second Row Atoms and Hydrides", *J. Chem. Phys.*, **101**, 6834.

Barone, V., Orlandini, L., Adamo, C., 1994a, "Density Functional Study of Diborane, Dialane, and Digallane", *J. Phys. Chem.*, **98**, 13185.

Barone, V., Orlandini, L., Adamo, C., 1994b, "Proton Transfer in Model Hydrogen-Bonded Systems by a Density Functional Approach", *Chem. Phys. Lett.*, **231**, 295.

Barone, V., 1994, "Validation of Self-Consistent Hybrid Approaches for the Study of Transition Metal Complexes. NiCO and CuCO as Case Studies", *Chem. Phys. Lett.*, **233**, 129.

Barone, V., 1995, "Structure, Magnetic Properties and Reactivities of Open-Shell Species from Density Functional and Self-Consistent Hybrid Methods", in *Recent Advances in Density Functional Methods, Part I*, Chong, D. P. (ed.), World Scientific, Singapore.

Barone, V., Adamo, C., Mele, F., 1996, "Comparison of Conventional and Hybrid Density Functional Approaches. Cationic Hydrides of First-Row Transition Metals as a Case Study", *Chem. Phys. Lett.*, **249**, 290.

Barone, V., Arnaud, 1996, "Study of Prototypical Diels-Alder Reactions by a Hybrid Density Functional/Hartree-Fock Approach", *Chem. Phys. Lett.*, **251**, 393.

Barone, V., Adamo, C., 1996, "Proton Transfer in the Ground and Lowest Excited States of Malonaldehyde: A Comparative Density Functional and Post-Hartree-Fock Study", *J. Chem. Phys.*, **105**, 11007.

Barone, V., Arnaud, 1997, "Diels-Alder Reactions: An Assessment of Quantum Chemical Procedures", *J. Chem. Phys.*, **106**, 8727.

Barone, V., Adamo, C., 1997a, "First Row Transition-Metal Hydrides: A Challenging Playground for New Theoretical Approaches", *J. Comput. Chem.*, **61**, 443.

Barone, V., Adamo, C., 1997b, "Toward a General Protocol for the Study of Static and Dynamic Properties of Hydrogen-Bonded Systems", *Int. J. Quant. Chem.*, **61**, 429.

Bartlett, R. J., Stanton, J. F., 1995, "Applications of Post-Hartree-Fock Methods: A Tutorial", *Rev. Comput. Chem.*, **5**, 65.

Basch, H., Mogi, K., Musaev, D. G., Morokuma, K., 1999, "Mechanism of the Methane → Methanol Conversion Reaction Catalyzed by Methane Monooxygenase: A Density Functional Study", *J. Am. Chem. Soc.*, **121**, 7249.

Bauernschmitt, R., Ahlrichs, R., 1996a, "Stability Analysis for Solutions of the Closed Shell Kohn-Sham Equation", *J. Chem. Phys.*, **104**, 9047.

Bauernschmitt, R., Ahlrichs, R., 1996b, "Treatment of Electronic Excitations Within the Adiabatic Approximation of Time Dependent Density Functional Theory", *Chem. Phys. Lett.*, **256**, 454.

Bauschlicher, C. W., Jr., 1995a, "A Comparison of the Accuracy of Different Functionals", *Chem. Phys. Lett.*, **246**, 40.

Bauschlicher, C. W., Jr., 1995b, "The Application of *Ab Initio* Electronic Structure Calculations to Molecules Containing Transition Metal Atoms" in *Modern Electronic Structure Theory, Part II*, Yarkony, D. R. (ed.), World Scientific, Singapore.

Bauschlicher, C. W., Jr., Partridge, H., 1995, "A Modification of the Gaussian-2 Approach Using Density Functional Theory", *J. Chem. Phys.*, **103**, 1788.

Bauschlicher, C. W., Jr., Maitre, P., 1995, "Structure of $Co(H_2)_n^+$ Clusters, for n = 1-6", *J. Phys. Chem.*, **99**, 3444.

Bauschlicher, C. W., Jr., Ricca, A., Partridge, H., Langhoff, S. R., 1997, "Chemistry by Density Functional Theory" in *Recent Advances in Density Functional Methods Part II*, Chong, D. P. (ed.), World Scientific, Singapore.

Bauschlicher, C. W., Jr., 1998, "Transition Metals: Applications" in *Encyclopedia of Computational Chemistry*, Schleyer, P. v. R. (Editor-in-Chief), Wiley, Chichester.

Bauschlicher, C. W., Jr., 1999, "A Further Study of Scandium and Dioxygen Reactions", *J. Phys. Chem. A*, **103**, 5463.

Bauschlicher, C. W., Jr., Hudgins, D. M., Allamandola, L. J., 1999, "The Infrared Spectra of Polycyclic Aromatic Hydrocarbons Containing a Five-Membered Ring: Symmetry Breaking and the B3LYP Functional", *Theor. Chem. Acc.*, **103**, 154.

Becke, A. D., 1986, "Density Functional Calculations of Molecular Bond Energies", *J. Chem. Phys.*, **84**, 4524.

Becke, A. D., 1988a, "Correlation Energy of an Inhomogeneous Electron Gas. A Coordinate Space Model", *J. Chem. Phys.*, **88**, 1053.

Becke, A. D., 1988b, "Density-Functional Exchange-Energy Approximation With Correct Asymptotic Behavior", *Phys. Rev. A*, **38**, 3098.

Becke, A. D., 1988c, "A Multicenter Numerical Integration Scheme for Polyatomic Molecules", *J. Chem. Phys.*, **88**, 2547.

Becke, A. D., Dickson, R. M., 1988, "Numerical Solution of Poisson's Equation in Polyatomic Molecules", *J. Chem. Phys.*, **89**, 2993.

Becke, A. D., 1989, "Basis-Set-Free Density-Functional Quantum Chemistry", *Int. J. Quant. Chem. Symp.*, **23**, 599.

Becke, A. D., 1992a, "Density Functional Thermochemistry. I. The Effect of Exchange-Only Gradient Correction", *J. Chem. Phys.*, **96**, 2155.

Becke, A. D., 1992b, "Density Functional Thermochemistry. II. The Effect of the Perdew-Wang Generalized-Gradient Correlation Correction", *J. Chem. Phys.*, **97**, 9173.

Becke, A. D., 1993a, "A New Mixing of Hartree-Fock and Local Density-Functional Theories", *J. Chem. Phys.*, **98**, 1372.

Becke, A. D., 1993b, "Density-Functional Thermochemistry. III. The Role of Exact Exchange", *J. Chem. Phys.*, **98**, 5648.

Becke, A. D., 1995, "Exchange-Correlation Approximations in Density-Functional Theory" in *Modern Electronic Structure Theory, Part II*, Yarkony, D. R., World Scientific, Singapore.

Becke, A. D., 1996a, "Density-Functional Thermochemistry. IV. A New Dynamical Correlation Functional and Implications for Exact-Exchange Mixing", *J. Chem. Phys.*, **104**, 1040.

Becke, A. D., 1996b, "Current-Density Dependent Exchange-Correlation Functionals", *Can. J. Chem.*, **74**, 995.

Becke, A. D., 1997, "Density-Functional Thermochemistry. V. Systematic Optimization of Exchange-Correlation Functionals", *J. Chem. Phys.*, **107**, 8554.

Becke, A. D., 1998, "A New Inhomogeneity Parameter in Density-Functional Theory", *J. Chem. Phys.*, **109**, 2092.

Becke, A. D., 1999, "Exploring the Limits of Gradient Corrections in Density Functional Theory", *J. Comput. Chem.*, **20**, 63.

Beno, B. R., Houk, K. N., Singleton, D. A., 1996, "Synchronous or Asynchronous? An 'Experimental' Transition State from a Direct Comparison of Experimental and Theoretical Kinetic Isotope Effects for a Diels-Alder Reaction", *J. Am. Chem. Soc.*, **118**, 9984.

Bérces, A., Ziegler, T., 1992, "The Harmonic Force Field of Benzene. A Local Density Functional Study", *J. Chem. Phys.*, **98**, 4793.

Bérces, A., 1996, "Harmonic Vibrational Frequencies and Force Constants of $M(CO)_5CX$ (M = Cr, Mo, W; X = O, S, Se). The Performance of Density Functional Theory and the Influence of Relativistic Effects", *J. Phys. Chem.*, **100**, 16538.

Berghold, G., Hutter, J., Parrinello, M.,1998, "Grid-Free DFT Implementation of Local and Gradient-Corrected XC Functionals", *Theor. Chem. Acc.*, **99**, 344.

Bergström, R., Lunell, S., Eriksson, L. A., 1996, "Comparative Study of DFT Methods Applied to Small Titanium/Oxygen Compounds", *Int. J. Quant. Chem.*, **59**, 427.

Bernardi, F., Bottoni, 1997, "Polar Effect in Hydrogen Abstraction Reactions from Halo-Substituted Methanes by Methyl Radical: A Comparison Between Hartree-Fock, Perturbation, and Density Functional Theories", *J. Phys. Chem.*, **101**, 1912.

Bertran, J., Branchadell, V., Oliva, A., Sodupe, M., 1998, "Pericyclic Reactions: The Diels-Alder Reaction" in *Encyclopedia of Computational Chemistry*, Schleyer, P. v. R. (Editor-in-Chief), Wiley, Chichester.

Bettinger, H. F., Schleyer, P. v. R., Schreiner, P. R., Schaefer, III, H. F., 1997, "Computational Analyses of Prototype Carbene Structures and Reactions" in *Modern Electronic Structure Theory and Applications in Organic Chemistry*, Davidson, E. R. (ed.), World Scientific, Singapore.

Bettinger, H. F., Schleyer, P. v. R., Schreiner, P. R., Schaefer, III, H. F., 1998, "Carbenes: A Testing Ground for Electronic Structure Methods" in *Encyclopedia of Computational Chemistry*, Schleyer, P. v. R. (Editor-in-Chief), Wiley, Chichester.

Bienati, M., Adamo, C., Barone, V., 1999, "Performance of a New Hybrid Hartree-Fock/Kohn-Sham Model (B98) in Predicting Vibrational Frequencies, Polarisabilities and NMR Chemical Shifts", *Chem. Phys. Lett.*, **311**, 69.

Blanchet, C., Duarte, H. A., Salahub, D. R., 1997, "Density Functional Study of Mononitrosyls of First-Row Transition-Metal Atoms", *J. Chem. Phys.*, **106**, 8778.

Bleiber, A., Sauer, J., 1995, "The Vibrational Frequency of the Donor OH Group in the H-Bonded Dimers of Water, Methanol, and Silanol. *Ab Initio* Calculations Including Anharmonicities", *Chem. Phys. Lett.*, **238**, 243.

Bloch, F., 1929, "The Electron Theory of Ferromagnetism and Electrical Conductivity", Z. *Physik*, **57**, 545.

Blöchl, P. E., Margl, P., Schwarz, K., 1996, "Ab Initio Molecular Dynamics With the Projector Augmented Wave Method" in *Chemical Applications of Density Functional Theory*, Laird, B. B., Ross, R. B., Ziegler, T. (eds.), American Chemical Society, Washington DC.

Blomberg, M. R. A., Siegbahn, P. E. M., Svensson, M., 1996, "Comparisons of Results From Parameterized Configuration Interaction (PCI-80) and From Hybrid Density Functional Theory With Experiments for First Row Transition Metal Compounds", *J. Chem. Phys.*, **104**, 9546.

Blomberg, M. R. A., Siegbahn, P. E. M., 1998, "Calculating Bond Strengths for Transition Metal Complexes" in *Computational Thermochemistry*, Irikura, K. K., Frurip, D. J. (eds.), *American Chemical Society Symposium Series* **677**, Washington, DC.

Blomberg, M. R. A., 1998, "Configuration Interaction: PCI-X and Applications" in *Encyclopedia of Computational Chemistry*, Schleyer, P. v. R. (Editor-in-Chief), Wiley, Chichester.

Blomberg, M. R. A., Yi, S. S., Noll, R. J., Weisshaar, J. C., 1999, "Gas Phase $Ni^+(^2D_{5/2})$ + n-C_4H_{10} Reaction Dynamics in Real Time: Experiment and Statistical Modeling Based on Density Functional Theory", *J. Phys. Chem. A*, **103**, 7254.

Boese, A. D., Doltsinis, N. L., Handy, N. C., Spriek, M., 2000, "New Generalized Gradient Approximation Functionals", *J. Chem. Phys.*, **112**, 1670.

Borowski, P., Jordan, K. D., Nichols, J., Nachtigall, P., 1998, "Investigation of a Hybrid TCSCF-DFT Procedure", *Theor. Chem. Acc.*, **99**, 135.

Bose, B., Zhao, S., Stenutz, R., Cloran, F., Bondo, P. B., Bondo, G., Hertz, B., Carmichael, I., Serianni, A. S., 1998, "Three-Bond C-O-C-C Spin-Coupling Constants in Carbohydrates: Development of a Karplus Relationship", *J. Am. Chem. Soc.*, **120**, 11158.

Botschwina, P., 1998, "The Saddle Point of the Nucleophilic Substitution Reaction Cl^- + CH_3Cl: Results of Large-Scale Coupled Cluster Calculations", *Theor. Chem. Acc.*, **99**, 426.

Boys, S. F., Bernardi, F., 1970, "The Calculation of Small Molecular Interactions by the Differences of Separate Total Energies. Some Procedures With Reduced Errors", *Mol. Phys.*, **19**, 553.

Bray, M. R., Deeth, R. J., Paget, V. J., Sheen, P. D., 1996, "The Relative Performance of the Local Density Approximation and Gradient Corrected Density Functional Theory for Computing Metal-Ligand Distances in Werner-Type and Organometallic Complexes", *Int. J. Quant. Chem.*, **61**, 85.

Brönstrup, M., Schröder, D., Kretzschmar, I., Schwarz, H., Harvey, J. N., 2001, "Platinum Dioxide Cation: Easy to Generate Experimentally but Difficult to Describe Theoretically", *J. Am. Chem. Soc.*, **123**, 142.

Brown, S. T., Rienstra-Kiracofe, J. C., Schaefer, H. F., 1999, "A Systematic Application of Density Functional Theory to Some Carbon-Containing Molecules and Their Anions", *J. Phys. Chem. A*, **103**, 4065.

Bühl, M., 1997, "Density Functional Calculations of Transition Metal NMR Chemical Shifts: Dramatic Effects of Hartree-Fock Exchange", *Chem. Phys. Lett.*, **267**, 251.

Bühl, M. 1998, "NMR Chemical Shift Computation: Structural Applications", in *Encyclopedia of Computational Chemistry*, Schleyer, P. v. R. (Editor-in-Chief), Wiley, Chichester.

Bühl, M., Kaupp, M., Malkina, O. L., Malkin, V. G., 1999, "The DFT Route to NMR Chemical Shifts", *J. Comput. Chem.*, **20**, 91.

Burke, K., Ernzerhof, M., Perdew, J. P., 1997, "The Adiabatic Connection Method: A Non-Empirical Hybrid", *Chem. Phys. Lett.*, **265**, 115.

Burke, K., Gross, E. K. U., 1998, "A Guided Tour of Time-Dependent Density Functional Theory" in *Density Functionals: Theory and Applications*, Lecture Notes in Physics, Vol. 500, Joubert, D. (ed.), Springer, Heidelberg.

Burke, K., Perdew, J. P., Ernzerhof, M., 1998, "Mixing Exact Exchange With GGA: When to Say When" in *Electronic Density Functional Theory. Recent Progress and New Directions*, Dobson, J. F., Vignale, G., Das, M. P. (eds.), Plenum Press, New York.

Burke, K., Perdew, J. P., Wang, Y., 1998, "Derivation of a Generalized Gradient Approximation: The PW91 Density Functional", in *Electronic Density Functional Theory. Recent Progress and New Directions*, Dobson, J. F., Vignale, G., Das, M. P. (eds.), Plenum Press, New York.

Calaminici, P., Jug, K., Köster, A. M., 1998, "Density Functional Calculations of Molecular Polarizabilities and Hyperpolarizabilities", *J. Chem. Phys.*, **109**, 7756.

Carrington, T., Jr., Miller, W. H., 1986, "Reaction Surface Description of Intramolecular Hydrogen Atom Transfer in Malonaldehyde", *J. Chem. Phys.*, **84**, 4364.

Carter, E. A., Goddard III, W. A., 1988, "Relationships between Bond Energies in Coordinatively Unsaturated and Coordinatively Saturated Transition Metal Complexes: A Quantitative Guide for Single, Double, and Triple Bonds", *J. Phys. Chem.*, **92**, 5679.

Casida, M. E., 1995, "Time Dependent Density Functional Response Theory for Molecules" in *Recent Advances in Density Functional Methods, Part I*, Chong, D. P. (ed.), World Scientific, Singapore.

Casida, M. E., Casida, K. C., Salahub, D. R., 1998, "Excited-State Potential Energy Curves from Time-Dependent Density-Functional Theory: A Cross Section of Formaldehyde's $^1A_1$ Manifold", *Int. J. Quant. Chem.*, **70**, 933.

Casida, M. E., Jamorski, C., Casida, K. C., Salahub, D. R., 1998, "Molecular Excitation Energies to High-Lying Bound States from Time-Dependent Density-Functional Response Theory: Characterization and Correction of the Time-Dependent Local Density Approximation Ionization Threshold", *J. Chem. Phys.*, **108**, 4439.

Ceperley, D. M., Alder, B. J., 1980, "Ground State of the Electron Gas by a Stochastic Method", *Phys. Rev. Lett.*, **45**, 566.

Chakravorty, S. J., Davidson, E. R., 1993, "The Water Dimer: Correlation Energy Calculations", *J. Phys. Chem.*, **97**, 6373.

Challacombe, M., Schwegler, E., Almlöf, J., 1996, "Fast Assembly of the Coulomb Matrix: A Quantum Chemical Tree Code", *J. Chem. Phys.*, **104**, 4685.

Chandra, A. K., Nguyen, M. T., 1998, "A Density Functional Study of Weakly Bound Hydrogen Bonded Complexes", *Chem. Phys.*, **232**, 299.

Cheeseman, J. R., Trucks, G. W., Keith, T. A., Frisch, M. J., 1996, "A Comparison of Models for Calculating Nuclear Magnetic Resonance Shielding Tensors", *J. Chem. Phys.*, **104**, 5497.

Chen, J., Houk, K. N., Foote, C. S., 1998, "Theoretical Study of the Concerted and Stepwise Mechanisms of Triazolinedione Diels-Alder Reactions", *J. Am. Chem. Soc.*, **120**, 12303.

Chermette, H., Razafinjanahary, H., Carrion, L., 1997, "A Density Functional Especially Designed for Hydrogen-Only Systems", *J. Chem. Phys.*, **107**, 10643.

Chermette, H., Lembarki, A., Razafinjanahary, H., Rogemond, 1998, "Gradient-Corrected Exchange Potential Functional With the Correct Asymptotic Behavior", *Adv. Quant. Chem.*, **33**, 105.

Citra, A., Andrews, L., 1999, "Reactions of Laser Ablated Rhodium Atoms With Nitrogen Atoms and Molecules. Infrared Spectra and Density Functional Calculations on Rhodium Nitrides and Dinitrogen Complexes", *J. Phys. Chem. A*, **103**, 3410.

Civalleri, B., Garrone, E., Ugliengo, P., 1997, "Density Functional Study of Hydrogen-Bonded Systems: Energetic and Vibrational Features of Some Gas-Phase Adducts of Hydrogen Fluoride", *J. Mol. Struct. (Theochem)*, **419**, 227.

Cloran, F., Carmichael, I., Serianni, A. S., 1999a, "^{13}C-^{1}H and ^{13}C-^{13}C Spin Coupling Behavior in Aldofuranosyl Rings from Density Functional Theory", *J. Phys. Chem. A*, **103**, 3783.

Cloran, F., Carmichael, I., Serianni, A. S., 1999b, "Density Functional Calculations on Disaccharide Mimics: Studies of Molecular Geometries and Trans-*O*-glycosidic $^{3}J_{COCH}$ and $^{3}J_{COCC}$ Spin Couplings", *J. Am. Chem. Soc.*, **121**, 9843.

Cohen, A. J., Tantirungrotechai, Y., 1999, "Molecular Electric Properties: An Assessment of Recently Developed Functionals", *Chem. Phys. Lett.*, **299**, 465.

Cohen, A. J., Handy, N. C., Tozer, D. J., 1999, "Density Functional Calculations of the Hyperpolarisabilities of Small Molecules", *Chem. Phys. Lett.*, **303**, 391.

Colwell, S. M., Murray, C. W., Handy, N. C., Amos, R. D., 1993, "The Determination of Hyperpolarisablities Using Density Functional Theory", *Chem. Phys. Lett.*, **210**, 261.

Colle, R., Salvetti, O, 1975, "Approximate Calculation of the Correlation Energy for the Closed Shells", *Theor. Chim. Acta*, **37**, 329.

Cook, M., Karplus, M., 1987, "Electron Correlation and Density-Functional Methods", *J. Phys. Chem.*, **91**, 31.

Cramer, C. J., Dulles, F. J., Giesen, D. J., Almlöf, J., 1995, "Density Functional Theory: Excited States and Spin Annihilation", *Chem. Phys. Lett.*, **245**, 165.

Csonka, G. I., Johnson, B. G., 1998, "Inclusion of Exact Exchange for Self-Interaction Corrected H_3 Density Functional Potential Energy Surface", *Theor. Chem. Acc.*, **99**, 158.

Cui, Q., Musaev, D. G., Svensson, M., Sieber, S., Morokuma, K., 1995, "N_2 Cleavage by Three-Coordinate Group 6 Complexes. W(III) Complexes Would Be Better Than Mo(III) Complexes", *J. Am. Chem. Soc.*, **117**, 12366.

Cui, Q., Musaev, D. G., Morokuma, K., 1998a, "Why do $Pt(PR_3)_2$ Complexes Catalyze the Alkyne Diboration Reaction, but Their Palladium Analogues Do Not? A Density Functional Study", *Organometallics*, **17**, 742.

Cui, Q., Musaev, D. G., Morokuma, K., 1998b, "Density Functional Study on the Mechanism of Palladium(0)-Catalyzed Thioboration Reaction of Alkynes. Differences between Pd(0) and Pt(o) Catalysts and between Thioboration and Diboration", *Organometallics*, **17**, 1383.

Cundari, T. R., Benson, M. T., Lutz, M. L., Sommerer, S. O., 1996, "Effective Core Potential Approaches to the Chemistry of the Heavier Elements", *Rev. Comput. Chem.*, **8**, 145.

Curtiss, L. A, Raghavachari, K., Trucks, G. W., Pople, J. A., 1991, "Gaussian-2 Theory for Molecular Energies of First- and Second-Row Compounds", *J. Chem. Phys.*, **94**, 7221.

Curtiss, L. A., Raghavachari, K., Redfern, P. C., Pople, J. A., 1997, "Assessment of Gaussian-2 and Density Functional Theories for the Computation of Enthalpies of Formation", *J. Chem. Phys.*, **106**, 1063.

Curtiss, L. A., Redfern, P. C., Raghavachari, K., Pople, J. A., 1998, "Assessment of Gaussian-2 and Density Functional Theories for the Computation of Ionization Potentials and Electron Affinities", *J. Chem. Phys.*, **109**, 42.

Dannenberg, J. J., Haskamp, J., Masunov, A., 1999, "Are Hydrogen Bonds Covalent of Electrostatic? A Molecular Orbital Comparison of Molecules in Electric Fields and H-Bonding Environments", *J. Chem. Phys. A* **103**, 7083.

Danovich, D., Shaik, S., 1997, "Spin-Orbit Coupling in the Oxidative Activation of H-H by FeO^+. Selection Rules and Reactivity Effects", *J. Am. Chem. Soc.*, **119**, 1773.

Dapprich, S., Ujaque, G., Maseras, F., Lledós, A., Musaev, D. G., Morokuma, K., 1996, "Theory Does Not Support an Osmaoxetane Intermediate in the Osmium-Catalyzed Dihydroxylation of Olefins", *J. Am. Chem. Soc.*, **118**, 11660.

Dargel, T. K., Hertwig, R. H., Koch, W., Horn, H., 1998, "Towards an Accurate Gold Carbonyl Binding Energy in $AuCO^+$: Basis Set Convergence and a Comparison between Density Functional and Conventional Methods", *J. Chem. Phys.*, **108**, 3876.

Daul, C., 1994, "Density Functional Theory Applied to the Excited States of Coordination Compounds", *Int. J. Quant. Chem.*, **52**, 867.

Daul, C. A., Doclo, K. G., Stückl, A. C., 1997, "On the Calculation of Multiplets" in *Recent Advances in Density Functional Methods, Part II*, Chong, D. P. (ed.), World Scientific, Singapore.

Davidson, E. R., 1976, *Reduced Density Matrices in Quantum Chemistry*, Academic Press, New York.

Dejaegere, A. P., Case, D. A., 1998, "Density Functional Study of Ribose and Deoxyribose Chemical Shifts", *J. Phys. Chem. A*, **102**, 5280.

Del Bene, J. E., Person, W. B., Szczepaniak, K., 1995, "Properties of Hydrogen-Bonded Complexes Obtained from the B3LYP Functional With 6-31G(d,p) and 6-31+G(d,p) Basis Sets: Comparison With MP2/6-31+G(d,p) Results and Experimental Data", *J. Phys. Chem.*, **99**, 10705.

Del Bene, J. E., Jordan, M. J. T., 1998, "A Comparative Study of Anharmonicity and Matrix Effects on the Complexes XH:NH_3, X=F, Cl, and Br", *J. Chem. Phys.*, **108**, 3205.

Del Bene, J. E., 1998, "Hydrogen Bonding: 1", in *Encyclopedia of Computational Chemistry*, Schleyer, P. v. R. (Editor-in-Chief), Wiley, Chichester.

Delley, B., 1990, "An All-Electron Numerical Method for Solving the Local Density Functional for Polyatomic Molecules", *J. Chem. Phys.*, **92**, 508.

Deng, L., Branchadell, V., Ziegler, T., 1994, "Potential Energy Surface of the Gas-Phase S_N2 Reactions X^- + CH_3X = XCH_3 + X^- (X = F, Cl, Br, I): A Comparative Study by Density Functional Theory and Ab Initio Methods", *J. Am. Chem. Soc.*, **116**, 10645.

Deng, L., Ziegler, T., 1997, "Theoretical Study of the Oxidation of Alcohols to Aldehyde by d^0 Transition-Metal-Oxo Complexes: Combined Approach Based on Density Functional Theory and the Intrinsic Reaction Coordinate Method", *Organometallics*, **16**, 716.

De Oliveira, G., Martin, J. M. L., De Proft, F., Geerlings, P., 1999, "Electron Affinities of the First and Second Row Atoms: Benchmark *Ab Initio* and Density Functional Calculations", *Phys. Rev. A*, **60**, 1034.

De Proft, F., Martin, J. M. L., Geerlings, P., 1996, "On the Performance of Density Functional Methods for Describing Atomic Populations, Dipole Moments and Infrared Intensities", *Chem. Phys. Lett.*, **250**, 393.

De Proft, F., Geerlings, P., 1997, "Calculation of Ionization Energies, Electron Affinities, Electronegativities and Hardnesses Using Density Functional Methods", *J. Chem. Phys.*, **106**, 3270.

Deubel, D. V., Frenking, G., 1999, "Are there Metal Oxides that Prefer a [2+2] Addition over a [3+2] Addition to Olefins? Theoretical Study of the Reaction Mechanism of $LReO_3$ Addition (L = O^-, Cl, Cp) to Ethylene", *J. Am. Chem. Soc.*, **121**, 2021.

Devlin, F. J., Stephens, P. J., 1999, "Ab Initio Density Functional Theory Study of the Structure and Vibrational Spectra of Cyclohexanone and Its Isotopomers", *J. Phys. Chem.*, **103**, 527.

Dickson, R. M., Becke, A. D., 1993, "Basis-set-free Local Density-Functional Calculations of Geometries of Polyatomic Molecules", *J. Chem. Phys.*, **99**, 3898.

Dickson, R. M., Becke, A. D., 1996, "Local Density-Functional Polarizabilities and Hyperpolarizabilities at the Basis-Set Limit", *J. Phys. Chem.*, **100**, 16105.

Dickson, R. M., Ziegler, T., 1996, "NMR Spin-Spin Coupling Constants from Density Functional Theory With Slater-Type Basis Functions", *J. Phys. Chem.*, **100**, 5286.

Dirac, P. A. M., 1930, "Note on Exchange Phenomena in the Thomas Atom", *Proc. Camb. Phil. Soc.*, **26**, 376.

Dirac, P. A. M, 1958, *The Principles of Quantum Mechanics*, 4th edition, Clarendon Press, Oxford.

Dkhissi, A., Alikhani, M. E., Bouteiller, Y., 1997, "Methodological Study of Becke3-LYP Density Functional Adapted to the Determination of Accurate Infrared Signature for Hydrogen-Bonded Complexes", *J. Mol. Struct. (Theochem)*, **416**, 1.

Dobson, J. F., Dinte, B. P., 1996, "Constraint Satisfaction in Local and Gradient Susceptibility Approximations: Application to a Van der Waals Density Functional", *Phys. Rev. Lett.*, **76**, 1780.

Dobson, J. F., 1998, "Prospects for a Van der Waals Density Functional", *Int. J. Quant. Chem.*, **69**, 615.

Dombroski, J. P., Taylor, S. W., Gill, P. M. W., 1996, "KWIK: Coulomb Energies in *O*(N) Work", *J. Phys. Chem.*, **100**, 6272.

Dreizler, R. M., Gross, E. K. U., 1995, *Density Functional Theory*, Springer, Heidelberg.

Dunlap B. I., Connolly, J. W. D., Sabin, J. R., 1979, "On Some Approximations in Applications of X_α Theory", *J. Chem. Phys.*, **71**, 3396.

Dunlap, B. I., 1987, "Symmetry and Degeneracy in X_α and Density Functional Theory" in *Ab Initio Methods in Quantum Chemistry - II*, Lawley, K. P. (ed.), Wiley, Chichester.

Dunning, T. H., 1989, "Gaussian Basis Sets for Use in Correlated Molecular Calculations. I. The Atoms Boron Through Neon and Hydrogen", *J. Chem. Phys.*, **90**, 1007.

Durant, J. L., 1996, "Evaluation of Transition State Properties by Density Functional Theory", *Chem. Phys. Lett.*, **256**, 595.

Ehlers, A. W., Ruiz-Morales, Y., Baerends, E. J., Ziegler, T., 1997, "Dissociation Energies, Vibrational Frequencies, and ^{13}C NMR Chemical Shifts of the 18-Electron Species $[M(CO)_6]^n$ (M = Hf-Ir, Mo, Tc, Ru, Ru, Cr, Mn, Fe). A Density Functional Study", *Inorg. Chem.*, **36**, 5031.

Eichkorn, K., Treutler, O., Öhm, H., Häser, M., Ahlrichs, R., 1995, "Auxiliary Basis Sets to Approximate Coulomb Potentials", *Chem. Phys. Lett.*, **240**, 283.

Eichkorn, K., Weigend, F., Treutler, O., Ahlrichs, R., 1997, "Auxiliary Basis Sets for Main Row Atoms and Transition Metals and their Use to Approximate Coulomb Potentials", *Theor. Chem. Acc.*, **97**, 119.

Eriksson, L. A., Malkina, O. L., Malkin, V. G., Salahub, D. R., 1994, "The Hyperfine Structures of Small Radicals from Density Functional Calculations", *J. Chem. Phys.*, **100**, 5066.

Eriksson, L. A., Pettersson, L. G. M., Siegbahn, P. E. M., Wahlgren, U., 1995, "On the Accuracy of Gradient Corrected Density Functional Methods for Transition Metal Complexes", *J. Chem. Phys.*, **102**, 872.

Ernzerhof, M., Perdew, J. P., Burke, K., 1996, "Density Functionals: Where Do They Come From, Why Do They Work?", *Top. Curr. Chem.*, **180**, 1.

Ernzerhof, M., Perdew, J. P., Burke, K., 1997, "Coupling-Constant Dependence of Atomization Energies", *Int. J. Quant. Chem.*, **64**, 285.

Ernzerhof, M., Scuseria, G. E., 1999a, "Assessment of the Perdew-Burke-Ernzerhof Exchange-Correlation Functional", *J. Chem. Phys.*, **110**, 5029.

Ernzerhof, M., Scuseria, G. E., 1999b, "Kinetic Energy Density Dependent Approximations to the Exchange Energy", *J. Chem. Phys.*, **111**, 911.

Eschrig, H., 1996, *"The Fundamentals of Density Functional Theory"*, Teubner, Stuttgart.

Estrin, D. A., Paglieri, G., Corongiu, G., Clementi, E., 1996, "Small Clusters of Water Molecules Using Density Functional Theory", *J. Phys. Chem.*, **100**, 8701.

Facelli, J. C., 1998, "Density Functional Theory Calculations of the Structure and the ^{15}N and ^{13}C Chemical Shifts of Methyl Bacteriopheophorbide a and Bacteriochlorophyll a", *J. Phys. Chem. B*, **102**, 2111.

Fan, L., Ziegler, T., 1991, "Optimization of Molecular Structures by Self-Consistent and Nonlocal Density-Functional Theory", *J. Chem. Phys.*, **95**, 7401.

Fan, L., Ziegler, T., 1992, "Application of Density Functional Theory to Infrared Absorption Intensity Calculations on Main Group Molecules", *J. Chem. Phys.*, **96**, 9005.

Fan, L., Ziegler, T., 1995, "The Application of Nonlocal and Self-Consistent Density Functional Theory to Molecular Problems" in *Density Functional Theory of Molecules, Clusters, and Solids*, Ellis, D. E. (ed.), Kluwer Academic Publishers, Dordrecht.

Feller, D., Davidson, E. R., 1990, "Basis Sets for Ab Initio Molecular Orbital Calculations and Intermolecular Interactions", *Rev. Comput. Chem.*, **1**, 1.

Feller, D., Schuchardt, K., Jones, D., 1998, Extensible Computational Chemistry Environment Basis Set Database, Version 1.0, as developed and distributed by the Molecular Science Computing Facility, Environmental and Molecular Sciences Laboratory which is part of the Pacific Northwest Laboratory, P. O. Box 999, Richland, Washington 99352, USA, and funded by the U. S. Department of Energy. The Pacific Northwest Laboratory is a multi-program laboratory operated by Battelle Memorial Institute for the U. S. Department of Energy under contract DE-AC06-76RLO 1830.

Fermi, E., 1927, "Un Metodo Statistice per la Determinazionedi Alcune Proprieta dell' Atomo", *Rend. Accad. Lincei*, **6**, 602.

Fiedler, A., Schröder, D., Shaik, S., Schwarz, H., 1994, "Electronic Structures and Gas-Phase Reactivities of Cationic Late-Transition-Metal Oxides", *J. Am. Chem. Soc.*, **116**, 10734.

Filatov, M., Thiel, W., 1997, "A New Gradient-Corrected Exchange-Correlation Density Functional", *Mol. Phys.*, **91**, 847.

Filatov, M., Thiel, W., 1998, "Exchange-Correlation Density Functional Beyond the Gradient Approximation", *Phys. Rev. A*, **57**, 189.

Filatov, M., Shaik, S., 1998a, "Spin-Restricted Density Functional Approach to the Open-Shell Problem", *Chem. Phys. Lett.*, **288**, 689.

Filatov, M., Shaik, S., 1998b, "Theoretical Investigation of Two-State-Reactivity Pathways of H-H Activation by FeO^+: Addition-Elimination, "Rebound", and Oxene-Insertion Mechanisms", *J. Phys. Chem. A*, **102**, 3835.

Filatov, M., Shaik, S., 1999, "Application of Spin-Restricted Open-Shell Kohn-Sham Method to Atomic and Molecular Multiplet States", *J. Chem. Phys.*, **110**, 116.

Filippi, C., Umrigar, C. J., Gonze, X., 1997, "Excitation Energies from Density Functional Perturbation Theory", *J. Chem. Phys.*, **107**, 9994.

Finley, J. W., Stephens, P. J., 1995, "Density Functional Theory Calculations of Molecular Structures and Harmonic Vibrational Frequencies Using Hybrid Density Functionals", *J. Mol. Struct. (Theochem)*, **357**, 225.

Florian, J., Johnson, B. G., 1994, "Comparison and Scaling of Hartree-Fock and Density Functional Harmonic Force Fields. 1. Formamide Monomer", *J. Phys. Chem.*, **98**, 3681.

Florian, J., Johnson, B. G., 1995, "Structure, Energetics, and Force Fields of the Cyclic Formamide Dimer: MP2, Hartree-Fock, and Density Functional Study", *J. Phys. Chem.*, **99**, 5899.

Fonseca Guerra, C., Snijders, J. G., te Velde, G., Baerends, E. J., 1998, "Towards an Order-*N* DFT Method", *Theor. Chem. Acc.*, **99**, 391.

Freiser, B. S. (ed.), 1996, *Organometallic Ion Chemistry*, Kluwer, Amsterdam.

Fournier, R., Andzelm, J., Salahub, D. R., 1989, "Analytical Gradient of the Linear Combination of Gaussian-Type Orbitals-Local Spin Density Energy", *J. Chem. Phys.*, **90**, 6371.

Fournier, R., 1993, "Theoretical Study of the Monocarbonyls of First-Row Transition Metal Atoms", *J. Chem. Phys.*, **99**, 1801.

Frankcombe, K. E., Cavell, K. J., Knott, R. B., Yates, B. F., 1997, "Competing Reaction Mechanisms for the Carbonylation of Neutral Palladium(II) Complexes Containing Bidentate Ligands: a Theoretical Study", *Organometallics*, **16**, 3199.

Frenking, G., Antes, I., Böhme, M., Dapprich, S., Ehlers, A. W., Jonas, V., Neuhaus, A., Otto, M., Stegmann, R., Veldkamp, A., Vyboishchikov, S. F., 1996, "Pseudopotential Calculations of Transition Metal Compounds - Scope and Limitations", *Rev. Comput. Chem.*, **8**, 63.

Frenking, G., Wagener, T., 1998, "Transition Metal Chemistry", in *Encyclopedia of Computational Chemistry*, Schleyer, P. v. R. (Editor-in-Chief), Wiley, Chichester.

Froese, R. D. J., Humbel, S., Svensson, M., Morokuma, K., 1997, "IMOMO(G2MS): A New High-Level G2-like Method for Large Molecules and Its Application the Diels-Alder Reactions", *J. Phys. Chem. A*, **101**, 227.

Fuentalba, P., Simón-Manso, Y., 1997, "Static Dipole Polarizabilities Through Density Functional Methods", *J. Phys. Chem. A*, **101**, 4231.

García, A., Cruz, E. M., Sarasola, C., Ugalde, J. M., 1997, "Properties of Some Weakly Bound Complexes Obtained with Various Density Functionals", *J. Mol. Struct. (Theochem)*, **397**, 191.

Garcia-Viloca, M., Gelabert, R., González-Lafont, À., Moreno, M., Lluch, J. M., 1998, "Temperature Dependence of Proton NMR Chemical Shift as a Criterion to Identify Low-Barrier Hydrogen Bonds", *J. Am. Chem. Soc.*, **120**, 10203.

Gauld, J. W., Eriksson, L. A., Radom, L., 1997, "Assessment of Procedures for Calculating Radical Hyperfine Structures", *J. Phys. Chem. A*, **101**, 1352.

Gauss, J., Stanton, J. F., 1996, "Perturbative Treatment of Triple Excitations in Coupled Cluster Calculations of Nuclear Magnetic Shielding Constants", *J. Chem. Phys.*, **104**, 2574.

Geerlings, P., De Proft, F., Martin, J. M. L., 1996, "Density Functional Theory Concepts and Techniques for Studying Molecular Charge Distributions and Related Properties", in *Recent Developments and Applications of Modern Density Functional*, Seminario, J. M. (ed.), Elsevier, Amsterdam.

Gill, P. M. W., Johnson, B. G., Pople, J. A., 1993, "A Standard Grid for Density Functional Calculations", *Chem. Phys. Lett.*, **209**, 506.

Gill, P. M. W., Adamson, R. D., 1996, "A Family of Attenuated Coulomb Operators", *Chem. Phys. Lett.*, **261**, 105.

Glaesemann, K. R., Gordon, M. S., 1998, "Investigation of a Grid-Free Density Functional Theory (DFT) Approach", *J. Chem. Phys.*, **108**, 9959.

Glaesemann, K. R., Gordon, M. S., 1999, "Evaluation of Gradient Corrections in Grid-Free Density Functional Theory", *J. Chem. Phys.*, **110**, 6580.

Glaesemann, K. R., Gordon, M. S., 2000, "Auxiliary Basis Sets for Grid Free Density Functional Theory", *J. Chem. Phys.*, **112**, 10738.

Glendening, E. D., Streitwieser, A., 1994, "Natural Energy Decomposition Analysis: An Energy Partitioning Procedure for Molecular Interactions With Application to Weak Hydrogen Bonding, Strong Ionic, and Moderate Donor-Acceptor Interactions", *J. Chem. Phys.*, **100**, 2900.

Glukhovtsev, M. N., Pross, A., Radom, L., 1995, "Gas Phase Identity S_N2 Reactions of Halide Anions With Methyl Halides: A High-Level Computational Study", *J. Am. Chem. Soc.*, **117**, 2024.

Glukhovtsev, M. N., Bach, R. D., Pross, A., Radom, L., 1996, "The Performance of B3-LYP Density Functional Theory in Describing S_N2 Reactions at Saturated Carbon", *Chem. Phys. Lett.*, **260**, 558.

Godbout, N., Salahub, D. R., Andzelm, J., Wimmer, E., 1992, "Optimization of Gaussian-Type Basis Sets for Local Spin Density Functional Calculations. Part I. Boron through Neon, Optimization Technique and Validation", *Can. J. Chem.*, **70**, 560.

Goddart, J. D., Orlova, G., 1999, "Density Functional Theory with Fractionally Occupied Frontier Orbitals and the Instabilities of the Kohn-Sham Solutions for Defining Diradical Transition States: Ring-Opening Reactions", *J. Chem. Phys.*, **111**, 7705.

Goedecker, S., Umrigar, C. J., 1997, "Critical Assessment of the Self-Interaction-Corrected–Local-Density-Functional Method and its Algorithmic Implementation", *Phys. Rev. A.*, **55**, 1765.

Goedecker, S., 1999, "Linear Scaling Electronic Structure Methods", *Rev. Mod. Phys.*, **71**, 1085.

Goh, S. K., Gallant, R. T., St-Amant, A., 1998, "Towards Linear Scaling for the Fits of the Exchange-Correlation Terms in the LCGTO-DF Method via a Divide-and-Conquer Approach", *Int. J. Quant. Chem.*, **69**, 405.

Goldstein, E., Beno, B., Houk, K. N., 1996, "Density Functional Theory Prediction of the Relative Energies and Isotope Effects for the Concerted and Stepwise Mechanism of the Diels-Alder Reaction of Butadiene and Ethylene", *J. Am. Chem. Soc.*, **118**, 6036.

Goldstein, E., Beno, B., Houk, K. N., 1999, "Transition Structures and Exo/Endo Stereoselectivities of Concerted [6+4] Cycloadditions with Density Functional Theory, *Theor. Chem. Acc.*, **103**, 81.

Gonzáles, L., Mó, O., Yáñez, M., 1996, "High Level *Ab Initio* Versus DFT Calculations on $(H_2O_2)_2$ and H_2O_2-H_2O Complexes as Prototypes of Multiple Hydrogen Bonded Systems", *J. Comput. Chem.*, **18**, 1124.

Gonzáles, L., Mó, O., Yáñez, M., Elguerdo, J., 1996, "Cooperative Effects in Water Trimers. The Performance of Density Functional Approaches", *J. Mol. Struct. (Theochem)*, **371**, 1.

Gonzáles, L., Mó, O., Yáñez, M., Elguerdo, J., 1998a, "High Level *Ab Initio* and Density Functional Theory Studies on Methanol-Water Dimers and Cyclic Methanol(Water)$_2$ Trimer", *J. Chem. Phys.*, **109**, 139.

Gonzáles, L., Mó, O., Yáñez, M., Elguerdo, J., 1998b, "Very Strong Hydrogen Bonds in Neutral Molecules: The Phosphonic Acid Dimers", *J. Chem. Phys.*, **109**, 2685.

Gonzáles, L., Mó, O., Yáñez, M., 1999, "Density Functional Theory Study on Ethanol Dimers and Cyclic Ethanol Trimers", *J. Chem. Phys.*, **111**, 3855.

Gonzalez, C., Restrepo-Cossio, A., Márquez, M., Wiberg, K. B., De Rosa, M., 1998, "Ab Initio Study of the Solvent Effects on the Singlet-Triplet Gap of Nitrenium Ions and Carbenes", *J. Phys. Chem. A*, **102**, 2732.

Görling, A., 1996, "Density Functional Theory for Excited States", *Phys. Rev. A*, **54**, 3912.

Görling, A., 1999, "Density-Functional Theory Beyond the Hohenberg-Kohn Theorem", *Phys. Rev. A*, **59**, 3359.

Gräfenstein, J., Kraka, E., Cremer, D., 1998, "Density Functional Theory for Open-Shell Singlet Biradicals", *Chem. Phys. Lett.*, **288**, 593.

Gräfenstein, J., Cremer, D., 2000, "The Combination of Density Functional Theory with Multi-Configuration Methods – CAS-DFT", *Chem. Phys. Lett.*, **316**, 569.

Greengard, L., 1987, *The Rapid Evaluation of Potential Fields in Particle Systems*, MIT Press, Cambridge, Mass.

Grimme, S., 1996, "Density Functional Calculations With Configuration Interaction for the Excited States of Molecules", *Chem. Phys. Lett.*, **259**, 128.

Grimme, S., Waletzke, M., 1999, "A Combination of Kohn-Sham Density Functional Theory and Multi-Reference Configuration Interaction Methods", *J. Chem. Phys.*, **111**, 5645.

Gritsenko, O. V., van Leuwen, R., Baerends, E. J., 1996, "On the Optimal Mixing of the Exchange Energy and the Electron-Electron Interaction Part of the Exchange-Correlation Energy", *Int. J. Quant. Chem.: Quant. Chem. Symp.*, **30**, 1375.

Gritsenko, O. V., Schipper, P. R. T., Baerends, E. J., 1997, "Exchange and Correlation Energy in Density Functional Theory. Comparison of Accurate DFT Quantities With Traditional Hartree-Fock Based Ones and Generalized Gradient Approximations for the Molecules Li_2, N_2, F_2", *J. Chem. Phys.*, **107**, 5007.

Gritsenko, O. V., Baerends, E. J., 1997, "Electron Correlation Effects on the Shape of the Kohn-Sham Molecular Orbitals", *Theor. Chem. Acc.*, **96**, 44.

Gross, E. K. U., Oliveira, L. N., Kohn, W., 1988a, "Rayleigh-Ritz Variational Principle for Ensembles of Fractionally Occupied States", *Phys. Rev. A*, **37**, 2805.

Gross, E. K. U., Oliveira, L. N., Kohn, W., 1988b, "Density-Functional Theory for Ensembles of Fractionally Occupied States. I. Basic Formalism", *Phys. Rev. A*, **37**, 2809.

Guan, J., Duffy, P., Carter, J. T., Chong, D. P., Casida, K., Casida, M. E., Wrinn, M., 1993, "Comparison of Local-Density and Hartree-Fock Calculations of Molecular Polarizabilities and Hyperpolarizabilities", *J. Chem. Phys.*, **98**, 4753.

Gunnarsson, O., Lundqvist, B. I., 1976, "Exchange and Correlation in Atoms, Molecules, and Solids by the Spin-Density-Functional Formalism", *Phys. Rev. B*, **13**, 4274.

Gunnarsson, O., Jones, R. O., 1985, "Total Energy Differences: Sources of Error in Local-Density Approximations", *Phys. Rev. B*, **31**, 7588.

Guo, H., Sirois, S., Proynov, E. I., Salahub, D. R., 1997, "Density Functional Theory and its Applications to Hydrogen-bonded Systems" in *Theoretical Treatments of Hydrogen Bonding*, Hadzi, D. (ed.), Wiley, New York.

Guo, H., Salahub, D. R., 1998, "Cooperative Hydrogen Bonding and Enzyme Catalysis", *Angew. Chem. Int. Ed. Engl.*, **37**, 2985.

Haase, F., Sauer, J., Hutter, J., 1997, "Ab Initio Molecular Dynamics Simulation of Methanol Adsorbed in Chabazite", *Chem. Phys. Lett.*, **266**, 397.

Hagemeister, F. C., Gruenloh, C. J., Zwier, T. S., 1998, "Density Functional Theory Calculations of the Structures, Binding Energies, and Infrared Spectra of Methanol Clusters", *J. Phys. Chem. A*, **102**, 82.

Halkier, A., Koch, H., Jørgensen, P., Christiansen, O., Nielsen, I. M. B., Helgaker, T., 1997, "A Systematic Ab Initio Study of the Water Dimer in Hierarchies of Basis Sets and Correlation Models", *Theor. Chem. Acc.*, **97**, 150.

Halls, M. D., Schlegel, H. B., 1998, "Comparison of the Performance of Local, Gradient-Corrected, and Hybrid Density Functional Models in Predicting Infrared Intensities", *J. Chem. Phys.*, **109**, 10587.

Halls, M. D., Schlegel, H. B., 1999, "Comparison Study of the Prediction of Raman Intensities Using Electronic Structure Methods", *J. Chem. Phys.*, **111**, 8819.

Hamprecht, F. A., Cohen, A. J., Tozer, D. J., Handy, N. C., 1998, "Development and Assessment of New Exchange-Correlation Functionals", *J. Chem. Phys.*, **109**, 6264.

Handy, N. C., Maslen, P. E., Amos, R. D., Andrews, J. S., Murray, C. W., Laming, G. J., 1992, "The Harmonic Frequencies of Benzene", *Chem. Phys. Lett.*, **197**, 506.

Handy, N. C., Tozer, D. J., Laming, G. J., Murray, C. W., Amos, R. D., 1993, "Analytic Second Derivatives of the Potential Energy Surface", *Isr. J. Chem.*, **33**, 331.

Handy, N. C., 1994, "Density Functional Theory" in *Lecture Notes in Quantum Chemistry II*, Roos, B. O. (ed.), Springer, Heidelberg.

Handy, N. C., Tozer, D. J., 1999, "Excitation Energies of Benzene from Kohn-Sham Theory", *J. Comput. Chem.*, **20**, 106.

Hariharan, P. C., Pople, J. A., 1973, "The Influence of Polarization Functions on Molecular Orbital Hydrogenation Energies", *Theor. Chim. Acta.*, **28**, 213.

Harris, J., 1984, "Adiabatic-Connection Approach to Kohn-Sham Theory", *Phys. Rev. A*, **29**, 1648.

Hay, P. J., 1977, "Gaussian Basis Sets for Molecular Calculations", *J. Chem. Phys.*, **66**, 4377.

Hay, P. J., Martin, R. L., 1998, "Theoretical Studies of the Structures and Vibrational Frequencies of Actinide Compounds Using Relativistic Effective Core Potentials With Hartree-Fock and Density Functional Methods: UF_6, NpF_6, and PuF_6", *J. Chem. Phys.*, **109**, 3875.

Hehre, W. J., Ditchfield, R., Pople, J. A., 1972, "Self-Consistent Molecular Orbital Methods. XII. Further Extensions of Gaussian-Type Basis Sets for Use in Molecular Orbital Studies of Organic Molecules" *J. Chem. Phys.*, **56**, 2257.

Hehre, W. J., Radom, L., Schleyer, P. v. R., Pople, J. A., 1986, *Ab Initio Molecular Orbital Theory*, Wiley, New York.

Helgaker, T., Taylor, P. R., 1995, "Gaussian Basis Sets and Molecular Integrals" in *Modern Electronic Structure Theory, Part II*, Yarkony, D. R., (ed.), World Scientific, Singapore.

Helgaker, T., Watson, M., Handy, N. C., 2000, "Analytical Calculation of Nuclear Magnetic Resonance Indirect Spin-Spin Coupling Constants at the Generalized Gradient Aproximation and Hybrid Levels of Density Functional Theory", *J. Chem. Phys.*, **113**, 9402.

Helgaker, T., Jaszunski, M., Ruud, K., 1999, "Ab Initio Methods for the Calculation of NMR Shielding and Indirect Spin-Spin Coupling Constants", *Chem. Rev.*, **99**, 293.

Hertwig, R. H., Koch, W., 1995, "On the Accuracy of Density Functionals and Their Basis Set Dependence: An Extensive Study on the Main Group Element Homonuclear Diatomic Molecules Li_2 to Br_2", *J. Comput. Chem*, **16**, 576.

Hertwig, R. H., Koch, W., Schröder, D., Schwarz, H., Hrušák, J., Schwerdtfeger, P., 1996, "A Comparative Computational Study of Cationic Coinage Metal-Ethylene Complexes $(C_2H_4)M^+$ (M = Cu, Ag, and Au)", *J. Phys. Chem.*, **100**, 12253.

Hertwig, R. H., Koch, W., 1997, "On the Parameterization of the Local Correlation Functional: What is Becke-3-LYP?", *Chem. Phys. Lett.*, **268**, 345.

Hertwig, R. H., Seemeyer, K., Schwarz, H., Koch, W., 1997, "The Origin of the Remarkable Regioselectivity of Fe^+-Mediated Dehydrogenation in Benzocycloalkenes", *Chem. Eur. J.*, **3**, 1214.

Hertwig, R. H., Koch, W., 1999, "A Theoretician's View of the C-F Bond Activation Mediated by the Lanthanide Cations Ce^+ and Ho^+", *Chem. Eur. J.*, **5**, 312.

Hill, J.-R., Freeman, C. M., Delley, B., 1999, "Bridging Hydroxyl Groups in Faujasite: Periodic vs Cluster Density Functional Calculations", *J. Phys. Chem. A*, **103**, 3772.

Hirata, S., Iwata, S., 1998, "*Ab Initio* Hartree-Fock and Density Functional Studies on the Structures and Vibrations of an Infinite Hydrogen Fluoride Polymer", *J. Phys. Chem A*, **102**, 8426.

Hirata, S., Head-Gordon, M., 1999, "Time-Dependent Density Functional Theory for Radicals. An Improved Description for Excited States With Substantial Double Excitation Character", *Chem. Phys. Lett.*, **302**, 375.

Hobza, P., Šponer, J., Reschel, T., 1995, "Density Functional Theory and Molecular Clusters", *J. Comput. Chem.*, **16**, 1315.

Hobza, P., Bludský, O. Suhai, S., 1999, "Reliable Theoretical Treatment of Molecular Clusters: Counterpoise-Corrected Potential Energy Surface and Anharmonic Vibrational Frequencies of the Water Dimer", *Phys. Chem. Chem. Phys.*, **1**, 3073.

Hohenberg, P., Kohn, W., 1964, "Inhomogeneous Electron Gas", *Phys. Rev.*, **136**, B864.

Holthausen, M. C., Heinemann, C., Cornehl, H. H., Koch, W., Schwarz, H., 1995, "The Performance of Density-Functional/Hartree-Fock Hybrid Methods: Cationic Transition-Metal Methyl Complexes MCH_3^+ (M = Sc - Cu, La, Hf - Au)", *J. Chem. Phys.*, **102**, 4931.

Holthausen, M. C., Mohr, M., Koch, W., 1995, "The Performance of Density Functional/Hartree-Fock Hybrid Methods: The Bonding in Cationic First-Row Transition Metal Methylene Complexes", *Chem. Phys. Lett.*, **240**, 245.

Holthausen, M. C., Koch, W., 1996a, "A Theoretical View on Co^+-Mediated C-C and C-H Bond Activations in Ethane", *J. Am. Chem. Soc.*, **118**, 9932.

Holthausen, M. C., Koch, W., 1996b, "Mechanistic Details of the Fe^+-Mediated C-C and C-H Bond Activations in Propane: A Theoretical Investigation", *Helv. Chim. Act.*, **79**, 1939.

Holthausen, M. C., Fiedler, A., Schwarz, H., Koch, W., 1996, "How Does Fe^+ Activate C-C and C-H Bonds in Ethane? A Theoretical Investigation Using Density Funcional Theory", *J. Phys. Chem.*, **100**, 6236.

Holthausen, M. C., Hornung, G., Schröder, D., Sen, S., Koch, W., Schwarz, H., 1997, "Synergy of Theory and Experiment in the Remote Functionalization of Aliphatic Nitriles by 'Bare' Fe(I) and Co(I) Cations in the Gas Phase", *Organometallics*, **16**, 3135.

Holthausen, M. C., Koch, W., Apeloig, Y., 1999, "Theory Predicts Triplet Ground-State Organic Silylenes", *J. Am. Chem. Soc.*, **121**, 2623.

Houk, K. N., Gonzalez, J., Li, Y., 1995, "Pericyclic Reaction Transition-States: Passions and Punctilios, 1935–1995", *Acc. Chem. Res.*, **28**, 81.

Hoyau, S., Ohanessian, G., 1997, "Absolute Affinities of α-Amino Acids for Cu^+ in the Gas Phase. A Theoretical Study", *J. Am. Chem. Soc.*, **119**, 2016.

Hricovíni, M., Malkina, O. L., Bízik, F., Nagy, L. T., Malkin, V. G., 1997, "Calculation of NMR Chemical Shifts and Spin-Spin Coupling Constants in the Monosaccharide Methyl-β-D-xylopyranoside Using a Density Functional Theory Approach", *J. Phys. Chem. A*, **101**, 9756.

Hutter, J., Lüthi, H. P., Diederich, F., 1994, "Structures and Vibrational Frequencies of the Carbon Molecules C_2-C_{18} Calculated by Density Functional Theory", *J. Am. Chem. Soc.*, **116**, 750.

Irigoras, A., Fowler, J. E., Ugalde, J. M., 1999, "Reactivity of $Cr^+(^6S, ^4D)$, $Mn^+(^7S, ^5S)$, and $Fe^+(^6D, ^4F)$: Reaction of Cr^+, Mn^+, and Fe^+ With Water", *J. Am. Chem. Soc.*, **121**, 8549.

Irikura, K. K., Frurip, D. J. (eds.), 1998, *Computational Thermochemistry, American Chemical Society Symposium Series* **677**, Washington, DC.

Jang, J. H., Lee, J. G., Lee, H., Xie, Y., Schaefer III, H. F., 1998, "Molecular Structures and Vibrational Frequencies of Iron Carbonyls: $Fe(CO)_5$, $Fe_2(CO)_9$, and $Fe_3(CO)_{12}$", *J. Chem. Phys. A*, **102**, 5298.

Jaramillo, J., Scuseria, G. E., 1999, "Performance of a Kinetic Energy Density Dependent Functional (VSXC) for Predicting Vibrational Frequencies", *Chem. Phys. Lett.*, **312**, 269.

Jarecki, A. A., Davidson, E. R., 1999, "Density Functional Theory Calculations for F^-", *Chem. Phys. Lett.*, **300**, 44.

Jensen, F., 1999, *Introduction to Computational Chemistry*, Wiley, Chichester.

Jeong, H. Y., Han, Y., "Comment on 'A Computational Study of the Structures of Van der Waals and Hydrogen Bonded Complexes of Ethene and Ethyne' ", *Chem. Phys. Lett.*, **263**, 345.

Johnson, B. G., Gill, P. M. W., Pople, J. A., 1993, "The Performance of a Family of Density Functional Methods", *J. Chem. Phys.*, **98**, 5612.

Johnson, B. G., Frisch, M. J., 1993, "Analytic Second Derivatives of the Gradient-Corrected Density Functional Energy. Effect of Quadrature Weight Derivatives", *Chem. Phys. Lett.*, **216**, 133.

Johnson, B. G., Frisch, M. J., 1994, "An Implementation of Analytic Second Derivatives of the Gradient-Corrected Density Functional Energy", *J. Chem. Phys.*, **100**, 7429.

Johnson, B. G., Gonzales, C. A., Gill, P. M. W., Pople, J. A., 1994, "A Density Functional Study of the Simplest Hydrogen Abstraction Reaction. Effect of Self-Interaction Correction", *Chem. Phys. Lett.*, **221**, 100.

Johnson, B. G., 1995, "Development, Implementation and Applications of Efficient Methodologies for Density Functional Calculations" in *Modern Density Functional Theory - A Tool for Chemistry*, Seminario, J. M., Politzer, P. (eds.), Elsevier, Amsterdam.

Johnson, B. G., White, C. A., Zhang, Q., Chen, B., Graham, R. L., Gill, P. M. W., Head-Gordon M., 1996, "Advances in Methodologies for Linear-Scaling Density Functional Calculations", in *Recent Developments and Applications of Modern Density Functional Theory*, J. M. Seminario, (ed.), Elsevier, Amsterdam.

Jonas, V., Thiel, W., 1995, "Theoretical Study of the Vibrational Spectra on the Transition Metal Carbonyls $M(CO)_6$ [M = Cr, Mo, W], $M(CO)_5$ [M=Fe, Ru, Os], and $M(CO)_4$ [M=Ni, Pd, Pt]", *J. Chem. Phys.*, **102**, 8474.

Jonas, V., Thiel, W., 1996, "Theoretical Study of the Vibrational Spectra of the Transition-Metal Carbonyl Hydrides $HM(CO)_5$ (M=Mn, Re), $H_2M(CO)_4$ (M=Fe, Ru, Os), and $HM(CO)_4$ (M=Co, Rh, Ir)", *J. Chem. Phys.*, **105**, 3636.

Jonas, V., Thiel, W., 1998, "Density Functional Study of the Vibrational Spectra of Octahedral Transition-Metal Hexacarbonyls: Neutral Molecules (M= Cr, Mo, W) and Isoelectronic Ions (M= V, Nb, Ta; Mn, Re; Fe, Ru, Os; Co, Rh, Ir; Pt; Au)", *Organometallics*, **17**, 353.

Jones, R. O., Gunnarsson, 1989, "The Density Functional Formalism, its Applications and Prospects", *Rev. Mod. Phys.*, **61**, 689.

Jursic, B. S., 1996, "Computing Transition State Structures With Density Functional Theory Methods" in *Recent Developments and Applications of Modern Density Functional Theory*, Seminario, J. M. (ed.), Elsevier, Amsterdam.

Jursic, B. S., 1999, "Study of the Water-Methanol Dimer With Gaussian and Complete Basis Set *Ab Initio*, and Density Functional Theory Methods", *J. Mol. Struct. (Theochem)*, **466**, 203.

Kang, H. C., 1996, "A Computational Study of the Structures of Van der Waals and Hydrogen Bonded Complexes of Ethene and Ethyne", *Chem. Phys. Lett.*, **254**, 135.

Kaschner, R., Seifert, G., 1994, "Investigations of Hydrogen-Bonded Systems: Local Density Approximation and Gradient Corrections", *Int. J. Quant. Chem.*, **52**, 957.

Kaupp, M., Malkina, O. L., Malkin, V. G., 1997, "The Calculation of ^{17}O Chemical Shielding in Transition Metal Oxo Complexes. I. Comparison of DFT and *Ab Initio* Approaches, and Mechanisms of Relativity-Induced Shielding", *J. Chem. Phys.*, **106**, 9201.

Kaupp, M., Malkin, V. G., Malkina, O. L., 1998, "NMR of Transition Metal Compounds", in *Encyclopedia of Computational Chemistry*, Schleyer, P. v. R. (Editor-in-Chief), Wiley, Chichester.

Kendall, R. A., Früchtl, H. A., 1997, "The Impact of the Resolution of the Identity Approximate Integral Method on Modern Ab Initio Algorithm Development", *Theor. Chem. Acc.*, **97**, 158.

Kesyczynski, J. Goodman, L., Kwiatkowski, J. S., 1997, "Density Functional Theory and Post-Hartree-Fock Studies on Molecular Structure and Harmonic Vibrational Spectrum of Formaldehyde", *Theor. Chem. Acc.*, **97**, 195.

Kieninger, M., Suhai, S., 1994, "Density Functional Studies on Hydrogen-Bonded Complexes", *Int. J. Quant. Chem.*, **52**, 465.

Kim, K., Jordan, K. D., 1994, "Comparison of Density Functional and MP2 Calculations on the Water Monomer and Dimer", *J. Phys. Chem.*, **98**, 10089.

Kim, J., Majumdar, D., Lee, H. M., Kim, K. S., 1999, "Structures and Energetics of the Water Heptamer: Comparison With the Water Hexamer and Octamer", *J. Chem. Phys.*, **110**, 9128.

Klopper, W., van Duijneveldt-van de Rijdt, J. G. C. M., van Duijneveldt, F. B., 2000, "Computational Determination of Equilibrium Geometry and Dissociation Energy of the Water Dimer", *Phys. Chem. Chem. Phys.*, **2**, 2227.

Knight, L. B., Kaup, J. G., Petzold, B., Ayyad, R., Ghanty, T. K., Davidson, E. R., 1999, "Electron Spin Resonance Studies of $^{45}Sc^{17}O$, $^{89}Y^{17}O$, and $^{139}La^{17}O$ in Rare Gas Matrices: Comparison With Ab Initio Electronic Structure and Nuclear Hyperfine Calculations", *J. Chem. Phys.*, **110**, 5658.

Koch, W., Hertwig, R. H., 1998, "Density Functional Theory Applications to Transition Metal Problems", in *Encyclopedia of Computational Chemistry*, Schleyer, P. v. R. (Editor-in-Chief), Wiley, Chichester.

Kohn, W., Sham, L. J., 1965, "Self Consistent Equations Including Exchange and Correlation Effects", *Phys. Rev.*, **140**, A1133.

Kohn, W., Becke, A. D., Parr, R. G., 1996, "Density Functional Theory of Electronic Structure", *J. Phys. Chem.*, **100**, 12974.

Kohn, W., Meir, Y., Makarov, D. E., 1998, "Van der Waals Energies in Density Functional Theory", *Phys. Rev. Lett.*, **80**, 4153.

Koopmans, T. A., 1934, "Über die Zuordnung von Wellenfunktionen und Eigenwerten zu den einzelnen Elektronen eines Atoms", *Physica*, **1**, 104.

Kozlowski, P. M., Rauhut, G., Pulay, P., 1995, "Potential Symmetry Breaking, Structure and Definite Vibrational Assignment for Azulene: Multiconfigurational and Density Functional Results", *J. Chem. Phys.*, **103**, 5650.

Kragten, D. D., van Santen, R. A., Lerou, 1999, "Density Functional Study of the Palladium Acetate Catalyzed Wacker Reaction in Acetic Acid", *J. Phys. Chem. A*, **103**, 80.

Krossner, M., Sauer, J., 1996, "Interaction of Water With Brønsted Acidic Sites of Zeolite Catalysts. *Ab Initio* Study of 1:1 and 2:1 Surface Complexes", *J. Phys. Chem.*, **100**, 6199.

Kryachko, E. S., Ludeña, E. V., 1990, *Energy Density Functional Theory of Many-Electron Systems*, Kluwer Academic Press, Dordrecht.

Kristyán, S., Pulay, P., 1994, "Can (Semi)Local Density Functional Theory Account for the London Dispersion Forces?", *Chem. Phys. Lett.*, **229**, 175.

Kumar, G. A., Pan, Y., Smallwood, C. J., McAllister, M. A., 1998, "Low-Barrier Hydrogen Bonds: *Ab Initio* and DFT Investigation", *J. Comput. Chem.*, **19**, 1345.

Kumar, G. A., McAllister, M. A., 1998, "Theoretical Investigation of the Relationship Between Proton NMR Chemical Shift and Hydrogen Bond Strength", *J. Org. Chem.*, **63**, 6968.

Kushto, G. P., Andrews, L., 1999, "Infrared Spectroscopic and Density Functional Theoretical Investigation of the Reaction Products of Laser-Ablated Zr, Hf, and Th Atoms With Nitric Oxide", *J. Phys. Chem. A*, **103**, 4836.

Kutzelnigg, W., Fleischer, U., Schindler, M., 1990, "The IGLO Method: Ab Initio Calculation and Interpretation of NMR Chemical Shifts and Magnetic Susceptibilities", in *NMR Basic Principles and Progress, Vol. 23*, Diehl, P., Fluck, E., Günther, H., Kosfield, R., Seeling, J., (eds.), Springer, Heidelberg.

Kwiatkowski, J. S., Leszczynski, J., 1997, "Density Functional Theory Study on Molecular Structure and Vibrational IR Spectra of Isocytosine", *Int. J. Quant. Chem.*, **61**, 453.

Laasonen, K., Parrinello, M., Car, R., Lee, C., Vanderbilt, D., 1993, "Structures of Small Water Clusters Using Gradient-Corrected Density Functional Theory", *Chem. Phys. Lett.*, **207**, 208.

Laasonen, K., Klein, M. K., 1994, "Structural Study of $(H_2O)_{20}$ and $(H_2O)_{21}H^+$ Using Density Functional Methods", *J. Phys. Chem.*, **98**, 10079.

Lacks, D. J., Gordon, R. G., 1993, "Pair Interactions of Rare-Gas Atoms as a Test of Exchange-Energy-Density Functionals in Regions of Large Density Gradients", *Phys. Rev. A*, **47**, 4681.

Laming, G. J., Termath, V., Handy, N, C., 1993, "A General Purpose Exchange-Correlation Energy Functional", *J. Chem. Phys.*, **99**, 8765.

Laming, G. J., Handy, N. C., Amos, R. D., 1993, "Kohn-Sham Calculations on Open-Shell Diatomic Molecules", *Mol. Phys.*, **80**, 1121.

Latajka, Z., Bouteiller, Y., 1994, "Application of Density Functional Methods for the Study of Hydrogen-Bonded Systems: The Hydrogen Fluoride Dimer", *J. Chem. Phys.*, **101**, 9793.

Lee, C., Yang, W., Parr, R. G., 1988, "Development of the Colle-Salvetti Correlation-Energy Formula into a Functional of the Electron Density", *Phys. Rev. B*, **37**, 785.

Lee, A. M., Colwell, S. M., 1994, "The Determination of Hyperpolarizabilities Using Density Functional Theory With Nonlocal Functionals", *J. Chem. Phys.*, **101**, 9704.

Lee, C., Chen, H., Fitzgerald, G., 1994, "Chemical Bonding in Water Clusters", *J. Chem. Phys.*, **102**, 1266.

Lee, A. M., Handy, N. C., Colwell, S. M., 1995, "The Density Functional Calculation of Nuclear Shielding Constants using London Atomic Orbitals", *J. Chem. Phys.*, **103**, 10095.

Lee, T. J., Bauschlicher, C. W., Jr., Jayatilaka, D., 1997, "A Challenge for Density Functional Theory: the XONO and XNO_2 (X = F, Cl, and Br) Molecules", *Theor. Chem. Acc.*, **97**, 185.

Lein, M., Dobson, J. F., Gross, E. K. U., 1999, "Toward the Description of Van der Waals Interactions Within Density Functional Theory", *J. Comput. Chem.*, **20**, 12.

Leininger, T., Stoll, H., Werner, H.-J., Savin, A., 1997, "Combining Long-Range Configuration Interaction With Short-Range Density Functionals", *Chem. Phys. Lett.*, **275**, 151.

Levy, M., 1979, "Universal Variational Functionals of Electron Densities, First Order Density Matrices, and Natural Spin Orbitals and Solution of the v-Representability Problem", *Proc. Natl. Acad. Sci. USA*, **76**, 6062.

Levine, I. N., 1991, *Quantum Chemistry*, 4th Edition, Prentice Hall, London.

Lii, J.-H., 1998, "Hydrogen Bonding: 2", in *Encyclopedia of Computational Chemistry*, Schleyer, P. v. R. (Editor-in-Chief), Wiley, Chichester.

London, F., 1937, "Quantum Theory of Interatomic Currents in Aromatic Compounds", *J. Phys. Radium*, **8**, 397.

Löwdin, P.-O., 1959, "Correlation Problem in Many-Electron Quantum Mechanics", *Adv. Chem. Phys.*, **2**, 207.

Luna, A., Amekraz, B., Morizur, J.-P., Tortajada, J., Mó, O., Yáñez, M., 1997, "Reactions Between Guanidine and Cu^+ in the Gas Phase: An Experimental and Theoretical Study", *J. Phys. Chem. A*, **101**, 5931.

Luna, A., Amekraz, B., Tortajada, J., Morizur, J.-P., Alcamí, M., Mó, O., Yáñez, M., 1998a, "Modeling the Interactions between Peptide Functions and Cu(I): Formamide-Cu^+ Reactions in the Gas Phase", *J. Am. Chem. Soc.*, **120**, 5411.

Luna, A., Morizur, J.-P., Tortajada, J., Alcamí, M., Mó, O., Yáñez, M., 1998b, "Role of Cu^+ Association on the Formamide $\rightarrow$ Formamidic Acid $\rightarrow$ (Aminohydroxy)carbene Isomerizations in the Gas Phase", *J. Phys. Chem.*, **102**, 4652.

Lundell, J., 1995, "A MPPT2 Investigation of the H_2O-CO Dimer. A Test of Geometries, Energetics, and Vibrational Spectra", *J. Phys. Chem.*, **99**, 14290.

Lundell, J., Latajka, Z., 1997, "Density Functional Study of Hydrogen-Bonded Systems: The Water-Carbon Monoxide Complex", *J. Phys. Chem. A*, **101**, 5004.

Lundqvist, B. I., Andersson, Y., Shao, H., Chan, S., Langreth, D. C., 1995, "Density Functional Theory Including Van der Waals Forces", *Int. J. Quant. Chem.*, **56**, 247.

Maerker, C., Schleyer, P. v. R., Liedl, K. R., Ha, T. K., Quack, M., Suhm, M. A., 1997, "A Critical Analysis of Electronic Density Functionals for Structural, Energetic, Dynamic, and Magnetic Properties of Hydrogen Fluoride Clusters", *J. Comput. Chem.*, **18**, 1695.

Malkin, V. G., Malkina, O. L., Salahub, D. R., 1993, "Calculations of NMR Shielding Constants by Uncoupled Density Functional Theory", *Chem. Phys. Lett.*, **204**, 80.

Malkin, V. G., Malkina, O. L., Salahub, D. R., 1994, "Calculation of Spin-Spin Coupling Constants Using Density Functional Theory", *Chem. Phys. Lett.*, **221**, 91.

Malkin, V. G., Malkina, O. L., Casida, M. E., Salahub, D. R., 1994, "Nuclear Magnetic Resonance Shielding Tensors Calculated With a Sum-Over-States Density Functional Perturbation Theory", *J. Am. Chem. Soc.*, **116**, 5898.

Malkin, V. G., Malkina, O. L., Eriksson, L. A., Salahub, D. R., 1995, "The Calculation of NMR and ESR Spectroscopy Parameters Using Density Functional Theory" in *Modern Density Functional Theory: A Tool for Chemistry*, Seminario, J. M., Politzer, P. (eds.), Elsevier, Amsterdam.

Malkina, O. L., Salahub, D. R., Malkin, V. G., 1996, "Nuclear Magnetic Resonance Spin-Spin Coupling Constants from Density Functional Theory: Problems and Results", *J. Chem. Phys.*, **105**, 8793.

March, N. H., 1975, *Self-Consistent Fields in Atoms*, Pergamon Press, Oxford.

March, N. H., 1992, *Electron Density Theory of Atoms and Molecules*, Academic Press, London.

Martell, J. M., Goddard, J. D., Eriksson, L. A., 1997, "Assessment of Basis Set and Functional Dependencies in Density Functional Theory: Studies of Atomization and Reaction Energies", *J. Phys. Chem. A*, **101**, 1927.

Martin, J. M. L., El-Yazal, J., François, J.-P., 1995a, "Basis Set Convergence and Performance of Density Functional Theory Including Exact Exchange Contributions for Geometries and Harmonic Frequencies", *Mol. Phys.*, **86**, 1437.

Martin, J. M. L., El-Yazal, J., François, J.-P., 1996, "Structure and Vibrational Spectrum of Some Polycyclic Aromatic Compounds Studied by Density Functional Theory. 1. Naphtalene, Azolene, Phenanthrene, and Anthracene", *J. Phys. Chem.*, **100**, 15358.

Martin, J. M. L., El-Yazal, J., François, J.-P., 1995b, "Structure and Vibrational Spectra of Carbon Clusters C_n (n = 2-10, 12, 14, 16, 18) Using Density Functional Theory Including Exact Exchange Contributions", *Chem. Phys. Lett.*, **242**, 570.

Martin, J. M. L., 2000, "Some Observations and Case Studies on Basis Set Convergence in Density Functional Theory" in *Density Functional Theory: A Bridge between Chemistry and Physics*, Geerlings, P., De Proft, F., Langenaeker, W. (eds.), VUB Press, Brussels.

McDowell, S. A. C., Amos, R. D., Handy, N. C., 1995, "Molecular Polarisabilities – A Comparison of Density Functional Theory with Standard Ab Initio Methods", *Chem. Phys. Lett.*, **235**, 1.

McMahon, M. T., deDios, A. D, Godbout, N., Salzmann, R., Laws, D. D., Le, H., Havlin, R. H., Oldfield, E., 1998, "An Experimental and Quantum Chemical Investigation of CO Binding to Heme Proteins and Model Systems: A Unified Model Based on ^{13}C, ^{17}O and ^{57}Fe Nuclear Magnetic Resonance and ^{57}Fe Mössbauer and Infrared Spectroscopies", *J. Am. Chem. Soc.*, **120**, 4784.

McWeeny, R., 1967, "The Nature of Electron Correlation in Molecules", *Int. J. Quant. Chem.*, **1S**, 351.

McWeeny, R., 1992, *Methods of Molecular Quantum Mechanics*, 2nd edition, Academic Press, London.

Mebel, A. M., Morokuma, K., Lin, M. C., 1995, "Modification of the GAUSSIAN-2 Theoretical Model: The Use of Coupled-Cluster Energies, Density Functional Geometries, and Frequencies", *J. Chem. Phys.*, **103**, 7414.

Meijer, E. J., Spriek, M., 1996, "A Density Functional Study of the Intermolecular Interactions of Benzene", *J. Chem. Phys.*, **105**, 8684.

Merkle, R., Savin, A., Preuss, H., 1992, "Singly Ionized First-Row Dimers and Hydrides Calculated With the Fully Numerical Density Program NUMOL", *J. Chem. Phys.*, **97**, 9216.

Michalska, D., Bienko, D. C., Abkowicz-Bienko, A. J., Latajka, Z., 1996, "Density Functional, Hartree-Fock, and MP2 Studies on the Vibrational Spectrum of Phenol", *J. Chem. Phys.*, **100**, 17786.

Milet, A., Korona, T., Moszynski, R., Kochanski, E., 1999, "Anisotropic Intermolecular Interactions in Van der Waals and Hydrogen-Bonded Complexes: What Can we Get from Density Functional Theory?", *J. Chem. Phys.*, **111**, 7727.

Millam, J. M., Scuseria, G. E., 1997, "Linear Scaling Conjugate Gradient Density Matrix Search as an Alternative to Diagonalization for First Principles Electronic Structure Calculations", *J. Chem. Phys.*, **106**, 5569.

Mó, O., Yáñez, M., Elguero, J., 1997, "Study of the Methanol Trimer Potential Energy Surface", *J. Chem. Phys.*, **107**, 3592.

Mó, O., Yáñez, M., 1998, "Density Functional Theory Calculations on Hydrogen-Bonded Tropolone-$(H_2O)_2$ Clusters", *J. Phys. Chem. A*, **102**, 8174.

Montgomery, J. A., Jr., Frisch, M. J., Ochterski, J. W., Petersson, G. A., 1999, "A Complete Basis Set Model Chemistry. VI. Use of Density Functional Geometries and Frequencies", *J. Chem. Phys.*, **110**, 2822.

Morokuma, K., 1977, "Why Do Molecules Interact? The Origin of Electron Donor-Acceptor Complexes, Hydrogen Bonding, and Proton Affinity", *Acc. Chem. Res.*, **10**, 294.

Morrison, R. C., Zhao, Q., 1995, "Solution to the Kohn-Sham Equations Using Reference Densities from Accurate, Correlated Wave Functions for the Neutral Atoms Helium Through Argon", *Phys. Rev. A*, **51**, 1980.

Munzarová, M., Kaupp, M., 1999, "A Critical Validation of Density Functional and Coupled Cluster Approaches for the Calculation for EPR Hyperfine Coupling Constants in Transition Metal Complexes", *J. Phys. Chem. A*, **103**, 9966.

Murcko, M. A., Castejon, H., Wiberg, K. B., 1996, "Carbon-Carbon Rotational Barriers in Butane, 1-Butene, and 1,3-Butadiene", *J. Phys. Chem.*, **100**, 16162.

Murray, C. W., Laming, G. J., Handy, N. C., Amos, R. D., 1992, "Kohn-Sham Bond Lengths and Frequencies Calculated With Accurate Quadrature and Large Basis Sets", *Chem. Phys. Lett.*, 199, 551.

Murray, C. W., Handy, N. C., Laming, G. J., 1993, "Quadrature Schemes for Integrals in Density Functional Theory", *Mol. Phys.*, **78**, 997.

Murray, C. W., Handy, N. C., Amos, R. D., 1993, "A Study of O_3, S_3, CH_2, and Be_2 Using Kohn-Sham Theory With Accurate Quadrature and Large Basis Sets", *J. Chem. Phys.*, **98**, 7145.

Nagy, Á., 1998a, "Excited States in Density Functional Theory", *Int. J. Quant. Chem.*, **70**, 681.

Nagy, Á., 1998b, "Kohn-Sham Equations for Multiplets", *Phys. Rev. A*, **57**, 1672.

Neuman, R., Handy, N. C., 1995, "Investigations Using the Becke-Roussel Exchange Functional", *Chem. Phys. Lett.*, **252**, 19.

Neuman, R., Handy, N. C., 1996, "Investigations Using the Becke95 Correlation Functional", *Chem. Phys. Lett.*, **246**, 381.

Neumann, R., Nobes, R. H., Handy, N. C., 1996, "Exchange Functionals and Potentials", *Mol. Phys.*, **87**, 1.

Nguyen, M. T., Creve, S., Vanquickenborne, L. G., 1996, "Difficulties of Density Functional Theory in Investigating Addition Reactions of the Hydrogen Atom", *J. Phys. Chem.*, **100**, 18422.

Nielsen, I. M. B., Seidl, E., Janssen, C. L., 1999, "Accurate Structures and Binding Energies for Small Water Clusters: The Water Trimer", *J. Chem. Phys.*, **110**, 9435.

Niu, S., Hall, M. B., 1997, "Comparison of Hartree-Fock, Density Functional, Møller-Plesset Perturbation, Coupled Cluster, and Configuration Interaction Methods for the Migratory Insertion of Nitric Oxide into a Cobalt-Carbon Bond", *J. Phys. Chem. A*, **101**, 1360.

Niu, S., Thomson, L. M., Hall, M. B., 1999, "Theoretical Characterization of the Reaction Intermediates in a Model of the Nickel-Iron Hydrogenase of *Desulfovibrio Gigas*", *J. Am. Chem. Soc.*, **121**, 4000.

Novak, A., 1974, "Hydrogen Bonding in Solids. Correlation of Spectroscopic and Crystallographic Data", in *Structure and Bonding*, Vol. 18, Dunitz, J. D., Hemmerich, P., Holm, R. H., Ibers, J. A., Jørgensen, C. K., Neilands, J. B., Reinen, D., Williams, R. J. P. (eds.), Springer, Heidelberg.

Novoa, J. J., Sosa, C., 1995, "Evaluation of Density Functional Approximation on the Computation of Hydrogen Bond Interactions", *J. Phys. Chem.*, **99**, 15837.

Ojamäe, L., Shavitt, I., Singer, S. J., 1995, "Potential Energy Surfaces and Vibrational Spectra of $H_5O_2^+$ and Larger Hydrated Proton Complexes", *Int. J. Quant. Chem. Symp.*, **29**, 657.

Oliveira, L. N., Gross, E. K. U., Kohn, W., 1988, "Density-Functional Theory for Ensembles of Fractionally Occupied States. II. Application to the He Atom", *Phys. Rev. A*, **37**, 2821.

Orendt, A. M., Facelli, J. C., Radziszewski, J. G., Horton, W. J., Grant, D. M., Michl, J., 1996, "^{13}C Dipolar NMR Spectrum of Matrix-Isolated *o*-Benzyne-*1,2*-$^{13}C_2$", *J. Am. Chem. Soc.*, **118**, 846.

Osinga, V. P., van Gisbergen, S. J. A., Snijders, J. G., Baerends, E. J., 1997, "Density Functional Results for Isotropic and Anisotropic Multipole Polarizabilities and C_6, C_7, and C_8 Van der Waals Dispersion Coefficients for Molecules", *J. Chem. Phys.*, **106**, 5091.

Paizs, B., Suhai, S., 1998, "Comparative Study of BSSE Correction Methods at DFT and MP2 Levels of Theory, *J. Comput. Chem.*, **19**, 575.

Parr, R. G., Yang, W., 1989, *Density-Functional Theory of Atoms and Molecules*, Oxford University Press, New York.

Parr, R. G., Yang, W., 1995, "Density-Functional Theory of the Electronic Structure of Molecules", *Annu. Rev. Phys. Chem.*, **46**, 701.

Patton, D. C., Pederson, M. R., 1997, "Application of the Generalized-Gradient Approximation to Rare-Gas Dimers", *Phys. Rev. A*, **56**, R2495.

Pavlov, M., Blomberg, M. R. A., Siegbahn, P. E. M., Wesendrup, R., Heinemann, C., Schwarz, H., 1997, "Pt^+-Catalyzed Oxidation of Methane: Theory and Experiment", *J. Phys. Chem. A*, **101**, 1567.

Pavlov, M., Blomberg, M. R. A., Siegbahn, P. E. M., 1999, "New Aspects of H_2 Activation by Nickel-Iron Hydrogenase", *Int. J. Quant. Chem.*, **73**, 197.

Perdew, J. P., Zunger, A., 1981, "Self-Interaction Correction to Density-Functional Approximations for Many-Electron Systems", *Phys. Rev. B*, **23**, 5048.

Perdew, J. P., Parr, R. G., Levy, M., Balduz, J. L., Jr., 1982, "Density Functional Theory for Fractional Particle Number: Derivative Discontinuities of the Energy", *Phys. Rev. Lett.*, **49**, 1691.

Perdew, J. P., 1986, "Density-Functional Approximation for the Correlation Energy of the Inhomogeneous Electron Gas", *Phys. Rev. B*, **33**, 8822.

Perdew, J. P., Wang, Y., 1986, "Accurate and Simple Density Functional for the Electronic Exchange Energy: Generalized Gradient Approximation", *Phys. Rev. B*, **33**, 8800.

Perdew, J. P., 1991, "Unified Theory of Exchange and Correlation Beyond the Local Density Approximation", in *Electronic Structure of Solids*, P. Ziesche, H. Eschrig (eds.), Akademie Verlag, Berlin.

Perdew, J. P., Wang, Y., 1992, "Accurate and Simple Analytic Representation of the Electron Gas Correlation Energy", *Phys. Rev. B.*, **45**, 13244.

Perdew, J. P., Burke, K., 1996, "Comparison Shopping for a Gradient-Corrected Density Functional", *Int. J. Quant. Chem.*, **57**, 309.

Perdew, J. P., Burke, K., Ernzerhof, M., 1996, "Generalized Gradient Approximation Made Simple", *Phys. Rev. Lett.*, **77**, 3865, Erratum: *Phys. Rev. Lett.*, **78**, 1396 (1997).

Perdew, J. P., Ernzerhof, M., Burke, K., 1996, "Rationale for Mixing Exact Exchange With Density Functional Approximations", *J. Chem. Phys.*, **105**. 9982.

Perdew, J. P., Ernzerhof, M., Burke, K., Savin, A., 1997, "On-Top Pair-Density Interpretation of Spin Density Functional Theory, With Applications to Magnetism", *Int. J. Quant. Chem.*, **61**, 197.

Perdew, J. P., Ernzerhof, M., 1998, "Driving Out the Self-Interaction Error" in *Electron Density Functional Theory. Recent Progress and New Directions*, Dobson, J. F., Vignale, G., Das, M. P. (eds.), Plenum Press, New York.

Perdew, J. P., Kurth, S., Zupan, A., Blaha, P., 1999, "Accurate Density Functional With Correct Formal Properties: A Step Beyond the Generalized Gradient Approximation", *Phys. Rev. Lett.*, **82**, 2544.

Pérez-Jordá, J. M., Becke, A. D., 1995, "A Density-Functional Study of Van der Waals Forces: Rare Gas Diatomics", *Chem. Phys. Lett.*, **233**, 134.

Pérez-Jordá, J. M., Yang, W., 1995, "An Algorithm for 3D Numerical Integration that Scales Linearly With the Size of the Molecule", *Chem. Phys. Lett.*, **241**, 469.

Pérez-Jordá, J. M., Yang, W., 1998, "On the Scaling of Multipole Methods for Particle-Particle Interactions", *Chem. Phys. Lett.*, **282**, 71.

Perrin, C. L., Kim, Y.-J., 1998, "Symmetry of the Hydrogen Bond in Malonaldehyde Enol in Solution", *J. Am. Chem. Soc.*, **120**, 12641.

Petersilka, M., Gossmann, U. J., Gross, E. K. U., 1998, "Time Dependent Optimized Effective Potential in the Linear Response Regime" in *Electronic Density Functional Theory. Recent Progress and New Directions*, Dobson, J. F., Vignale, G., Das, M. P. (eds.), Plenum Press, New York.

Petitjean, L., Pattou, D., Ruiz-López, M. F., 1999, "Theoretical Study of the Mechanisms of Ethylene Polymerization with Metallocene-Type Catalysts", *J. Phys. Chem. B*, **103**, 27.

Pople, J. A., Gill, P. M. W., Handy, N. C., 1995, "Spin-Unrestricted Character of Kohn-Sham Orbitals for Open-Shell Systems", *Int. J. Quant. Chem.*, **56**, 303.

Pribble, R. N., Hagemeister, F. C., Zwier, T. S., 1997, "Resonant Ion-Dip Infrared Spectroscopy of Benzene-(Methanol)$_m$ Clusters With m = 1-6", *J. Chem. Phys.*, **106**, 2145.

Proynov, E. I., Vela, A., Salahub, D. R., 1994, "Nonlocal Correlation Functional Involving the Laplacian of the Density", *Chem. Phys. Lett.*, **230**, 419.

Proynov, E., Chermette, H., Salahub, D. R., 2000, "New τ-Dependent Correlation Functional Combined with a Modified Becke Exchange", *J. Chem. Phys.*, **113**, 10013.

Rablen, P. R., Lockman, J. W., Jorgensen, W. L., 1998, "Ab Initio Study of Hydrogen-Bonded Complexes of Small Organic Molecules With Water", *J. Phys. Chem. A*, **102**, 3782.

Rablen, P. R., Pearlman, S. A., Finkbiner, J., 1999, "A Comparison of Density Functional Methods for the Estimation of Proton Chemical Shifts With Chemical Accuracy", *J. Phys. Chem. A*, **103**, 7357.

Rabuck, A. D., Scuseria, G. E., 1999, "Assessment of Recently Developed Density Functionals for the Calculation of Enthalpies of Formation in Challenging Cases", *Chem. Phys. Lett.*, **309**, 450.

Raghavachari, K., Trucks, G. W., 1989a, "Highly Correlated Systems. Excitation Energies of First Row Transition Metals Sc-Cu", *J. Chem. Phys.*, **91**, 1062.

Raghavachari, K., Trucks, G. W., 1989b, "Highly Correlated Systems. Ionization Energies of First Row Transition Metals Sc-Cu ", *J. Chem. Phys.*, **91**, 2457.

Raghavachari, K., 2000, "Perspective on 'Density Functional Thermochemistry. III. The Role of Exact Exchange'", *Theor. Chem. Acc.*, **103**, 361.

Rappé, A. K., Bernstein, E. R., 2000, "Ab Initio Calculation of Nonbonded Interactions: Are We There Yet?", *J. Phys. Chem. A*, **104**, 6117.

Rauhut, G., Pulay, P., 1995, "Transferable Scaling Factors for Density Functional Derived Vibrational Force Fields", *J. Phys. Chem.*, **99**, 3093.

Rauhut, G., Puyear, S., Wolinski, K., Pulay, P., 1996, "Comparison of NMR Shielding Calculated from Hartree-Fock and Density Functional Wave Functions Using Gauge-Including Atomic Orbitals", *J. Phys. Chem.*, **100**, 6310.

Raymond, K. S., Wheeler, R. A., 1999, "Compatibility of Correlation-Consistent Basis Sets With a Hybrid Hartree-Fock/Density Functional Method", *J. Comput. Chem.*, **20**, 207.

Redfern, P. C., Blaudeau, J.-P., Curtiss, L. A., 1997, "Assessment of Modified Gaussian-2 (G2) and Density Functional Theories for Molecules Containing Third-Row Atoms Ga-Kr", *J. Phys. Chem. A*, **101**, 8701.

Reed, A. E., Weinhold, F., Curtiss, L. A., Pochatko, D. J., 1986, "Natural Bond Orbital Analysis of Molecular Interactions: Theoretical Studies of Binary Complexes of HF, H_2O, NH_3, N_2, O_2, F_2, CO, and CO_2 With HF, H_2O, and NH_3", *J. Chem. Phys.*, **84**, 5687.

Remer, L. C., Jensen, J. H., 2000, "Toward a General Theory of Hydrogen Bonding: The Short, Strong Hydrogen Bond [HOH ··· OH]$^-$", *J. Phys. Chem. A*, **104**, 9266.

Ricca, A., Bauschlicher, C. W., Jr., 1994, "Successive Binding Energies for $Fe(CO)_5^+$ ", *J. Phys. Chem.*, **98**, 12899.

Ricca, A., Bauschlicher, C. W., Jr., 1995a, "The MCH_2^+ Systems: Do $ScCH_2^+$ and $TiCH_2^+$ have C_s or C_{2v} Symmetry and a Comparison of the B3LYP Method to Other Approaches", *Chem. Phys. Lett.*, **245**, 150.

Ricca, A., Bauschlicher, C. W., Jr., 1995b, "Theoretical Study of $Fe(CO)_n^-$ ", *J. Phys. Chem.*, **99**, 5922.

Ricca, A., Bauschlicher, C. W., Jr., 1995c, "Successive Binding Energies for $Fe(H_2O)_n^+$ ", *J. Phys. Chem.*, **99**, 9003.

Rodriguez, J. H., Wheeler, D. E., McCusker, J. K., 1998, "Density Functional Studies of a Heisenberg Spin Coupled Chromium-Semiquinone Complex and its Chromium-Catechol Analog", *J. Am. Chem. Soc.*, **120**, 12051.

Roos, B. O., Andersson, K., Fülscher, M. P., Malmqvist, P.-Å., Serrano-Andrés, L., 1996, "Multiconfigurational Perturbation Theory: Applications in Electronic Spectroscopy" in *Advances in Chemical Physics* **XCIII**, Prigogine, I. and Rice, S. A. (eds.), Wiley, Chichester.

Roothaan, C. C. J., 1951, "New Developments in Molecular Orbitals Theory", *Rev. Mod. Phys.*, **23**, 69.

Rosa, A. Ehlers, A. W., Baerends, E. J., Snijders, J. G., te Velde, G., 1996, "Basis Set Effects in Density Functional Calculations on the Metal-Ligand and Metal-Metal Bonds of $Cr(CO)_5$-CO and $(CO)_5$Mn-Mn$(CO)_5$", *J. Phys. Chem.*, **100**, 5690.

Rösch, N., Trickey, S. B., 1997, "Comment on 'Concerning the Applicability of Density Functional Methods to Atomic and Molecular Negative Ions' [J. Chem. Phys., 105, 862 (1996)]", *J. Chem. Phys.*, **106**, 8940.

Ruiz, E., Salahub, D. R., Vela, A., 1996, "Charge-Transfer Complexes: Stringent Tests for Widely Used Density Functionals" *J. Phys. Chem.*, **100**, 12265.

Russo, T. V., Martin, R. L., Hay, P. J., 1994, "Density Functional Calculations on First-Row Transition Metals", *J. Chem. Phys.*, **101**, 7729.

Russo, T. V., Martin, R. L., Hay, P. J., 1995, "Effective Core Potentials for DFT Calculations", *J. Phys. Chem.*, **99**, 17085.

Sadhukhan, S., Muñoz, D., Adamo, C., Scuseria, G. E., 1999, "Predicting Proton Transfer Barriers With Density Functional Methods", *Chem. Phys. Lett.*, **306**, 83.

Salahub, D. R., 1987, "Transition Metal Atoms and Dimers" in *Ab Initio Methods in Quantum Chemistry – II*, Lawley, K. P. (ed.), Wiley, Chichester.

Salahub, D. R., Chrétien, S., Milet, A., Proynov, E. I., 1999, "Performance of Density Functionals for Transition States" in *Transition State Modeling for Catalysis*, Truhlar, D. G., Morokuma, K. (eds.), *ACS Symp. Ser.*, **721**, American Chemical Society, Washington, D. C.

Sambe, H, Felton, R. H., 1975, "A New Computational Approach to Slater's SCF X_α Equation", *J. Chem. Phys.*, **62**, 1122.

Sändig, N., Koch, W., 1997, "On the Mechanism of the Ta^+ Mediated Activation of the C-H Bond in Methane", *Organometallics*, **16**, 5244.

Sändig, N., Koch, W., 1998, "A Quantum Chemical View on the Mechanism of the Ta^+ Mediated Cooupling of Carbon Dioxide With Methane", *Organometallics*, **17**, 2344.

Santamaria, R., Charro, E., Zacarías, A., Castro, M., 1999, "Vibrational Spectra of Nucleic Acid Bases and Their Watson-Crick Pair Complexes", *J. Comput. Chem.*, **20**, 511.

Sauer, J., Ugliengo, P., Garrone, E., Saunders, V. R., 1994, "Theoretical Study of Van der Waals Complexes at Surface Sites in Comparison With the Experiment", *Chem. Rev.*, **94**, 2095.

Sauer, J., 1998, "Zeolites: Applications of Computational Methods" in *Encyclopedia of Computational Chemistry*, Schleyer, P. v. R. (Editor-in-Chief), Wiley, Chichester.

Savin, A., 1995, "Beyond the Kohn-Sham Determinant", in *Recent Advances in Density Functional Methods. Part I*, Chong, D. P. (ed.), World Scientific, Singapore.

Savin, A., 1996, "On Degeneracy, Near Degeneracy and Density Functional Theory", in *Recent Developments of Modern Density Functional Theory*, Seminario, J. M. (ed.), Elsevier, Amsterdam.

Savin, A., Umrigar, C. J., Gonze, X., 1998, "Relationship of Kohn-Sham Eigenvalues to Excitation Energies", *Chem. Phys. Lett.*, **288**, 391.

Schäfer, A., Horn, H., Ahlrichs, R., 1992, "Fully Optimized Contracted Gaussian Basis Sets for Atoms Li to Kr", *J. Chem. Phys.*, **97**, 2571.

Scheiner, S., 1991, "Calculating the Properties of Hydrogen Bonds by Ab Initio Methods", *Rev. Comput. Chem.*, **2**, 165.

Scheiner, S., 1997, *Hydrogen Bonding*, Oxford University Press, Oxford.

Scheiner, A. C., Baker, J., Andzelm, J. W., 1997, "Molecular Energies and Properties from Density Functional Theory: Exploring Basis Set Dependence of Kohn-Sham Equation Using Several Density Functionals", *J. Comput. Chem.*, **18**, 775.

Schipper, P. R. T., Gritsenko, O. V., Baerends, E. J., 1998a, "One-Determinantal Pure State versus Ensemble Kohn-Sham Solutions in the Case of Strong Electron Correlation: CH_2 and C_2", *Theor. Chem. Acc.*, **99**, 329.

Schipper, P. R. T., Gritsenko, O. V., Baerends, E. J., 1998b, "Kohn-Sham Potentials and Exchange and Correlation Energy Densities From One- and Two-Electron Density Matrices for Li_2, N_2 and F_2", *Phys. Rev. A.*, **57**, 1729.

Schipper, P. R. T., Gritsenko, O. V., Baerends, E. J., 1999, "Benchmark Calculations of Chemical Reactions in Density Functional Theory: Comparison of the Accurate Kohn-Sham Solution With Generalized Gradient Approximations for the H_2+H and H_2+H_2 Reactions", *J. Chem. Phys.*, **111**, 4056.

Schmider, H. L., Becke, A. D., 1998a, "Optimized Density Functionals from the Extended G2 Test Set", *J. Chem. Phys.*, **108**, 9624.

Schmider, H. L., Becke, A. D., 1998b, "Density Functionals from the Extended G2 Test Set: Second-Order Gradient Corrections", *J. Chem. Phys.*, **109**, 8188.

Schreckenbach, G., Ziegler, T., 1997a, "Calculation of NMR Shielding Tensors Based on Density Functional Theory and a Scalar Relativistic Pauli-Type Hamiltonian. Application to Transition Metal Complexes", *Int. J. Quant. Chem.*, **61**, 899.

Schreckenbach, G., Ziegler, T., 1997b, "Calculation of the *g*-Tensor of Electron Paramagnetic Resonance Spectroscopy Using Gauge-Including Atomic Orbitals and Density Functional Theory", *J. Phys. Chem. A*, **101**, 3388.

Schreckenbach, G., Ziegler, T., 1998, "Density Functional Calculations of NMR Chemical Shifts and ESR *g*-Tensors", *Theor. Chem. Acc.*, **99**, 71.

Schreckenbach, G., 1999, "The ^{57}Fe NMR Shielding in Ferrocene Revisited. A Density-Functional Study of Orbital Energies, Shielding Mechanisms, and the Influence of the Exchange-Correlation Functional", *J. Chem. Phys.*, **110**, 11936.

Schröder, D., Schwarz, H., 1995, "C-H and C-C Bond Activation by Bare Transition-Metal Oxide Cations in the Gas Phase", *Angew. Chem. Int. Ed. Engl.*, **34**, 1973.

Schröder, D., Heinemann, C., Koch, W., Schwarz, H., 1997, "Perspectives and Challenges in Physical Organic Chemistry", *Pure Appl. Chem.*, **69**, 273.

Schütz, M., Brdarski, S., Widmark, P.-O., Lindh, R., Karlström, G., 1997, "The Water Dimer Interaction Energy: Convergence to the Basis Set Limit at the Correlated Level", *J. Chem. Phys.*, **107**, 4597.

Scott, A. P., Radom, L., 1996, "Harmonic Vibrational Frequencies: An Evaluation of Hartree-Fock, Møller-Plesset, Quadratic Configuration Interaction, Density Functional Theory, and Semiempirical Scale Factors", *J. Phys. Chem.*, **100**, 16502.

Scuseria, G. E., 1999, "Linear Scaling Density Functional Calculations With Gaussian Orbitals", *J. Phys. Chem. A*, **103**, 4782.

Seifert, G., Krüger, K., 1995, "Density Functional Theory, Calculations of Potential Energy Surfaces and Reaction Paths" in *The Reaction Path in Chemistry: Current Approaches and Perspectives*, Heidrich, D. (ed.), Kluwer, Amsterdam.

Serrano-Andrés, L., Merchán, M., Nebot-Gil, I., Lindh, R., Roos, B. O., 1993, "Towards an Accurate Molecular Orbital Theory for Excited States: Ethene, Butadiene, and Hexatriene", *J. Chem. Phys.*, **98**, 3151.

Shaik, S., Schlegel, H. B., Wolfe, S., 1992, *Theoretical Aspects of Physical Organic Chemistry: The S_N2 Mechanism*, Wiley, New York.

Shaik, S., Danovich, D., Fiedler, A., Schröder, D., Schwarz, H., 1995, "Two-State Reactivity in Organometallic Gas-Phase Ion Chemistry", *Helv. Chim. Act.*, **78**, 1393.

Shaik, S., Filatov, M., Schröder, D., Schwarz, H., 1998, "Electronic Structure Makes a Difference: Cytochrome P-450 Mediated Hydroxylations of Hydrocarbons as a Two-State Reactivity Paradigm", *Chem. Eur. J.*, **4**, 193.

Sherrill, C. D., Lee, M. S., Head-Gordon, M., 1999, "On the Performance of Density Functional Theory for Symmetry-Breaking Problems", *Chem. Phys. Lett.*, **302**, 425.

Shida, N., Barbara, P. F., Almlöf, J. E., 1989, "A Theoretical Study of Multidimensional Nuclear Tunneling in Malonaldehyde", *J. Chem. Phys.*, **91**, 4061.

Siegbahn, P. E. M., 1996a, "Electronic Structure Calculations for Molecules Containing Transition Metals" in *Advances in Chemical Physics* **XCIII**, Prigogine, I. and Rice, S. A. (eds.), Wiley, Chichester.

Siegbahn, P. E. M., 1996b, "Two, Three, and Four Water Chain Models for the Nucelophilic Addition Step in the Wacker Process", *J. Phys. Chem.*, **100**, 14672.

Siegbahn, P. E. M., Crabtree, R. H., 1997, "Mechanism of C–H Activation by Diiron Methane Monooxygenase: Quantum Chemical Studies", *J. Am. Chem. Soc.*, **119**, 3103.

Sim, F., St-Amant, A., Papai, I., Salahub, D., 1992, "Gaussian Density Functional Calculations on Hydrogen Bonded Systems", *J. Am. Chem. Soc.*, **114**, 4391.

Simon, S., Duran, M., Dannenberg, J. J., 1999, "Effect of Basis Set Superposition Error on the Water Dimer Surface Calculated at Hartree-Fock, Møller-Plesset, and Density Functional Theory Levels", *J. Phys. Chem. A*, **103**, 1640.

Sirois, S., Proynov, E. I., Nguyen, D. T., Salahub, D. R., 1997, "Hydrogen Bonding in Glycine and Malonaldehyde: Performance of the Lap1 Correlation Functional", *J. Chem. Phys.*, **107**, 6770.

Slater, J. C., 1951, "A Simplification of the Hartree-Fock Method", *Phys. Rev.*, **81**, 385.

Smallwood, C. J., McAllister, M. A., 1997, "Characterization of Low-Barrier Hydrogen Bonds. 7. Relationship Between Strength and Geometry of Short-Strong Hydrogen Bonds. The Formic Acid-Formate Anion Model System. An Ab Initio and DFT Investigation", *J. Am. Chem. Soc.*, **119**, 11277.

Sodupe, M., Branchadell, V., Rosi, M., Bauschlicher, C. W., Jr., 1997a, "Theoretical Study of M^+-CO_2 and OM^+CO Systems for First Transition Row Metal Atoms", *J. Phys. Chem. A*, **1997**, 7854.

Sodupe, M., Rios, R., Branchadell, V., Nicholas, T., Oliva, A., Dannenberg, J. J., 1997b, "A Theoretical Study of the Endo/Exo Selectivity of the Diels-Alder Reaction between Cyclopropene and Butadiene", *J. Am. Chem. Soc.*, **119**, 4232.

Sodupe, M., Bertran, J., Rodríguez-Santiago, L., Baerends, E. J., 1999, "Ground State of the $(H_2O)_2^+$ Radical Cation: DFT versus Post-Hartree-Fock Methods", *J. Phys. Chem. A*, **103**, 166.

Sokolov, N. D., Savel'ev, V. A., 1977, "Dynamics of the Hydrogen Bond: Two-Dimensional Model and Isotopic Effects", *Chem. Phys.*, **22**, 383.

Sokolov, N. D., Vener, M. V., Savel'ev, V. A., 1990, "Tentative Study of Strong Hydrogen Bond Dynamics. II. Vibrational Frequency Consideration", *J. Mol. Struct. (Theochem)*, **222**, 365.

Spears, K. G., 1997, "Density Functional Study of Geometry and Vibrational Spectra for the Isoelectronic $V(CO)_6^-$ and $Cr(CO)_6$ Molecules", *J. Phys. Chem. A*, **101**, 6273.

Šponer, J., Hobza, P., 1998, "DNA Bases and Base Pairs: *Ab Initio* Calculations" in *Encyclopedia of Computational Chemistry*, Schleyer, P. v. R. (Editor-in-Chief), Wiley, Chichester.

Springborg, M., 1997, "Some Recent Density-Functional Studies of Molecular Systems", in *Density Functional Methods in Chemistry and Materials Science*, Springborg, M. (ed.), Wiley, Chichester.

St. Amant, A., 1996, "Practical Density Functional Approaches in Chemistry and Biochemistry" in *Quantum Mechanical Simulation Methods for Studying Biological Systems*, Bicout, D., Field, M. (eds.), Springer, Heidelberg.

Stahl, M., Schopfer, U., Frenking, G., Hoffmann, R. W., 1997, "Conformational Analysis with Carbon-Carbon Coupling Constants. A Density Functional and Molecular Mechanics Study", *J. Org. Chem.*, **62**, 3702.

Stanton, R. V., Merz, K. M., Jr., 1994, "Density Functional Transition States of Organic and Organometallic Reactions", *J. Chem. Phys.*, **100**, 434.

Stein, M., Sauer, J., 1997, "Formic Acid Tetramers: Structure Isomers in the Gas Phase", *Chem. Phys. Lett.*, **267**, 111.

Stepanian, S. G., Reva, I. D., Rosado, M. T. S., Duarte, M. L. T. S., Fausto, R., Radchenko, E. D., Adamowicz, L., 1998a, "Matrix-Isolation Infrared and Theoretical Studies of the Glycine Conformers", *J. Phys. Chem. A*, **102**, 1041.

Stepanian, S. G., Reva, D., Radchenko, E. D., Adamowicz, L., 1998b, "Conformational Behavior of Alanine. Matrix-Isolation Infrared and Theoretical DFT and ab Initio Study", *J. Phys. Chem. A*, **102**, 4623.

Stepanian, S. G., Reva, I. D., Radchenko, E. D., Adamowicz, 1999, "Combined Matrix-Isolation Infrared and Theoretical DFT and Ab Initio Study of the Nonionized Valine Conformers", *J. Phys. Chem. A*, **103**, 4404.

Stephens, P. J., Devlin, J. F., Chabalowski, C. F., Frisch, M. J, 1994, "Ab Initio Calculations of Vibrational Absorption and Circular Dichroism Spectra Using SCF, MP2, and Density Functional Theory Force Fields", *J. Phys. Chem.*, **98**, 11623.

Stewart, P. A., Gill, P. M. W., 1995, "Becke-Wigner: A Simple but Powerful Density Functional", *J. Chem. Soc. Faraday Trans.*, **91**, 4337.

Stirling, A., 1996, "Raman Intensities from Kohn-Sham Calculations", *J. Chem. Phys.*, **104**, 1254.

Stowasser, R., Hoffmann, R., 1999, "What Do the Kohn-Sham Orbitals and Eigenvalues Mean?", *J. Am. Chem. Soc.*, **121**, 3414.

Strain, M. C., Scuseria, G. E., Frisch, M. J., 1996, "Achieving Linear Scaling for the Electronic Quantum Coulomb Problem", *Science*, **271**, 51.

Stratmann, R. E., Scuseria, G. E., Frisch, M. J., 1996, "Achieving Linear Scaling in Exchange-Correlation Density Functional Quadratures", *Chem. Phys. Lett.*, **257**, 213.

Stratmann, R. E., Burant, J. C., Scuseria, G. E., Frisch, M. J., 1997, "Improving Harmonic Frequency Calculations in Density Functional Theory", *J. Chem. Phys.*, **106**, 10175.

Stratmann, R. E., Scuseria, G. E., Frisch, M. J., 1998, "An Efficient Implementation of Time-Dependent Density-Functional Theory for the Calculation of Excitation Energies of Large Molecules", *J. Chem. Phys.*, **109**, 8218.

Süle, P., Nagy, Á., 1996, "Density Functional Study of Strong Hydrogen-Bonded Systems: The Hydrogen Diformiate Complex", *J. Chem. Phys.*, **104**, 8524.

Sychrovský, V., Gräfenstein, J., Cremer, D., 2000, "Nuclear Magnetic Resonance Spin-Spin Coupling Constants from Coupled Perturbed Density Functional Theory", *J. Chem. Phys.*, **113**, 3530.

Szabo, A., Ostlund, N. S., 1982, *Modern Quantum Chemistry: Introduction to Advanced Electronic Structure Theory*, MacMillan Publishing Co., New York.

Szilagyi, R. K., Frenking, G., 1997, "Structure and Bonding of the Isoelectronic Hexacarbonyls $[HF(CO)_6]^{2-}$, $[Ta(CO)_6]^-$, $W(CO)_6$, $[Re(CO)_6]^+$, $[Os(CO)_6]^{2+}$, and $[Ir(CO)_6]^{3+}$: A Theoretical Study", *Organometallics*, **16**, 4807.

Taylor, P., 1992, "Accurate Calculations and Calibration" in *Lecture Notes in Quantum Chemistry*, Roos, B. O. (ed.), Springer, Heidelberg.

Termath, V., Sauer, J., 1997, "Ab Initio Molecular Dynamics Simulation of $H_5O_2^+$ and $H_7O_3^+$ Gas Phase Clusters Based on Density Functional Theory", *Mol. Phys.*, **91**, 963.

Theophilou, A., 1979, "The Energy Density Functional Formalism for Excited States", *J. Phys. C*, **12**, 5419.

Thomas, L. H., 1927, "The Calculation of Atomic Fields", *Proc. Camb. Phil. Soc.*, **23**, 542.

Thomas, J. L. C., Bauschlicher, C. W., Jr., Hall, M. B., 1997, "Binding of Nitric Oxide to First-Transition-Metal-Row Metal Cations. An *Ab Initio* Study", *J. Phys. Chem. A*, **101**, 8530.

Tietze, L. F., Pfeiffer, T., Schuffenhauer, A., 1998, "Stereoselective Intramolecular Hetero Diels-Alder Reactions of Cyclic Benzylidenesulfoxides and DFT Calculations on the Transition Structures", *Eur. J. Org. Chem.*, 2733.

Topol, I. A., Burt, S. K., Rashin, A. A., 1995, "Can Contemporary Density Functional Theory Yield Accurate Thermodynamics for Hydrogen Bonding?", *Chem. Phys. Lett.*, **247**, 112.

Torrent, M., Deng, L., Ziegler, T., 1998, "A Density Functional Study of [2+3] versus [2+2] Addition of Ethylene to Chromium-Oxygen Bonds in Chromyl Cloride", *Inorg. Chem.*, **37**, 1307.

Tozer, D. J., Handy, N. C., 1998, "Improving Virtual Kohn-Sham Orbitals and Eigenvalues: Application to Excitation Energies and Static Polarizabilities", *J. Chem. Phys.*, **109**, 10180.

Tozer, D. J., Amos, R. D., Handy, N. C., Roos, B. O., Serrano-Andrés, L., 1999, "Does Density Functional Theory Contribute to the Understanding of Excited States of Unsaturated Organic Compounds?", *Mol. Phys.*, **97**, 859.

Treutler, O., Ahlrichs, R., 1995, "Efficient Molecular Numerical Integration Schemes", *J. Chem. Phys.*, **102**, 346.

Truhlar, D. G., Morokuma, K., (eds.), 1999, *Transition State Modelling for Catalysis, ACS Symposium Series*, **721**, American Chemical Society, Washington, D. C.

Tschumper, G. S., Schaefer, H. F., 1997, "Predicting Electron Affinities With Density Functional Theory: Some Positive Results for Negative Ions", *J. Chem. Phys.*, **107**, 2529.

Tuma, C., Boese, A. D., Handy, N. C., 1999, "Predicting the Binding Energies of H-Bonded Complexes: A Comparative DFT Study", *Phys. Chem. Chem. Phys.*, **1**, 3939.

Umeyama, H., Morokuma, K., 1977, "The Origin of Hydrogen Bonding. An Energy Decomposition Study" *J. Am. Chem. Soc.*, **99**, 1316.

Valeev, E. F., Schaefer III, H. F., 1998, "The Protonated Water Dimer: Brueckner Methods Remove the Spurious C_1 Symmetry Minimum", *J. Chem. Phys.*, **108**, 7197.

Valiron, P., Vibók, Á., Mayer, I., 1993, "Comparision of a Posteriori and a Priori BSSE Correction Schemes for SCF Intermolecular Energies", *J. Comput. Chem.*, **14**, 401.

Van Caillie, C., Amos, R. D., 1998, "Static and Dynamic Polarisabilities, Cauchy Coefficients and Their Anisotropies: A Comparison of Standard Methods", *Chem. Phys. Lett.*, **291**, 71.

Van Caillie, C., Amos, R. D., 2000, "Raman Intensities Using Time Dependent Density Functional Theory", *Phys. Chem. Chem. Phys.*, **2**, 2123.

van Duijneveldt-van de Rijdt, J. G. C. M., van Duijneveldt, F. B., 1992, "Convergence to the Basis Set Limit in Ab Initio Calculations at the Correlated Level on the Water Dimer", *J. Chem. Phys.*, **97**, 5019.

van Duijneveldt, F. B., van Duijneveldt-van de Rijdt, J. G. C. M., van Lenthe, J. H., 1994, "State of the Art in Counterpoise Theory", *Chem. Rev.*, **94**, 1873.

van Duijneveldt-van de Rijdt, J. G. C. M., van Duijneveldt, F. B., 1999, "Interaction Optimized Basis Sets for Correlated Ab Initio Calculations on the Water Dimer", *J. Chem. Phys.*, **111**, 3812.

van Gisbergen, S. J. A., Osinga, V. P., Gritsenko, O. V., van Leeuwen, R., Snijders, J. G., Baerends, E. J., 1996, "Improved Density Functional Theory Results for Frequency-Dependent Polarizabilities, by Use of an Exchange-Correlation Potential With Correct Asymptotic Behavior", *J. Chem. Phys.*, **105**, 3142.

van Gisbergen, S. J. A., Snijders, J. G., Baerends, E. J., 1998a, "Calculating Frequency-Dependent Hyperpolarizabilities Using Time-Dependent Density Functional Theory", *J. Chem. Phys.*, **109**, 10644. Erratum: *J. Chem. Phys.*, **111**, 6652 (1999).

van Gisbergen, S. J. A., Snijders, J. G., Baerends, E. J., 1998b, "Accurate Density Functional Calculations on Frequency-Dependent Hyperpolarizabilities of Small Molecules", *J. Chem. Phys.*, **109**, 10657.

van Gisbergen, S. J. A., Kootstra, F., Schipper, P. R. T., Gritsenko, O. V., Snijders, J. G., Baerends, E. J., 1998, "Density-Functional-Theory Response-Property Calculations With Accurate Exchange-Correlation Potentials", *Phys. Rev. A*, **57**, 2556.

van Leeuwen, R., Baerends, E. J., 1994, "Exchange-Correlation Potential With Correct Asymptotic Behavior", *Phys. Rev. A*, **49**, 2421.

van Leeuwen, R., Gritsenko, O. V., Baerends, E. J., 1996, "Analysis and Modeling of Atomic and Molecular Kohn-Sham Potentials", *Top. Curr. Chem.*, **180**, 107.

van Lenthe, E., Wormer, P. E. S., van der Avoird, A., 1997, "Density-Functional Calculations of Molecular g-Tensors in the Zero-Order Regular Approximation for Relativistic Effects", *J. Chem. Phys.*, **107**, 2488.

van Voorhis, T., Scuseria, G. E., 1998, "A Novel Form for the Exchange-Correlation Energy Functional", *J. Chem. Phys.*, **109**, 400.

van Wüllen, C., 1996, "On the Use of Common Effective Core Potentials in Density Functional Calculations. I. Test Calculations on Transition-Metal Carbonyls", *Int. J. Quant. Chem.*, **58**, 147.

Vargas, R., Galván, M., Vela, A., 1998, "Singlet-Triplet Gaps and Spin Potentials", *J. Phys. Chem. A*, **102**, 3134.

Ventura, O. N., 1997, "Density Functional Studies of Open-Shell Species" in *The Molecular Modeling e-Conference (TMMeC)*, **1**, 57, available at www.tmmec.org.uy, and online at fcindy5.ncifcrf.gov/tmmec/current/.

Venturini, A., Joglar, J., Fustero, S., Gonzalez, J., 1997, "Diels-Alder Reactions of 2-Azabutadienes With Aldehydes: Ab Initio and Density Functional Theoretical Study of the Reaction Mechanism, Regioselectivity, Acid Catalysis, and Stereoselectivity", *J. Org. Chem.*, **62**, 3919.

Versluis, L., Ziegler, T., 1988, "The Determination of Molecular Structures by Density Functional Theory. The Evaluation of Analytical Energy Gradients by Numerical Integration", *J. Chem. Phys.*, **88**, 322.

Vignale, G., Rasolt, M., Geldart, D. J. W., 1990, "Magnetic Fields and Density Functional Theory", *Adv. Quantum Chem.*, **21**, 235.

von Arnim, M., Ahlrichs, R., 1998, "Performance of Parallel Turbomole for Density Functional Calculations", *J. Comput. Chem.*, **15**, 1746.

von Barth, U., 1979, "Local-Density Theory of Multiplet Structure", *Phys. Rev. A*, **20**, 1693.

Vosko, S. J., Wilk, L., Nusair, M., 1980, "Accurate Spin-Dependent Electron Liquid Correlation Energies for Local Spin Density Calculations: A Critical Analysis", *Can. J. Phys.*, **58**, 1200.

Wachters, A. J. H., 1970, "Gaussian Basis Set for Molecular Wavefunctions Containing Third-Row Atoms", *J. Chem. Phys.*, **52**, 1033.

Wang, S. G., Schwarz, W. H. E., 1996, "Simulation of Nondynamical Correlation in Density Functional Calculations by the Optimized Fractional Occupation Approach: Application to the Potential Energy Surfaces of O_3 and SO_2", *J. Chem. Phys.*, **105**, 4641.

Wang, S. G., Schwarz, W. H. E., 1998, "Density Functional Study of First Row Transition Metal Dihalides", *J. Chem. Phys.*, **109**, 7252.

Wei, D., Salahub, D. R., 1994, "Hydrated Proton Clusters and Solvent Effects on the Proton Transfer Barrier: A Density Functional Study", *J. Chem. Phys.*, **101**, 7633.

Wei, D., Salahub, D. R., 1997, "Hydrated Proton Clusters: Ab Initio Molecular Dynamics Simulation and Simulated Annealing", *J. Chem. Phys.*, **106**, 6086.

Wheeless, C. J. M., Zhou, X., Liu, R., 1995, "Density Functional Theory Study of Vibrational Spectra. 2. Assignment of Fundamental Vibrational Frequencies of Fulvene", *J. Phys. Chem.*, **99**, 12488.

Werpetinski, K. S., Cook, M., 1997, "A New Grid-Free Density Functional Technique: Application to the Torsional Energy Surfaces of Ethane, Hydrazine, and Hydrogen Peroxide", *J. Chem. Phys.*, **106**, 7124.

Wesolowski, T. A., Parisel, O., Ellinger, Y., Weber, J., 1997, "Comparative Study of Benzene···X (X = O_2, N_2, CO) Complexes Using Density Functional Theory: The Importance of an Accurate Exchange-Correlation Energy Density at High Reduced Density Gradients", *J. Phys. Chem. A*, **101**, 7818.

Westerberg, J., Blomberg, M. R. A., 1998, "Methane Activation by Naked Rh^+ Atoms. A Theoretical Study", *J. Phys. Chem. A*, **102**, 7303.

White, C. A., Johnson, B. G., Gill, P. M. W, Head-Gordon, M., 1994, "The Continuous Fast Multipole Method", *Chem. Phys. Lett.*, **230**, 8.

White, C. A., Head-Gordon, M., 1996, "A *J*-Matrix Engine for Density Functional Theory Calculations", *J. Chem. Phys.*, **104**, 2620.

Wiberg, K. B., 1999, "Comparison of Density Functional Theory Models' Ability to Reproduce Experimental ^{13}C-NMR Shielding Values", *J. Comput. Chem.*, **20**, 1299.

Wiest, O., Houk, K. N., 1996, "Density Functional Theory Calculations of Pericyclic Reaction Transition Structures" *Top. Curr. Chem.*, **182**, 1.

Wiest, O., 1998, "Transition States in Organic Chemistry: *Ab Initio*" in *Encyclopedia of Computational Chemistry*, Schleyer, P. v. R. (Editor-in-Chief), Wiley, Chichester.

Wiest, O., 1999, "Structure and [2+2] Cycloreversion of the Cyclobutane Radical Cation", *J. Phys. Chem. A*, **103**, 7907.

Wittborn, A. M., Costas, M., Blomberg, M. R. A., Siegbahn, P. E. M., 1997, "The C-H Activation Reaction of Methane for all Transition Metal Atoms from the Three Transition Rows", *J. Chem. Phys.*, **107**, 4318.

Wittbrodt, J. M., Schlegel, H. B., 1996, "Some Reasons Not to Use Spin Projected Density Functional Theory", *J. Chem. Phys.*, **105**, 6574.

Wolff, S. K., Ziegler, T., van Lenthe, E., Baerends, E. J., 1999, "Density Functional Calculations of Nuclear Magnetic Shieldings Using the Zeroth-Order Regular Approximation (ZORA) for Relativistic Effects: ZORA Nuclear Magnetic Resonance", *J. Chem. Phys.*, **110**, 7689.

Wong, M. W., 1996, "Vibrational Frequency Prediction Using Density Functional Theory", *Chem. Phys. Lett.*, **256**, 391.

Woodward, R. B., Hoffmann, R., 1970, *The Conservation of Orbital Symmetry*, Academic Press, New York. (The text of this book also appears in *Angew. Chem. Int. Ed. Engl.*, **8**, 781, (1969)).

Worthington, S. E., Cramer, C. J., 1997, "Density Functional Calculations of the Influence of Substitution on Singlet-Triplet Gaps in Carbenes and Vinylidenes", *J. Phys. Org. Chem.*, **10**, 755.

Xantheas, S. S., Dunning, T. H., Jr., 1993, "Ab Initio Studies of Cyclic Water Clusters $(H_2O)_n$, n = 1–6. I. Optimal Structures and Vibrational Spectra", *J. Chem. Phys.*, **99**, 8774.

Yang, W., 1992, "Electron Density as the Basic Variable: A Divide-and-Conquer Approach to the Ab Initio Computation of Large Molecules", *J. Mol. Struct. (Theochem)*, **255**, 461.

Yoshizawa, K., Shiota, Y., Yamabe, T., 1998, "Methane-Methanol Conversion by MnO^+, FeO^+, and CoO^+: A Theoretical Study of Catalytic Selectivity", *J. Am. Chem. Soc.*, **120**, 564.

Yoshizawa, K., Shiota, Y., Yamabe, T., 1999, "Intrinsic Reaction Coordinate Analysis of the Conversion of Methane to Methanol by an Iron-Oxo Species: A Study of Crossing Seams of Potential Energy Surfaces", *J. Chem. Phys.*, **111**, 538.

Zhang, Q., Bell, R., Truong, T. N., 1994, "*Ab Initio* and Density Funcional Theory Studies of Proton Transfer Reactions in Multiple Hydrogen Bond Systems", *J. Phys. Chem.*, **99**, 592.

Zhang, Y., Pan, W., Yang, W., 1997, "Describing Van der Waals Interaction in Diatomic Molecules With Generalized Gradient Approximations: The Role of the Exchange Functional", *J. Chem. Phys.*, **107**, 7921.

Zhang, Y., Yang, W., 1998, "A Challenge for Density Functionals: Self-Interaction Error Increases for Systems With a Noninteger Number of Electrons", *J. Chem. Phys.*, **109**, 2604.

Zhang, Y., Yang, W., 2000, "Perspective on 'Density Functional Theory for Fractional Particle Number: Derivative Discontinuities of the Energy'", *Theor. Chem. Acc.*, **103**, 346.

Zheng, Y. C., Almlöf, J., 1993, "Density Functionals Without Meshes and Grids", *Chem. Phys. Lett.*, **214**, 397.

Zhou, M., Andrews, L., 1999a, "Infrared Spectra and Density Functional Calculations of $Cu(CO)^+_{1\text{-}4}$, $Cu(CO)_{1\text{-}3}$, and $Cu(CO)^-_{1\text{-}3}$ in Solid Neon", *J. Chem. Phys.*, **111**, 4548.

Zhou, M., Andrews, L., 1999b, "Infrared Spectra and Density Functional Calculations on $RuCO^+$, $OsCO^+$, $Ru(CO)_x$, $Os(CO)_x$, $Ru(CO)^-_x$ and $Os(CO)^-_x$ (x = 1-4) in Solid Neon", *J. Phys. Chem. A*, **103**, 6956.

Zhou, M., Andrews, L., 1999c, "Infrared Spectra and Density Functional Calculations of the CrO_2^-, MoO_2^-, and WO_2^- Molecular Anions in Solid Argon", *J. Chem. Phys.*, **111**, 4230.

Zhou, X., Wheeless, C. J. M., Liu, R., 1996, "Density Functional Theory Study of Vibrational Spectra. 1. Performance of Several Density Functional Methods in Predicting Vibrational Frequencies", *Vib. Spectrosc.*, **12**, 53.

Ziegler, T., Rauk, A., Baerends, E. J., 1977, "On the Calculation of Multiplet Energies by the Hartree-Fock-Slater Method", *Theor. Chim. Acta*, **43**, 261.

Ziegler, T., 1991, "Approximate Density Functional Theory as a Practical Tool in Molecular Energetics and Dynamics", *Chem. Rev.*, **91**, 651.

Ziegler, T., Li, J., 1994, "Bond Energies for Cationic Bare Metal Hydrides of the First Transition Series: A Challenge to Density Functional Theory", *Can. J. Chem.*, **72**, 783.

Ziegler, T., 1995, "Density Functional Theory as a Practical Tool in Studies of Organometallic Energetics and Kinetics. Beating the Heavy Metal Blues With DFT", *Can. J. Chem.*, **73**, 743.

Ziegler, T., 1997, "Density-Functional Theory as a Practical Tool in Studies of Transition Metal Chemistry and Catalysis", in *Density Functional Methods in Chemistry and Materials Science*, Springborg, M. (ed.), Wiley, Chichester.

Zygmunt, S. A., Mueller, R. M., Curtiss, L. A., Iton, L. E., 1998, "An Assessment of Density Functional Methods for Studying Molecular Adsorption in Cluster Models of Zeolites", *J. Mol. Struct. (Theochem)*, **430**, 9.

Index

Note that references to ubiquitously used terms (e. g., B3LYP) are limited to those pages where definitions or other key information can be found.

Lightning Source UK Ltd.
Milton Keynes UK
05 January 2011

165178UK00001B/9/P

9 783527 303724